Gallium Arsenide and related compounds, 1974

Gallium Arsenide and related compounds, 1974

Papers from the Fifth International Symposium on
Gallium Arsenide and related compounds held in
Deauville, 24–26 September 1974

Conference Series Number 24
The Institute of Physics
London and Bristol

ISBN 0 85498 114 4
ISSN 0 305 2346

The Fifth International Symposium on Gallium Arsenide and related
compounds was jointly sponsored by The Institute of Physics, la Société
Française de Physique, la Société des Electriciens, des Electroniciens et des
Radioélectriciens and the United States Air Force Avionics Laboratory.

International Organizing Committee
  J Bok *(France, Chairman)*
  C Hilsum *(UK)*
  R Runnels *(USA)*
  H Strack *(Germany)*
  R Veilex *(France, Secretary)*

Honorary Editor
  J Bok

Published by The Institute of Physics, 47 Belgrave Square, London SW1X
8QX, England, and Techno House, Redcliffe Way, Bristol BS1 6NX, England,
in association with the American Institute of Physics, 335 East 45th Street,
New York, NY 10017, USA.

Set in 10/12 Press Roman, and printed in Great Britain by Adlard and Son
Ltd, Dorking, Surrey.

Preface

The impact of semiconductor devices on modern industry and even on everyday life is well known. Among all semiconductors, silicon has emerged as the only one to be used extensively at the moment. The success of this material is due to the fundamental knowledge of its physical properties, to the almost perfect control of its preparation, and to the device technology.

For many years gallium arsenide and the related III–V compounds have been actively studied for two reasons. Some of their properties allow important improvements on the performances of classical devices: higher mobilities and larger gaps, allowing higher frequency and temperature operation. Some characteristic features of III–V compounds give rise to new physical phenomena, such as high frequency electrical instabilities (Gunn effect) and light emission by carrier recombination.

These phenomena have recently been used in such types of devices as microwave oscillators and amplifiers, and light-emitting diodes and lasers. The future of these new compounds seems very promising especially for their application to optical telecommunications. This is the reason why physicists and metallurgists actively engaged in research on gallium arsenide and related compounds have met every two years since 1966 to discuss all aspects of device research and applications of the larger gap III–V compounds. The fifth international symposium of this series was held in Deauville, France, in September 1974. Two hundred and twenty participants attended: over ninety papers were submitted to the programme committee who had to select thirty six of these to be presented at the conference. Scientists from all the main universities, government or industrial laboratories throughout the world (with the exception of the Soviet Union), came to discuss the latest progress made on materials preparation, characterization and device technology. Topics discussed at the conference covered material analysis, crystal growth by liquid phase, vapour phase or molecular beam epitaxy, microwave devices, optoelectronic devices and light-emitting diodes, and device processing.

Some important progress was reported, the most significant being the large increase of operating life of (AlGa)As heterojunction lasers. It is well established that metallurgical factors are important in the control of gradual degradation of electroluminescent and laser diodes. New methods of processing and careful junction area definition give cw operating lifetimes of over 8000 hours. One of the conclusions of the conference is that there is definitely an industrial future for GaAs and related compounds in microwave and optical telecommunications.

Acknowledgments are due to our colleagues of the programme committee who were remarkably efficient in preparing the scientific programme and in selecting the papers presented at the conference, to the secretary of the meeting, Dr R Veilex, who took care of all the administration and to Dr J Lefèvre for local organization in Deauville.

It is my pleasure to thank our sponsors, the Délégation Générale à la Recherche Scientifique et Technique (DGRST), Ministère de l'Industrie et de la Recherche and the Direction de la Recherche et des Moyens d'Essais (DRME), Ministère de la Défense, for their financial support which made this conference possible.

J Bok

Contents

x   *Contents*

Acceptor incorporation in GaAs grown by beam epitaxy

M Ilegems and R Dingle

Bell Laboratories, Murray Hill, New Jersey 07974, USA

Abstract. Undoped GaAs grown from Ga and As_4 beams is generally 10^{15}–10^{16} cm^{-3} p-type with mobilities in the range 380–400 cm^2 V^{-1} s^{-1} (296 K) and 4200–5100 cm^2 V^{-1} s^{-1} (77 K). C (27 meV) and Mn (113 meV) are identified as persistent residual acceptors by low-temperature photoluminescence and Hall measurements. Under Ga-rich growth conditions the concentration of C acceptors (on As sites) increases, while the concentration of Mn acceptors (on Ga sites) decreases. These observations are consistent with reduced As-coverage of the growth surface and parallel the behaviour previously reported for Ge. Under As-rich growth conditions 296 K hole concentrations up to about 1×10^{18} cm^{-3} may be achieved with Mn doping at a growth temperature of about 580 °C and with surface morphology equivalent to that obtained in undoped layers.

1. Introduction

Incorporation of dopants in GaAs grown by molecular beam epitaxy (MBE) is quite different from that in material grown by the more usual liquid phase epitaxy (LPE) or chemical vapour deposition (CVD) techniques. The difference is especially marked for the p-type dopants. Thus the traditional group II dopants, Zn and Cd, are very difficult to incorporate during MBE growth because of their high vapour pressures and concomitant low sticking probabilities to the growth surface, whereas the amphoteric group IV dopants Si, Ge and Sn are incorporated predominantly as donors on Ga sites under the usual MBE growth conditions.

Serious efforts are presently being made to exploit the unique capabilities of MBE for device applications and fundamental studies. Among the most significant recent advances in this respect are the demonstration by Cho and Casey (1974) of pulsed lasing action in AlGaAs–GaAs–AlGaAs double heterostructure lasers and the observation by Dingle *et al* (1974) of quantum effects in extremely thin AlGaAs–GaAs–AlGaAs heterostructures.

In view of the emerging device applications it is increasingly important to study impurity incorporation and basic material parameters in further detail. In this context, we report in the present paper first, on the nature of the residual impurities in beam epitaxial GaAs and their incorporation under different growth conditions, and secondly, on the use of Mn as a possible dopant for achieving high-conductivity p-type layers with beam epitaxy.

2. Crystal growth

Crystals from two different growth systems were studied. The growth configuration is similar to that described by Cho (1971) and a schematic cross-section of one of the systems used is illustrated in figure 1. The apparatus consists essentially of an ion pumped high-vacuum station which contains a substrate mounted on a heating block and several individually heated pyrolytic BN evaporation cells enclosed in a liquid-nitrogen cooled shroud which serves to collimate the beams from the ovens as well as to pump condensible gases. The associated instrumentation includes a quadrupole mass spectrometer for

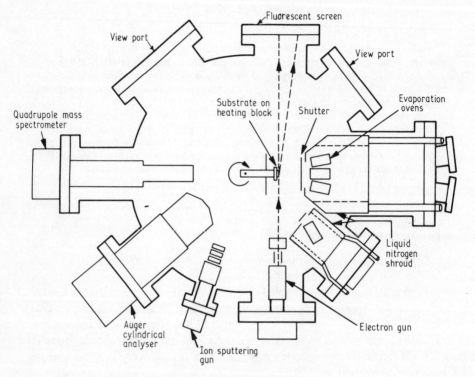

Figure 1. Cross-section of vacuum deposition system.

monitoring the molecular-beam constituents and background gases in the system, an Auger spectrometer for analysis of the chemical composition of the surface and a 3–5 kV electron gun for the observation of surface structures by glancing-incidence electron diffraction during growth.

Layers were grown on ⟨100⟩-oriented Cr-doped GaAs substrates which had been chemically polished in bromine-methanol and cleaned *in situ* by ion sputtering. Substrate temperatures were maintained in the range of 560–600 °C during epitaxial deposition and growth rates were generally around $1 \mu m\,h^{-1}$. Elemental α-As and Ga, both with nominal total impurity concentrations below 1×10^{-6} (mole fraction) were used as source materials.

Oven temperatures during growth are about 900 °C for Ga and about 300 °C for As with As_4 as the dominant species effusing from the As oven. Temperatures are measured with chromel–alumel thermocouples embedded in a graphite slug which either supports the crucible or, for more accurate measurements, is placed at the bottom of the crucible in direct contact with the melt. Background pressures prior to growth are in the low

10^{-9} Torr range. During growth the system pressure is around 2×10^{-7} Torr and is due mainly to As_4.

3. As-rich versus Ga-rich growth conditions

At any given substrate temperature a minimum As-flux is needed to maintain the so-called (Cho 1970) As-stabilized surface structure. This minimum flux corresponds to the equilibrium vapour pressure of As over a stoichiometric GaAs surface at that temperature.

When the substrate is heated in vacuum prior to growth, the amount of As evaporating from the surface will initially exceed that of Ga so that the surface of the crystal becomes depleted in As or Ga-rich (Arthur 1974). The electron diffraction pattern obtained under these conditions has been referred to by Cho (1970) as originating from a Ga-stabilized surface. Since under these conditions the surface and the bulk are of different composition, one would expect that the Ga-stabilized surface structure will be very difficult to maintain during growth except perhaps at very low growth rates. In practice, when the As/Ga flux ratio during growth is below that required to maintain an As-rich condition, excess Ga will build up at the surface. In extreme cases droplets of Ga appear, while under only slight excess-Ga conditions the growth surface remains smooth and mirror-like but shows a characteristic haziness. The properties of layers grown under these conditions may be expected to approach those of crystals grown by LPE. Our results, discussed below, substantiate this expectation.

When the As/Ga flux ratio is larger than that required to grow stoichiometric GaAs, the excess As simply evaporates from the surface. Structurally perfect layers can thus be grown under a wide range of As fluxes. However, one may expect that the As pressure during growth will influence both impurity incorporation and native defect concentrations, and hence the electrical and optical properties of the layers.

4. Residual impurities in GaAs

4.1. Luminescence

Comparison of the low-temperature luminescence spectra of non-intentionally doped MBE films with those observed in very detailed studies of high-purity CVD and LPE layers, summarized by Dean (1974), leads to the identification of C, Si, and Ge as residual impurities. Luminescence spectra of two not advertently doped crystals grown in the same system and under similar conditions except for the As/Ga flux ratio are shown in figure 2. The bands marked C, Si, and Ge, arise from donor–acceptor (DA) and free electron–bound hole (BA) recombination involving unidentified shallow (~ 6 meV) donors and respectively a carbon (~ 27 meV), silicon (~ 35 meV) and germanium (~ 41 meV) acceptor. The higher energy features relate to free exciton (1·515 eV), exciton–donor (1·5138 eV), and exciton–acceptor (1·5120 eV) processes. Finally, the characteristic feature near 1·407 eV, previously observed in MBE crystals by Cho and Hayashi (1971a), is attributed to the presence of approximately 113 meV deep manganese acceptors.

There are several characteristic differences between the spectra of crystals grown under respectively Ga-rich and As-rich conditions. First, we find that in crystals grown under Ga-rich conditions the exciton region (1·515−1·511 eV) is dominated by the strong sharp structure near 1·5120 eV which has been previously identified with excitons decaying at neutral acceptors. In these crystals one sees only very weak evidence for excitons decaying at neutral donor sites. In the corresponding As-rich crystals the inverse situation is observed with dominating neutral donor−exciton recombination at 1·5138 eV and very weak or no acceptor−exciton emission at 1·5120 eV. These observations demonstrate that growth under Ga-rich conditions leads to an increased incorporation of amphoteric dopants on acceptor (As) sites, and are consistent with the Ge-doping results of Cho and Hayashi (1971b). Secondly, the 1·407 eV spectrum, which we associate with an isolated manganese acceptor on a Ga site, is quite strong in crystals grown under As-rich conditions but invariably much weaker in layers grown under Ga-rich conditions. Again this is consistent with the conclusion that growth under Ga-rich conditions favours the incorporation of impurities on As-sites.

Of the four elements, C, Si, Ge, and Mn, only C should be considered as a residual impurity generally associated with the MBE growth process. In the spectra of figure 2 we attribute the presence of Si and Ge to contamination resulting from previous evaporation of these elements in the system, while the Mn is believed to originate from inadvertent heating of the stainless steel rods that support the evaporation ovens. Crystals grown subsequently in a newly assembled growth system in which molybdenum rods were used to support the ovens showed only carbon associated features. For the present discussion, however, we choose to show spectra from a weakly contaminated system which most clearly illustrate the effects of changing growth conditions.

The identification of the bands labelled C in figure 2 with BA and DA pair transitions

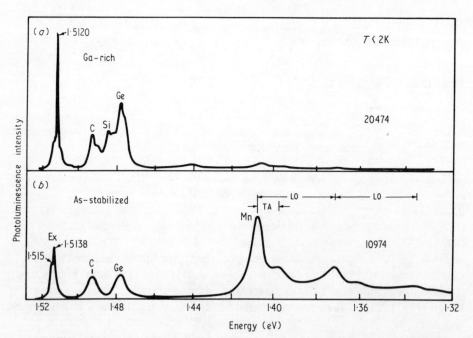

Figure 2. Photoluminescence of nominally undoped GaAs films grown under Ga-rich (*a*) or As-rich (*b*) conditions.

involving carbon acceptors is based on a comparison with the ionization energy for C reported by Dean (1974). In addition, we have also been able, in some cases, to resolve the characteristic (~ 0·2 meV) doublet splitting of the exciton peak at approximately 1·5120 eV. The presence of carbon in the grown layers is consistent with the fact that carbon is systematically detected by Auger analysis as a contaminant on initially clean GaAs surfaces after long exposure to the vacuum ambient.

Preliminary attempts were made to increase the concentration of C in the crystals by deliberate introductions of CO and C_2H_4 in the growth zone during deposition. These experiments proved inconclusive as only rather weak increases in the strength of the pair and exciton transitions involving C respective to that of the other spectral features was observed. In addition no change in the electrical behaviour of the layers was apparent. The result is not unexpected in view of the very low sticking coefficients of CO and C_2H_4 on GaAs at the growth temperature. The precise source of carbon contamination during growth is therefore still unidentified.

4.2. Electrical properties

Undoped layers grown from elemental As and Ga sources are generally p-type. This is in contrast with the results reported (Cho 1971) for films grown using undoped GaAs as a source for As where the residual conductivity is n-type, presumably as a result of Si contamination originating from the GaAs source. We find residual room temperature hole concentrations from about 1×10^{15} to about $1 \times 10^{16} \, cm^{-3}$. In one growth system and for crystals grown under As-rich conditions the p-type conductivity is invariably controlled by a deep acceptor at about 110 meV which we associate with the presence of Mn in the layers. In another growth system Mn is present in the crystals at much lower levels and the Hall data do not show the 110 meV controlled freeze-out at low temperatures. Plots of carrier concentration $p = (eR_H)^{-1}$ and mobilities $\mu_h = R_H/\rho$ against temperature are shown in figures 3 and 4 for two nominally undoped samples that are typical of the two growth systems. Where indicated the Hall curves have been fitted with the usual one acceptor-level expression:

$$p(p + N_D)/(N_A - N_D - p) = (m^*/g) \exp(-E_A/kT)$$

using the parameters listed in the figures.

In both 'undoped' samples of figure 3, as well as in all other undoped crystals measured, the carrier type changes from p to n at temperatures in the range from 100 to 200 °C. This transition to n-type suggests the presence of a moderately deep and presently unidentified electron trap. All data reported here have been limited to the temperature range where the influence of two carrier effects on the transport properties can be neglected.

The acceptor ionization energy $E_A = 97$ meV deduced from the fit to the Hall data in figure 3 corresponds closely to the binding energy of the centre responsible for the deep 1·407 eV band observed in luminescence of crystals from the same growth system. This supports the DA origin of the 1·407 eV band observed in luminescence. In crystal 61273 which does not show the characteristic freeze-out the Hall data suggest the presence of a shallower level with energy around 30 meV, consistent with the presence of C in these crystals.

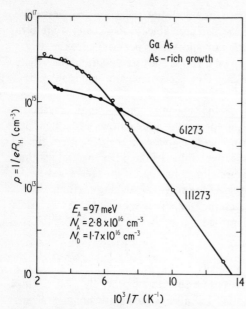

Figure 3. Hole concentration as a function of $1/T$ for crystals from a Mn-contaminated (111273) or uncontaminated (61273) system.

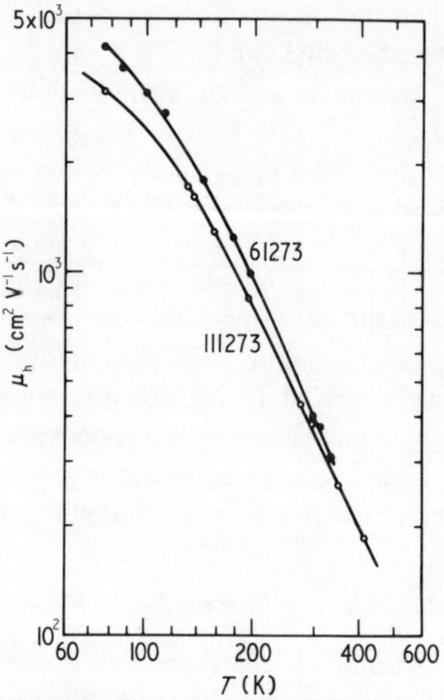

Figure 4. Mobility as a function of T for samples of figure 4.

The temperature dependence of the mobility shown in figure 4 for the same crystals as in figure 3 is typical of good-quality GaAs. The highest 77 K mobility measured in crystals grown under As-rich conditions is $4200\ \mathrm{cm^2\,V^{-1}\,s^{-1}}$; for comparison the highest 77 K mobility reported to data for high purity $p \sim 6 \times 10^{13}\ \mathrm{cm^{-3}}$ GaAs by Zschauer (1973) is $9700\ \mathrm{cm^2\,V^{-1}\,s^{-1}}$.

Electrical measurements confirm the dependence of Mn-acceptor incorporation upon the As/Ga ratio as observed in luminescence. Clear evidence of this is given in figure 5 where we compare two crystals grown in successive runs in the same system and under identical conditions except for a change in As-flux. The As-rich sample shows the usual dominance of the Mn-acceptor typical for this growth system. The Ga-rich crystal on the other hand shows essentially no carrier freeze-out associated with the acceptor level at about 113 meV. Mobilities of the Ga-rich crystals are generally slightly higher than those of the As-rich samples and their temperature dependence follows the behaviour shown in figure 4.

5. Mn doping

The initial identification of Mn as the impurity responsible for the 113 meV deep luminescence band and the corresponding carrier freeze out was based on the similarity between the luminescent spectra of the present crystals and those reported for Mn-diffused (Lee and Anderson 1964) samples as well as by comparison with the known activation energy for Mn (Vieland 1962). Other possibilities were considered, including the notion that a Ga-vacancy defect might be involved as was originally suggested by Cho and Hayashi (1971a). The difference observed between samples from different

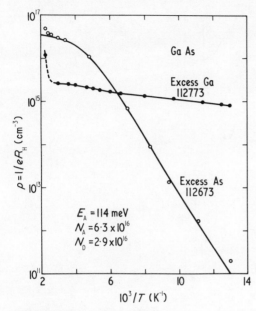

Figure 5. Hole concentration as a function of $1/T$ for crystals grown in successive runs and under identical conditions except for a change in the As/Ga flux ratio.

growth systems however argues convincingly against a simple vacancy assignment although some forms of intrinsic defect–impurity complex cannot be ruled out on that basis.

The variation of the shape and intensity of the 1·407 eV luminescence band with excitation energy is consistent with the behaviour expected for transitions involving a deep isolated acceptor. Figure 6 shows how the high energy band of the spectrum resolves into several components as the excitation energy is reduced and gives evidence for the participation of conduction band–acceptor (1·4065 eV) and donor–acceptor

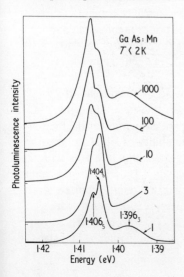

Figure 6. Photoluminescence of Mn-associated feature as a function of excitation level.

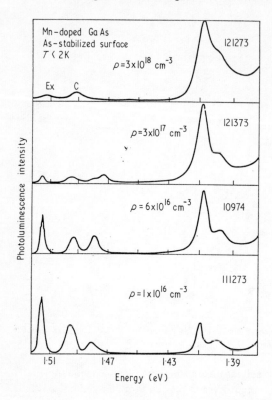

Mn–doped Ga As
As–stabilized surface
$T < 2K$

$\rho = 3 \times 10^{18}$ cm^{-3} 121273

Ex   C

$\rho = 3 \times 10^{17}$ cm^{-3} 121373

$\rho = 6 \times 10^{16}$ cm^{-3} 10974

111273
$\rho = 1 \times 10^{16}$ cm^{-3}

Photoluminescence intensity

1·51   1·47   1·43   1·39

Energy (eV)

Figure 7. Photoluminescence of Mn-doped GaAs layers arranged in order of increasing doping level.

pair (1·4046 eV) recombination. The analysis of these data results in an acceptor binding energy of $112·7 \pm 0·5$ meV.

In order to further characterize the Mn-acceptor as well as to investigate its possible use for p-type doping as an alternative to the use of Mg (Cho and Panish 1972) a number of growth runs were made using elemental Mn as a dopant. For all runs made under As-rich conditions, hole concentrations and 1·407 eV luminescence band intensities increased with increasing Mn addition. The luminescence results are illustrated in figure 7 for four layers with 296 K hole concentrations ranging from 1×10^{16} cm^{-3} to 3×10^{18} cm^{-3}. For crystals grown under Ga-rich conditions the Mn-associated luminescence remained very weak throughout.

The highest 296 K carrier concentration that could be achieved without affecting the growth morphology was $p \simeq 1 \times 10^{18}$ cm^{-3} in the present system. For higher Mn fluxes the surface appearance deteriorated rapidly. It is not clear whether the limit at approximately 1×10^{18} cm^{-3} represents the maximum solubility of Mn at the growth temperature of about 580 °C or whether the relatively high Mn flux interferes with the growth process. Hole activation energies in the Mn-doped crystals decrease with increasing doping levels above the 10^{17} cm^{-3} range, consistent with the results reported by Blakemore *et al* (1973). Room temperature mobilities measured for crystals with $p = 2 \times 10^{17}$ cm^{-3} and $1·2 \times 10^{18}$ cm^{-3} are 270 and 180 cm^2 V^{-1} s^{-1} respectively, as compared to mobilities in the range 380–400 cm^2 V^{-1} s^{-1} found in undoped crystals (figure 4).

6. Conclusions

The properties of undoped GaAs layers grown from Ga and As_4 beams appear to be comparable to good quality GaAs grown by liquid phase or vapour phase epitaxy techniques. The major uncontrolled impurity in the crystals is C; the combined electrical and luminescence results suggest that, even under As-rich growth conditions, C is principally incorporated as an acceptor on As-sites. Finally, it has been demonstrated that hole concentrations up to about $1 \times 10^{18} \, cm^{-3}$ can be realized with Mn doping while maintaining excellent growth morphology.

Acknowledgments

We thank J W Robinson and L Kopf for their expert technical assistance.

References

Arthur J R 1974 *Surface Sci.* **43** 449
Blakemore J S, Brown W J, Stass M L and Woodbury D A 1973 *J. Appl. Phys.* **44** 3352
Cho A Y 1970 *J. Appl. Phys.* **41** 2780
—— 1971 *J. Vac. Sci. Technol.* **8** S31
Cho A Y and Casey H C 1974 *Appl. Phys. Lett.* **25** 288
Cho A Y and Hayashi I 1971a *Solid St. Electron.* **14** 125
—— 1971b *J. Appl. Phys.* **42** 4422
Cho A Y and Panish M B 1972 *J. Appl. Phys.* **43** 5118
Dean P J 1974 *J. Phys. (Paris)* **35** C127
Dingle R, Wiegmann W and Henry C H 1974 *Phys. Rev. Lett.* **33** 827
Lee T C and Anderson W W 1964 *Solid St. Commun.* **2** 265
Vieland L J 1962 *J. Appl. Phys.* **33** 2007
Zschauer K-H 1973 *Proc. 4th Int. Symp. Gallium Arsenide and Related Compounds* (London and Bristol: Institute of Physics) p3

Liquid phase epitaxy apparatus for thin layers and multiple layers

E Bauser, M Frik, K S Loechner and L Schmidt

Max-Planck-Institut für Festkörperforschung, Stuttgart, Germany

Abstract. A liquid phase epitaxy apparatus, designed for the growth of thin layers and multiple layers of compound semiconductors, is presented. The apparatus utilizes thin melts. Sliding parts in the boat arrangement are avoided as the melts are transported to the substrate and removed from the substrate by centrifugal forces. For growing multiple layers, two or more melts of different composition can be applied. The principle of operation of the apparatus and the properties of GaAs layers grown in the system are described.

1. Introduction

In the 15 years since the technique of liquid phase epitaxy was started, three basic types of growth apparatus have been developed: the 'tipping' furnace (Ditric and Nelson 1960, Nelson 1963, Hicks and Manley 1969, Hicks and Greene 1971, Kressel and Nelson 1973 and references therein), the 'dipping' system (Woodall *et al* 1969, Kressel and Nelson 1973) and the 'multibin' system (Alferov *et al* 1969, Panish *et al* 1970, Kressel *et al* 1970, Miller *et al* 1972, Dawson 1973, Casey *et al* 1974). These basic types of growth equipment, and of course modifications thereof (Lien and Bestel 1973) are in use in many countries. As liquid phase epitaxy offers unique advantages, for instance in the preparation of materials for light emitting devices, the epitaxial techniques have become more and more adapted to the special requirements of device technology and there are trends towards the high capacity liquid phase apparatus now (Saul *et al* 1973).

All three systems have proved themselves in many applications. However, there are still a few problems which have not been completely solved. The most important are:

(i) the complete removal of the melt after the termination of growth of an epitaxial layer is difficult
(ii) the surfaces are scratched while the solutions are removed
(iii) the production of very thin layers.

We have looked for an alternative possibility to overcome these difficulties while maintaining the technique of liquid phase epitaxy. The idea is that the crucible should be without any moving parts. The forces to be exerted on the solution should exceed gravity in order to secure rapid supply and removal of the solution. Any forces can be used to move the solutions in the crucible, for example, excess pressure† or centrifugal

† Suggested by C Hilsum (private communication).

forces. A number of possibilities exists for realizing such a system. Among these we chose one which applies centrifugal forces, as this was regarded as the easiest to set up in a short time.

2. Description of the apparatus

The growth system is a vertical one (figure 1). It has the usual constituents: a furnace and a quartz tube. The quartz tube can be evacuated or filled with hydrogen or any other inert gas. The quartz tube can be disconnected beneath and above the furnace.

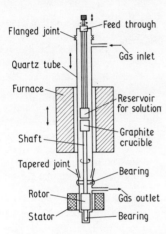

Figure 1. Schematic cross section of the liquid phase epitaxy system. The crucible and the reservoir for solutions are adjacent in the centre of the furnace. For loading and unloading the crucible, the furnace is pulled up. Then the quartz tube is disconnected at the tapered joint and the upper parts of the quartz tube containing the reservoir is pulled up. A photograph of the apparatus with the furnace and the upper parts of the quartz tube removed is shown in figure 2.

A graphite crucible is fixed at the top of a shaft which extends into the centre of the furnace. The lower end of the shaft is connected to a rotor. The rotor is held between two bearings inside the quartz tube. A stator, which is outside the quartz tube, drives the rotor with variable frequencies between 0 and 3000 cycles per minute. The furnace and the quartz tube can be pulled up for loading and unloading of the crucible. Figure 2 shows a photograph of the apparatus; the furnace and the upper part of the quartz tube have been removed.

A schematic cross section and a top view of the crucible are shown in figure 3. The crucible is symmetrical with respect to the rotational axis. It consists of an inner and an outer part. In the inner part there are small oblong stockbins for the solutions. These stockbins are adjacent to the centre of rotation and have slotted openings towards a narrow gap turned away from the axis. Substrates can be fixed on one or on both sides of the gap. The outer part of the crucible is cylindrical in shape and is used as a big container for used solution. The cover which is used for the crucible is omitted in figure 3.

Figure 2. The boat is cylindrical in shape and fastened at the upper end of a rotatable shaft. The surrounding quartz tube and the furnace have been pulled up.

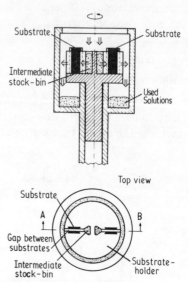

Figure 3. Schematic cross section of the crucible without covers (sectional view and top view). The arrows in the upper part of the sketch indicate the motion of the solution.

3. Typical run

In the growth of an epitaxial layer, for instance GaAs on a GaAs substrate, the first step is to place the gallium and gallium arsenide into the small stockbins and to heat the solution in a hydrogen atmosphere for about two hours. The high surface tension of the liquid gallium prevents the solutions from escaping through the slotted openings of the stockbins. After cooling to room temperature, the substrates are inserted into the inner crucible. The substrates in this special case are bevelled so that they fit into dovetailed grooves in the boundaries of the gaps. After the substrates are inserted, the cover is put on top of the crucible. Then, the unit is filled with hydrogen and heated to the desired growth temperature.

When the crucible assembly has reached thermal equilibrium, the system is rotated at a frequency sufficiently high for the centrifugal forces to exceed the resultant force originating from the surface tension and the interfacial tension. The solution then enters the gap and passes the substrate surfaces. While traversing the substrates, epitaxial layers are deposited from these 'thin melts'. Having passed the substrates, the solution leaves the inner crucible and enters the outer one. This acts as a collector for the solutions used. When the stock bins are empty, the rotational speed of the crucible is raised to remove melt residues from the surfaces of the grown layers. After this step the crucible is stopped. While it is decelerated, the used solution drops into the container at the bottom of the outer crucible.

The stock bins can be filled manually, but also they can be filled with a preheated solution from a reservoir, which is held above the crucible as is shown in figure 1. This reservoir may have containers for different solutions. By a mechanism, which can be actuated with a milled head at the top end of the quartz tube, the stock bins can be filled sequentially with different solutions as indicated by the upper arrows in figure 3. Every time, after the stock bins have been filled, the crucible is rotated. In this way it is possible to grow multiple layers using this kind of crucible. Figure 4 shows a photograph of the crucible with its cover removed. It has two stock bins and two substrates can be inserted. The width of the two gaps can be adjusted for experiments. Gap widths between 0·3 and 0·6 mm were used for the experiments. Figure 5 shows the crucible disassembled.

Figure 4. Graphite crucible with two triangle-shaped stock bins for solutions. The cover of the crucible has been removed. In this special case provisions are made for only one substrate per gap. The width of the gap can be adjusted (0−2 mm) and fixed by screws.

Figure 5. Crucible taken apart.

The time during which the layers are allowed to grow is only a fraction of a second. As the interface kinetics are probably very fast (Small and Barnes 1969, Minden 1970, Rode 1973), the growth time is still long compared to the time constants of the atomic processes taking place at the interface. The growth rates are expected to be much higher than in all cases where unstirred solutions are used. The growth will probably not be diffusion limited (Small and Barnes 1969, Rode 1973). It might be expected that the doping profiles of layers grown in a system like this differ from the profiles of layers grown out of unstirred solutions. Parameters with which the growth can be varied are the temperature, the width of the gap, and the speed of rotation of the crucible.

4. Results

We have grown a variety of layers in our first series of experiments. The thicknesses of the epitaxial layers were below 1 μm. For one of the single layers, the thickness has been determined by SEM to be 0·65 μm using a cleavage plane. For a double layer, the thickness was found out to be 1·95 μm by the same method†. We have also grown much thinner layers. As proof of the existence of these thin layers see figures.

Figure 6. SEM photograph showing the surface structure of the epitaxial layer grown by the centrifugal methŏd.

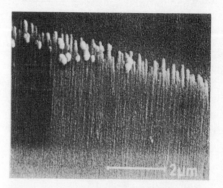

Figure 7. SEM photograph of the edge of the layer shown in figure 6. The white region is the crossing over from the epitaxial layer to the substrate.

Figure 6 shows a SEM photograph of the surface of such a thin epitaxial layer. The edge of this layer is shown in figure 7. The white region is the step between the surface of the epitaxial layer and the substrate surface. The upper part of the photograph (beyond the white region) shows the substrate surface without a layer. From this photograph, the height of the step can be estimated to be much less than 0·5 μm. Figure 8 shows the same phenomenon (the passing from the layer to the substrate) but for a very rough and nonhomogeneous layer.

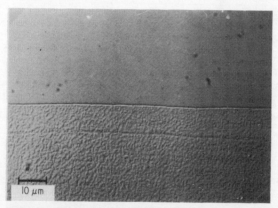

Figure 8. The lower part of the photograph shows an example of a very rough and nonhomogeneous layer grown by the centrifugal method. The upper part shows the substrate surface without a grown layer.

† We are grateful to W Gramann for taking these measurements.

5. Conclusions

Preliminary experiments with the apparatus described have shown that, applying centrifugal forces, thin layers and multiple layers of compound semiconductors can be grown without any sliding parts in the boat arrangement. The thin melts used can be removed uniformly over the entire substrate after the termination of layer growth. Unwanted impurity transport and scratching of the surfaces by a slider are avoided.

Acknowledgments

We would like to thank H J Queisser for continued encouragement and valuable advice and E Schoenherr for helpful cooperation. Many colleagues of the MPI deserve our thanks for active and quick help during the sometimes difficult and hectic phases of this work.

References

Alferov Zh I, Andreev M V, Korolkov E I, Portnoi E L, Tretyakov D N 1969 *Sov. Phys.—Semicond.* 2 1289
Casey H C, Panish M B, Schlosser W D and Paoli T L 1974 *J. Appl. Phys.* 45 322
Dawson L R 1973 *143rd Electrochem. Soc. Meeting, Chicago* RNP 325
Ditric N H and Nelson H 1960 *RCA Engineer* 6 19
Hicks H G B and Manley D F 1969 *Solid St. Commun.* 7 1463
Hicks H G B and Greene P D 1970 *Proc. 3rd Int. Symp. Gallium Arsenide and Related Compounds* (London and Bristol: Institute of Physics) p92
Kressel H and Nelson H 1973 *Physics of Thin Films* (New York and London: Academic) 7 115
Kressel H, Nelson H and Hawrylo F Z 1970 *J. Appl. Phys.* 41 2019
Lien S Y and Bestel J L 1973 *J. Electrochem. Soc.* 120 1571
Miller B I, Pikas E, Hayashi I and Capik R L 1972 *J. Appl. Phys.* 43 2817
Minden H G 1970 *J. Crystal Growth* 6 228
Nelson H 1963 *RCA Rev.* 24 603
Panish M B, Hayashi I and Sumski S 1970 *Appl. Phys. Lett.* 16 326
Rode D L 1973 *J. Crystal Growth* 20 13
Saul H R H, Lorimor O G, Dawson L R and Paola C R 1973 *143rd Electrochem. Soc. Meeting, Chicago* RNP 327
Small M B and Barnes J F 1969 *J. Crystal Growth* 5 9
Woodall J M, Rupprecht H and Reuter W 1969 *J. Electrochem. Soc.* 116 899

Growth of $In_{1-x}Ga_xSb$ by liquid phase epitaxy

Hidejiro Miki, Kazuaki Segawa, Mutsuyuki Otsubo, Kiyoshi Shirahata and Keiji Fujibayashi

Central Research Laboratory, Mitsubishi Electric Corporation, Itami, Hyogo, Japan

Abstract. High purity n-type GaSb and $In_{1-x}Ga_xSb$ epitaxial layers have been grown by liquid phase epitaxy. The influence of such growth conditions as growth temperature, substrate orientation, heat treatment of the solution and the presence of dopants on the carrier concentration of the $In_{1-x}Ga_xSb$ (mainly the GaSb) have been investigated. Outside of the InSb rich range, where the contribution due to thermal excitation dominates, an epitaxial layer in the order of $10^{15}\,cm^{-3}$ was obtained in the whole range of $In_{1-x}Ga_xSb$. Current oscillation was observed and the corresponding threshold field was estimated to be roughly $500-800\,V^{-1}\,cm^{-1}$.

1. Introduction

Several mixed III—V semiconductor compounds have certain advantages over pure binary compounds. For example, $In_{1-x}Ga_xSb$ is a promising microwave oscillator material because of its band structure in a particular range of composition x. Hilsum and Rees (1970) pointed out that $In_{1-x}Ga_xSb$ would show the possibility of increasing the peak—valley ratio in the drift velocity versus electric field characteristics. However, obtaining the high purity n-type $In_{1-x}Ga_xSb$ that is necessary to cause the Gunn effect has proved very difficult.

GaSb and InSb are basic binary compounds for the study of the $In_{1-x}Ga_xSb$ ternary compound. High purity n-type InSb has been obtained, but high purity n-type GaSb has not. According to reports presented in the past, all the undoped GaSb grown without intentional doping with donor impurity at time of crystal growth has been of the p-type. This is the result, it is thought, of a lattice defect, such as an antimony vacancy or an antistructure with Ga on the Sb site, but whether or not it is the only mechanism that produces the p-type carrier is still open to question.

This paper deals with investigations of substrate orientation, heat treatment of the solution and the presence of dopants on the carrier concentration of the $In_{1-x}Ga_xSb$ epitaxial layer (mainly the GaSb epitaxial layer), and growth of high purity n-type $In_{1-x}Ga_xSb$ ternary compounds by liquid phase epitaxy.

2. Experimental method

The equipment used to produce epitaxial growth is a tilting tube furnace similar to that used for growing high purity GaAs crystals (Otsubo *et al* 1973), since that system is very simple. The reaction tube was made of high purity quartz. The ambient gas in the furnace was flowing palladium-diffused hydrogen. The ternary phase diagram for

In–Ga–Sb was determined experimentally by the method reported by Antypas (1972). The solidus data was obtained by growing an In$_{1-x}$Ga$_x$Sb epitaxial layer on GaSb and InSb substrates. The composition of each growth layer was determined with an electron probe microanalyser. Lattice constant was measured by x-ray diffraction, assuming that the alloy system obeys Vegard's law. When the composition of the growth layer was inhomogeneous, the part near the interface of the growth layer and the substrate was considered for analytical purposes, as the epitaxial layer.

Epitaxial growth of GaSb was tried on an undoped GaSb substrate by using Ga–Sb solution with a surplus of Ga. Epitaxial growth of In$_{1-x}$Ga$_x$Sb was done on GaSb and InSb substrates using an In–Ga–Sb solution. In case of need, Te was added to the solution as a donor impurity. The electrical properties of the epitaxial layer were obtained by measuring the Hall effect after removing the substrate. Ohmic contacts to the epitaxial layer were obtained by soldering an In dot containing 1% Te for an n-type sample and pure In for a p-type sample. The current–voltage characteristics of the n-type layer were measured using the pulse method. The samples were fabricated in a planar structure with a thickness of 100 μm, width 300 μm and a distance of 400 μm between the ohmic contacts. The current was measured by a 1 Ω resistor connected in series with the sample, using 100 Hz voltage pulses of 100 ns duration.

3. Results and discussion

Figure 1 shows the liquidus data of a ternary phase diagram of In–Ga–Sb, the lines indicating the result calculated by Blom and Plaskett (1971) on the basis of regular solution theory, and circles representing the data obtained from the present experiments.

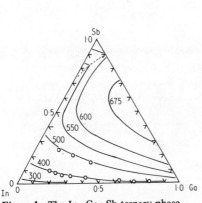

Figure 1. The In–Ga–Sb ternary phase diagram.

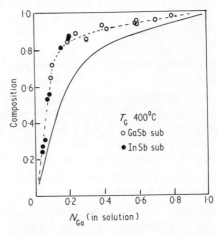

Figure 2. The 400 °C solidus isotherm of the In–Ga–Sb system.

Close agreement is seen between theory and experiment. Figure 2 shows a solidus data extrapolated from experimental data (broken curve) and the calculated solidus isotherm of Blom and Plaskett (solid curve), but here, the agreement between theory and experiment is not as satisfactory as that for the liquidus data. Similar discrepancies have been reported by Antypas (1972). It is supposed that regular solution theory has its limits because of the large difference in the lattice constants of InSb and GaSb. Figure 3 shows the temperature dependence of the hole concentration of a GaSb epitaxial layer for three differently oriented substrates. As the starting temperature of growth decreases, the hole concentrations of the epitaxial layer decrease in the order (111)B > (111)A > (100) plane. Hojo and Kuru (1973) reported similar temperature dependence of the hole concentration of Ga–Sb epitaxial layer grown on (111) plane. They have obtained

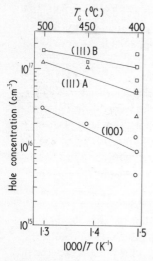

Figure 3. The temperature dependence of the hole concentration of a GaSb epitaxial layer for three differently oriented substrates.

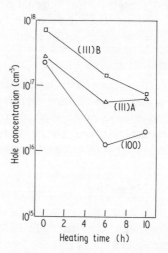

Figure 4. Heating effect of Ga–Sb solution.

a formation energy of the defect as $1 \cdot 1$ eV. In the present experiments, however, the hole concentration appears to be affected by an impurity judging from the scattered values of hole concentration about the same growth temperature. Figure 4 shows the influence of GaSb solution on the hole concentration of epitaxial layer. The first layer was grown from an unbaked solution before epitaxial growth, and subsequent layers from the same solution after heating at 800 °C in a palladium-diffused hydrogen gas flow. When heat treatment was carried out, the hole concentration of the epitaxial layer decreased. The effect of the heat treatment is also confirmed by the cooling rate dependence of the hole concentration of the epitaxial layer. The hole concentration decreased with an increase of the cooling rate when an epitaxial growth was obtained without preheating of solution. However, in heat-treated solutions, the hole concentration of the epitaxial layer showed a lower dependence on the cooling rate. These results suggest that a foreign impurity was more readily incorporated into the solution when heat treatment was not done.

When the growth starting temperature was 400 °C and the substrate orientation (100), and a special ingot of GaSb was used as the Sb source, the use of preheated solutions made it possible to obtain, on occasion, an n-type GaSb epitaxial layer without intentional doping of a donor impurity. The maximum purity obtained for n-type epitaxial layers was $3\cdot8 \times 10^{15}\,\mathrm{cm}^{-3}$ in electron concentration and $7700\,\mathrm{cm}^2\,\mathrm{V}^{-1}\,\mathrm{s}^{-1}$ in electron mobility at room temperature. The formation of an n-type layer is attributed to the following causes:

(i) The vacancy concentration becomes lower with a decrease in growth temperature.
(ii) The concentration of acceptor impurity incorporated into the epitaxial layer is lowest on the (100) oriented substrate.
(iii) The acceptor impurity was evaporated by the heat treatment of the solution.
(iv) An undoped n-type layer could only be obtained when the special GaSb ingot was used as Sb source, suggesting that the concentration of donor impurity in that ingot was higher than in others. However, the difference of concentration of donor impurity should not be so large, given the fact that in the n-type layer electron concentration is low and mobility is large.

Figure 5 shows the doping characteristics of silicon and oxygen dopants. Oxygen doping was achieved by adding Ga$_2$O$_3$ to the growth solution. The gradient of the line is about 2 for silicon and 0·4 for oxygen. This contrasts with usual liquid epitaxy doping behaviour where the gradient is approximately one. In the doping experiments on silicon and oxygen, there are several practical difficulties. One is the rather low solubility of oxygen in Ga–Sb solution in the operating temperature range and the other is due to the narrow range of doping density since the background density of impurity is high.

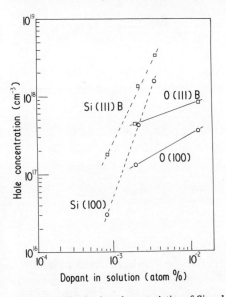

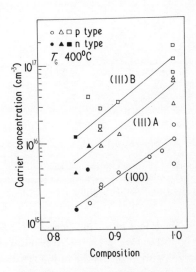

Figure 5. The doping characteristics of Si and oxygen. (Growth temperature 400 °C.)

Figure 6. The carrier concentration of In$_{1-x}$Ga$_x$Sb grown on a GaSb substrate.

Rather erratic results were obtained as shown in figure 5, but it emerged that silicon and oxygen were acceptor impurities in GaSb layers. A Si concentration as high as 47 ppm was detected in the undoped impure p-type GaSb epitaxial layer having hole concentration of $8 \times 10^{17}\,\mathrm{cm}^{-3}$ indicating that one origin for acceptor is silicon.

Epitaxial growth of $In_{1-x}Ga_xSb$ was carried out on GaSb and InSb substrates. At alloy compositions x where $0.84 \leqslant x \leqslant 1$, it was possible to grow single crystalline $In_{1-x}Ga_xSb$ epitaxial layers on a GaSb substrate and figure 6 shows the carrier concentration, which decreases with a decrease in x. The orientation of the substrate dependence of carrier concentration is similar to that case of GaSb layers grown on a GaSb substrate and becomes lower in the order of $(111)B > (111)A > (100)$ plane. High purity n-type layers with electron concentrations in the range $1-5 \times 10^{15} cm^{-3}$ and electron mobility between 10 000 and 14 000 $cm^2 V^{-1} s^{-1}$ at 300 K were obtained without intentional doping at a composition x of 0.84. At alloy compositions where $0 \leqslant x \leqslant 0.86$, single crystalline $In_{1-x}Ga_xSb$ epitaxial layers were obtained on an InSb substrate. When an InSb substrate was used, epitaxial growth for a wide range of compositions was possible in spite of the troublesome problem of the difference in lattice parameter between overgrowth and substrate. Figure 7 shows how the carrier concentration of an $In_{1-x}Ga_xSb$ epitaxial layer grown on an InSb substrate decreases with an increase in x. The effect of substrate orientation on epitaxial layer carrier concentration is not so clear as that of the GaSb substrate. Outside the InSb rich range,

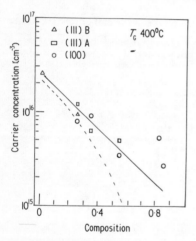

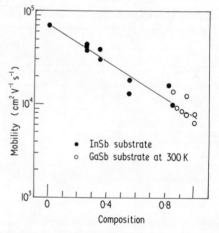

Figure 7. The carrier concentration of $In_{1-x}Ga_xSb$ grown on an InSb substrate. (Broken curve, intrinsic carrier.)

Figure 8. The electron mobility of $In_{1-x}Ga_xSb$ grown on GaSb and InSb substrates (at 300 K).

where the effect of intrinsic carrier due to thermal excitation must have been taken into consideration, an epitaxial layer in the order of $10^{15} cm^{-3}$ was obtained without doping. Figure 8 shows the mobility of the whole range of $In_{1-x}Ga_xSb$ layers grown on GaSb and InSb substrates, some of which was grown doped with Te at the epitaxial growth. In figure 8, electron mobility is seen to vary linearly with x. Similar results have been obtained by Coderre and Woolley (1969) in $In_{1-x}Ga_xSb$ crystals grown by the horizontal Bridgman technique. This would indicate that alloy scattering has negligible effects in these materials.

    The current–voltage characteristics of n-type layers with composition ranging from 0.91 to 0.55 were measured. Current oscillation was observed and was postulated to be caused by negative differential mobility. The corresponding threshold field was estimated to be roughly 500–800 V cm^{-1}, as shown in table 1. These values are close to that of previous reports (McGroddy *et al* 1969, Hojo and Kuru 1974), however, the composition dependence of the threshold field that was calculated theoretically by Sakai *et al* (1974) was not confirmed. The scatter of observations is due to the inhomogeneous layers.

Table 1. Threshold field for current oscillation

x	n (cm^{-3})	μ (cm^2 V^{-1} s^{-1})	E_{th} (V cm^{-1})
0·91	3.2×10^{15}	4900	600
0·87	4.7×10^{15}	13800	800
0·87	4.7×10^{15}	13800	500
0·84	1.4×10^{15}	10300	800
0·82	5.6×10^{15}	15800	550
0·65	7.1×10^{15}	7780	600
0·55	2.0×10^{15}	12200	800

4. Conclusion

High purity n-type GaSb and In$_{1-x}$Ga$_x$Sb epitaxial layers have been grown by the liquid phase epitaxy. An epitaxial layer in the order of 10^{15} cm^{-3} was obtained in the whole range of In$_{1-x}$Ga$_x$Sb outside the InSb rich range. Current oscillation was observed and the corresponding threshold field was 500–800 V cm^{-1}.

Acknowledgments

This research has been sponsored by the Japanese Science and Technology Agency. The authors wish to express their appreciation to Dr Mitsui and Messrs Kotani and Horiuchi for performing the measurements of current–voltage characteristics.

References

Antypas G A 1972 *J. Crystal Growth* **16** 181
Blom G M and Plaskett T S 1971 *J. Electrochem. Soc.* **118** 1831
Coderre W M and Woolley J C 1969 *Can. J. Phys.* **47** 2553
Hilsum C and Rees H D 1970 *Electron. Lett.* **6** 277
Hojo A and Kuru I 1973 *J. Japan. Soc. Appl. Phys.* **43** suppl 226
—— 1974 *Electron. Lett.* **10** 61
McGroddy J C, Lorenz M R and Plaskett T S 1969 *Solid St. Commun.* **7** 901
Otsubo M, Segawa K and Miki H 1973 *Japan. J. Appl. Phys.* **12** 797
Sakai K, Adachi Y and Ikoma T 1974 *Nat. Conv. Rec. IECE Japan* 446

Vapour phase epitaxial growth of GaAs for multiple applications

L Hollan

Laboratoires d'Electronique et de Physique Appliquée, 3, Avenue Descartes, 94450 Limeil-Brevannes, France

Abstract. An improved $Ga-AsCl_3-H_2$ VPE method for growing multi-layer structures including n^- layers ($n \leqslant 10^{14} cm^{-3}$) is described. The advantages of a heat pipe furnace are discussed. The influence of large variations of the $AsCl_3$ molar fraction on the doping level and the growth rate are studied under different growth conditions; it allows a good control of the doping level between 5×10^{12} and $2 \times 10^{16} cm^{-3}$. This process, together with a sulphur doping source, allows the growth of multilayer structures for millimetre Gunn diodes, nuclear detectors, Gunn amplifiers (TEA) and FET. Some device performances are given.

1. Introduction

The necessity has arisen of growing multilayer structures for the realization of several GaAs devices, including n^- doped layers (with $n^- < 10^{14}$). For instance, for transferred electron Gunn amplifiers (TEA) one needs a well-controlled notch in the $n \simeq 10^{14}$ range at the anode, and for FET application a buffer layer with $n < 10^{14} cm^{-3}$. Such doping levels have already been obtained in VPE (Di Lorenzo 1972, Wolfe *et al* 1970) by reducing the residual doping level using high $AsCl_3$ molar fractions (MF). Nevertheless, the variation of n and of the growth rate as a function of $AsCl_3$ MF under different growth conditions is not yet very well known, and the reproducibility in the $n \leqslant 10^{14}$ range does not seem to be good. On the other hand, the growth of multilayer structures with $n < 10^{14}$ has not been reported up to now.

Our previously described method (Hollan and Mircea 1973) is well fitted to the growth of such structures in good conditions, and we have used it with some improvements for the realization of various devices.

2. Experimental

2.1. Utility of heat pipe furnaces

The principle of our method is to use, in a classical $AsCl_3-H_2-Ga$ system, a long deposition zone divided into successive differently doped zones. The multilayer structures are obtained in stationary conditions by moving the substrates from one of these zones to the other (Hollan and Mircea 1973). Thus, in addition to the constant temperature gallium source zone (the necessity of which has been demonstrated by Shaw 1971), we need a long constant temperature deposition zone (ie 40 cm). Such a temperature profile is advantageously obtained by using two heat-pipe furnace elements joined end to end. These elements are heated by wound coaxial resistances each one being driven by a classical PID temperature controller. The main advantages of such a furnace compared to a classical one are the following:

(i) The desired temperature gradient is obtained without any calibration and for any values of the source and deposition temperatures (T_S, T_D). Furthermore, the gradient is insensitive to the efficiency of heat insulation. As an example, we have shown in figure 1 the temperature gradient obtained by two heat pipe furnaces containing sodium (manufactured by Sodern†), of 60 mm internal diameter and respectively 35 and 50 cm long. The temperatures are measured in a small quartz tube near the walls of the furnace.

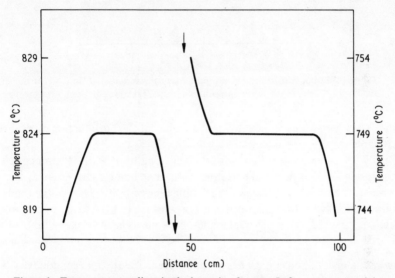

Figure 1. Temperature gradient in the heat pipe furnace. Left, source zone; right, deposition zone.

Owing to the solid angle of radiation, the gradient is, of course, slightly different (less sharp) in the axis of the epitaxial reactor.

(ii) At the end of the heating cycle, the temperature gradient is immediately reached without any possibility of over or underheating at any point in the reactor (the edges, or the transition between T_S and T_D zones etc). Experiments can start without a stabilization period, as soon as the temperature regulation begins. This is illustrated by figure 2(a) on which is shown a recording of the temperature of six regularly located thermocouples plotted against the time at the end of the heating cycle. (The distance between the first and the last thermocouple is 15 cm. In order to make the figure clear, we have artificially introduced a shift in the six recorded temperatures. Each thermocouple recorded exactly the same temperature.)

(iii) Any operation introducing a local temperature perturbation in the reactor (like introduction of the substrates) reacts on the whole furnace and as soon as the temperature regulator is set, the normal profile is re-established. This is illustrated on the right

† Sodern, Limeil-Brévannes, France.

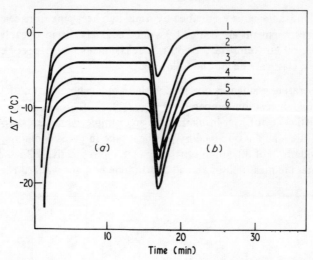

Figure 2. Temperature in the heat pipe furnace as a function of time: (*a*) at the end of the heating cycle, (*b*) after the introduction of the substrates in the hot reactor.

of figure 2(*b*) where we have recorded the influence of the introduction of a quite large substrate holder.

(iv) Finally, we can point out that, with this type of a furnace, the thermocouple is set in a tube inside the heat pipe and therefore has a very good thermal coupling with it. In this way we can easily get the best regulation allowed by the class of the temperature controller especially because the furnace is less sensitive to local external perturbations (convections, draught). These characteristics make this type of furnace also profitable in LPE growth.

However, a major disadvantage has been observed. The thermal conductivity being very poor below about 450 °C, during a normally quick heating cycle, important over-heating appears at the lower temperatures, resulting in stresses which, after hundreds of cycles, may crack the heat pipe elements. This problem has now been solved by Sodern and a good lifetime can be expected.

2.2. Experimental conditions and results

In all the experiments, the deposition zone is divided into two regions, the first one near the source for the n (or n⁻) layer, and the other one 25 cm downstream, just behind the additional doping source. The total flow rates were about $24 \, \mathrm{l \, h^{-1}}$ (except for very low $AsCl_3$ MF for which it was set at $50 \, \mathrm{l \, h^{-1}}$). The inner tube diameter was 40 mm. Conditions were maintained in such a way that no (or at any rate negligible) crystallization took place on the walls of the reactor. In the experiments performed to determine the influence of the $AsCl_3$ MF, we have previously grown an approximately 10 μm thick n⁺ buffer layer with MF $= 4 \times 10^{-3}$. The substrates (about 2×2 cm) are [100] oriented with 3° of misorientation (in the [110] Ga direction) and horizontally disposed in the tube. Then we observe typical thickness variations as a function of the position of the substrate in the direction of the flow, as shown in figure 3; the perpendicular variation is less pronounced, a slight increase being only observed near the edges.

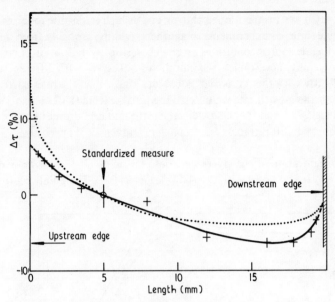

Figure 3. Typical growth rate variations on the substrates in the gas flow direction.

These variations are more important for MF $< 10^{-3}$, and they can be as large as 50% or more in this case. In order to be in the nearly constant thickness range, the measurements have always been taken at a point 5 mm from the upstream sample edge and far enough from the sides. Figure 4 shows the thickness variation plotted against AsCl$_3$ MF for different deposition temperatures; 770, 750 and 720. The corresponding source temperatures were 825, 825 and 790 or 770 °C. These curves correspond to an average over a large number of experiments. For the curve related to 770 °C, we have reported the observed spread of the measurements which is due to the thickness variation on the

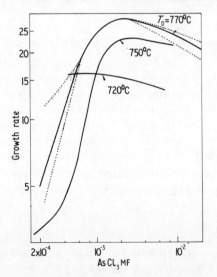

Figure 4. Standardized growth rate plotted against AsCl$_3$ MF at different temperatures.

substrate at low AsCl₃ MF (as mentioned above), and at high MF is due to growth rate variations from one series of experiments to another, resulting probably from not absolutely reproducible source conditions or small parasitic crystallization. The epitaxial surfaces are always shiny, nevertheless the morphology observed with differential contrast varies slightly with the AsCl₃ MF. For values higher than the ones corresponding to the maximum of the growth rate we observe the usual texture (Di Lorenzo 1972, Hollan and Schiller 1974), and for the lower values, the surface becomes textureless. We have never observed dull surfaces (due to pitting, Minden 1971) even at MF = 2×10^{-4}.

The carrier concentration of the layers were determined by $C-V$ measurements. Figure 5 shows the variation of n with the AsCl₃ MF at 750 °C. The points plotted in the

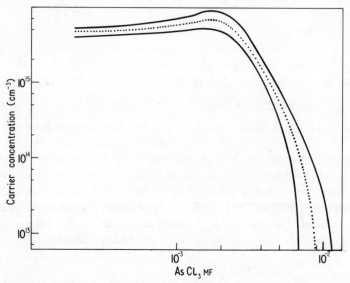

Figure 5. Carrier concentration against AsCl₃ MF at 750 °C. Spread of a large number of experiments.

figure correspond to a spread of experiments carried on over about two years especially in the $2 \times 10^{-3} < \text{MF} < 6 \times 10^{-3}$ range, and gives an idea of the reproducibility which is, of course, better in a single series of experiments. Taking only the mean value (this corresponds to the broken curve in figure 5), we have represented in figure 6 the results corresponding to the conditions of figure 4.

The compensation and the deep trap concentration plotted against AsCl₃ MF is studied by appropriate characterization techniques and will be published later.

3. Discussion

In order to allow comparisons between our results and previously published data (Di Lorenzo 1972, Cairns and Fairman 1968, Aoki and Yamaguchi 1971) we have represented these data on figure 7 as well as a theoretical curve representing silicon

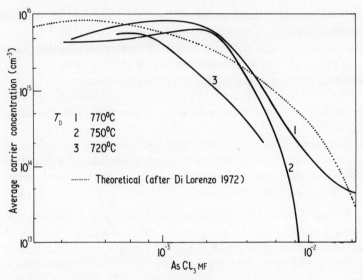

Figure 6. Average carrier concentration against $AsCl_3$ MF at different temperatures. Comparison with Di Lorenzo's theoretical silicon activity curve which has been arbitrarily translated (dotted line).

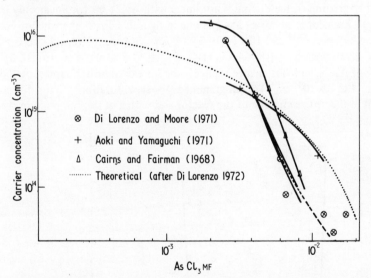

Figure 7. Published carrier concentration as a function of $AsCl_3$ MF data with the same theoretical curve.

activity in such a growing system assessed by Di Lorenzo (1972). This same curve is also shown in figure 6. It must be noticed that owing to the different source and substrate temperatures, this curve is only defined within an arbitrary translation along both axes. The experimental study and examination of the different figures calls for the following remarks:

(i) The growth rate τ varies markedly with $AsCl_3$ MF which seems to be an important growth parameter.

(ii) The slight decrease of τ with $AsCl_3$ MF under the usual conditions must be taken into account for precise thickness determination.

(iii) τ decreases rapidly at low values of $AsCl_3$ MF. The morphology then becomes textureless.

(iv) The value of $AsCl_3$ MF for which τ is maximum depends upon the growth conditions (T_S, T_D).

(v) The variation of n with $AsCl_3$ MF is not linear and depends on the growth parameters (growth temperatures, figure 6) and on the experimental conditions (different authors, figure 7). On the other hand, our results could be explained taking into account variations of the silicon activity pointed out by Di Lorenzo and Moore (1971).

However, points 1 and 3 show quite clearly that there is a change in the growth limiting kinetics at the value of $AsCl_3$ MF where τ is maximum. This value corresponds more or less to the value at which the doping level changes. So there is no evidence whether the change in the doping level comes from the kinetics of growth or from the change in silicon activity.

4. Device applications

Even if the variation of the doping is not linear with $AsCl_3$ MF (or linear only in a narrow range), knowledge of curves like the one in figure 5 allows reproducible control of the doping between 10^{14} and 10^{16} (with $2 < AsCl_3$ MF $< 7 \times 10^{-3}$) and also the growth of n^- layers with $n^- \leqslant 10^{14}$ ($AsCl_3$ MF $> 7 \times 10^{-3}$). The MF is controlled by the temperature of $AsCl_3$, which is changed quickly (as for the transition represented on figure 8 from 4 to 8×10^{14} cm^{-3}) or programmed (for slow variations). With this doping level control in the upstream zone of the reactor and with a n^+ sulphur dope in the

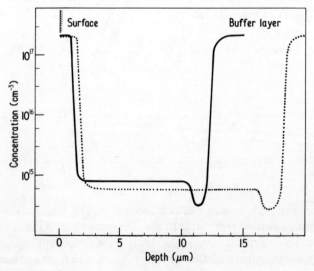

Figure 8. Typical doping profiles for TEA.

downstream zone (Hollan and Mircea 1973) we can obtain in a same reactor nearly all the epitaxial multilayer structures realized thus far.

In addition to the avalanche diodes (Farrayre *et al* 1975) the following devices have been produced in our laboratory.

4.1. Nuclear detectors on 400 μm thick layers with $n \simeq 10^{13} cm^{-3}$

These layers are grown on n^+ buffer layers in order to avoid the p^- interface otherwise observed at such low doping levels. The resolution obtained with a 70 μm thick n^- layer is 1·9 keV at 60 keV and 2·1 keV at 122 keV, at 300 K, which is the best result reported for a GaAs detector.

4.2. High-frequency Gunn diodes on $n^+/n/n^+/n^+$ substrate structures

These have been realized (Michel 1974). Here again, for the control of the n layer, the residual dope can be used, near the maximum of the curve represented on figure 6 ($n = 5 \times 10^{15}$). The results obtained on devices (7% efficiency at 18 GHz, and 4% at 30 GHz with 210 mW) are slightly lower than those previously reported (Omori 1974) even though the $n^+ \rightarrow n$ and $n \rightarrow n^+$ transitions obtained by this method are good (Hollan and Mircea 1973).

4.3. Gunn amplifiers (TEA)

The problem is to obtain a good control of the doping notch near the cathode (Spitalnik *et al* 1973). Low noise devices need a low (*nl*) product and a carefully controlled notch to maintain a uniform electric field throughout the sample. The best result obtained is a noise figure of 10·5 dB at 11·0 GHz (Magarshack *et al* 1974). Standard 7 dB gain modules have been made with bandwidths of 7—11 GHz and saturation power levels of 20 mW at 1 dB compression. A typical doping profile of such a device is shown in figure 8. These doping profiles can be obtained reproducibly.

4.4. Field effect transistors

Finally we point out that for FET n^+/n^- buffer/semiconductor/insulator/substrate structures can be easily obtained using high AsCl$_3$ MF for the n^- buffer layer and sulphur doping for the 0·2—0·3 μm thick active layer (grown in the second zone of the reactor). We have already realized some of these structures with $n^- < 10^{14}$ and about 10 μm thick. It has been observed that the buffer layer decreases the NF_{min} and increases the associated gain (Baudet *et al* 1974). As a first result, we have obtained at 7·5 GHz an NF_{min} of 3·3 dB and an associated gain $G_{NF} = 5$ dB.

Acknowledgments

The author would like to express his acknowledgments to Mme Mennechez for her collaboration in the epitaxial growth, to C Schemali for setting the heat pipe up and for the measurements, to Mme Baudet for the carrier concentration measurements and to MM Hallais, Chané and Fabre for their helpful discussions.

References

Aoki T and Yamaguchi M 1971 *Japan. J. Appl. Phys.* **10** 953

Baudet P, Binet M, Boccon-Gibod D and Parquet P 1974 *Report* 73 CNES 0653

Cairns B and Fairman R 1968 *J. Electrochem. Soc.* **115** 327C

Di Lorenzo J V 1972 *J. Crystal Growth* **17** 189

Di Lorenzo J V and Moore G E 1971 *J. Electrochem. Soc.* **118** 1823

Farrayre A, Kramer B and Mircea A 1975 this volume

Hollan L and Mircea A 1973 *Proc. 4th Int. Symp. Gallium Arsenide and Related Compounds* (London and Bristol: Institute of Physics) p217

Hollan L and Schiller C 1974 *J. Crystal Growth* **22** 175

Magarshack J, Rabier A and Spitalnik R 1974 *IEEE Trans. Electron Dev.* October

Michel J 1974 *Report DGRST 73 7 1377*

Minden H 1971 *J. Crystal Growth* **8** 37

Omori M 1974 *Microwave J.* **17** 57

Shaw D W 1971 *J. Crystal Growth* **8** 117

Spitalnik R, Shaw M P, Rabier A and Magarshack J 1973 *Appl. Phys. Lett.* **22** 162

Wolfe C M, Stillman G E and Owen E B 1970 *J. Electrochem. Soc.* **117** 129

Diffusion-limited LPE growth of InP for microwave devices

V L Wrick III† and L F Eastman

Electrical Engineering School, Cornell University, Ithaca, NY 14850, USA

Abstract. Liquid phase epitaxial InP has been grown using a multiple-well, bottom-of-the-melt, sliding substrate apparatus.

The diffusion coefficient of phosphorus in an Indium-rich solution D_P was determined for the first time, as a function of temperature, from LPE data acquired over the temperature range 750 °C to 715 °C. D_P varies similarly to the diffusion coefficient D_A of As in Ga, but $D_P/D_A \simeq 2\cdot3$ at 720 °C. The diffusion model used follows the work of D Rode (1973 *J. Crystal Growth* **20** 13 and private communication) at Bell Telephone Laboratories for As in a Ga-rich solution.

By applying the diffusion model to various growth schemes, we were able to conclude that a step-cooled growth technique was the optimal method for achieving microwave device quality InP epi-layers. Layer thicknesses between 20–60 μm can be reproducibly grown with a 5% variation of surface morphology. Hall samples produced from the same melt before and after device layers show no significant variation in material parameters from run to run. Liquid nitrogen Hall mobilities in the range 20 000 cm² V⁻¹ s⁻¹ were achieved for $N_D - N_A \simeq 2 \times 10^{15}$ cm⁻³. Preliminary analysis at Royal Radar Establishment indicates Zn and C are present as acceptors (P Dean, A M White and M Webb private communication) and there is a strong implication that Si is the primary donor after pre-growth baking of the melt.

Because of the complex device geometry of the Gunn and Read avalanche diodes, horizontal slider systems have recently become very popular for growing microwave semiconductor wafers. Last year Wood *et al* (1973) reported the use of such an apparatus to grow multiple layer InP using a uniform cooling thermal schedule. Recently Rode (1973) has demonstrated that such a system is diffusion limited in the transport of the melt constituents to the seed crystal. In order to better characterize our epitaxial growth procedure, we have applied this diffusion model to our data to determine the diffusion coefficient of phosphorus in an Indium-rich melt.

Figure 1 is a view of the graphite boat used for LPE growth of InP. Essential features of the apparatus are the source crystals at the bottom of the melt used to maintain local equilibrium during the pre-growth cycle. Also, the substrate is protected under graphite during the heating and cooling cycles to minimize phosphorus depletion.

Figure 2 is a representation of the melt geometry. It is important to recognize that only the lower half of the melt contributes to growth on the substrate. The top source crystal was employed to maintain liquidus equilibrium despite phosphorus loss from the melt. Since regrowth will occur on any excess material present in the melt, it is also necessary to provide a planar geometry which lends itself to analytical solution.

† Presently with Electrical Engineering School, Princeton University, Princeton, NJ 08540, USA.

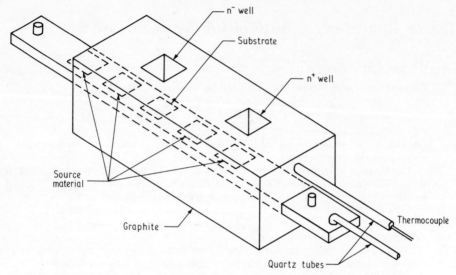

Figure 1. Multiple-well, sliding-substrate graphite boat for LPE growth of InP.

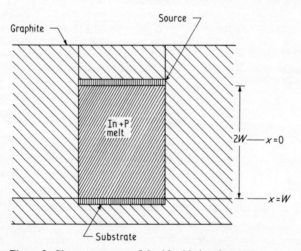

Figure 2. Planar geometry of double-sided melt system.

The transport of the phosphorus to the seed crystal is described by the following equation:

$$D(T) \frac{\partial^2 n(x, t)}{\partial x^2} = \frac{\partial C}{\partial t} - \frac{\partial n(x, t)}{\partial t}$$

where n is the excess concentration of phosphorus in the melt, C is the liquidus concentration of phosphorus and D is the diffusion coefficient of phosphorus. The solution to the equation, given by Rode (1973) is:

$$n(x, t) = \sum_{i=0}^{\infty} n_i(t) \cos \left[\frac{(2i+1)\pi x}{2W} \right].$$

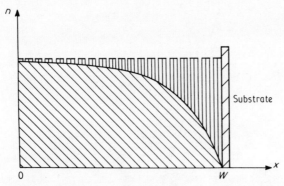

Figure 3. Plot of excess phosphorus density in lower half of melt before and after an arbitrary length of growth time.

Figure 3 gives an arbitrary spatial plot of n for an arbitrary time. The area shaded with the vertical lines represents the epitaxially deposited material, and the ratio of this area to the total supersaturation of the run defines the growth efficiency achieved.

Since a computer is required to generate the $n_i(t)$, it is convenient to gain a more intuitive approach to the problem by applying the semi-infinite melt approximation. In essence, this requires the constituents to diffuse a finite distance W from a non-depletable source located at $x = 0$. Because of the nature of the approximation, crystal growth can never re-establish melt equilibrium, hence its results are only valid for times less than a diffusion time W^2/D. The growth formulae are:

Uniform Cool
$$\frac{d}{d_1} = \frac{4}{3W} \left(\frac{Dt}{\pi}\right)^{1/2}$$

Step Cool
$$\frac{d}{d_1} = \frac{2}{W} \left(\frac{Dt}{\pi}\right)^{1/2}$$

where d is the measured epitaxial layer thickness, d_1 is the epitaxial thickness predicted from the liquidus curve for a given temperature, and t is the growth time. By applying the approximation for the uniform cool to our growth data, we were able to infer the diffusion coefficient for phosphorus in an indium-rich melt as shown in figure 4. A similar curve germane to GaAs growth is included for comparison. It may be seen that D for phosphorus follows an activation behaviour with markedly greater temperature variation than D for arsenic.

In our attempts to grow thick layers of InP, we considered the steady-state, or Sorét, method of inducing growth. The following formula for growth thickness in diffusion-limited steady state growth was given by Long *et al* (1974):

$$d = \frac{D}{W} \frac{C_0 T'}{T_0^2} \frac{(T_w - T_0)t}{[C_0 - C_s \exp(T'/T_w)]}$$

where C_s is the phosphorus concentration in the InP crystal, T_w is the temperature of the source crystal, T_0 is the temperature of the seed crystal, and C_0 and T' are defined

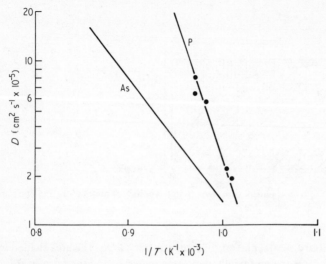

Figure 4. Plot of the logarithm of diffusion constant of P in In, and As in Ga, against the reciprocal of absolute temperature.

in the liquidus relationship: $C(T) = C_0 \exp - (T'/T)$. By applying this formula to typical steady state parameters, we conclude that diffusion-limited steady state growth of InP in the 730 °C region will yield about 1 μm per hour. Since the InP crystals pit severely with long term exposure to heat, we have rejected this method of growth in favour of the step cool.

The step cool relies upon an abrupt supersaturation of the solution and growth proceeds in an isothermal environment. Maximum layer thickness of course depends on the size of the supercooling step; however, we have realized nearly 100% growth efficiencies using this method for layers 60 μm thick. These epitaxial layers were achieved for growth times of 1 h at 730 °C for a 30 °C cooling step to 700 °C.

The two fundamental differences between the uniform cool and the step cool are the thermal environment during growth and the growth efficiency. It has recently been suggested by J J Hsieh (personal communication) that supersaturation of several degrees is necessary prior to growth to ensure uniform nucleation on the substrate. This is automatically ensured in the step cool, yet nearly 100% growth efficiency may still be achieved for sufficient growth time. In the uniform cool, any practical cooling rate will greatly outrun the diffusion rate, hence prohibiting high growth efficiencies in reasonable times, and requiring large temperature drops for thick layers.

In addition, since the growth proceeds in a changing temperature, doping gradients will be induced by the variation of impurity segregation coefficients. These doping gradients can degrade device performance substantially. Thus, we feel that the step cool may be the optimal method of growing LPE InP.

We would now like to address the problem of crystal purity. Figure 5 shows our interpretation of published work of Astles and Williams (1972) in deciding a growth temperature. The curve is meant to imply that for long term heating of the melt (16 h) prior to growth, there are two competing classes of donors, one which is 'volatile' and another which increases with temperature. Our data, depending on suppliers of In and

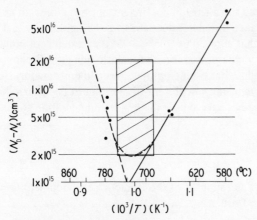

Figure 5. Plot of the logarithm of net donor density against the reciprocal of absolute temperature of pre-growth bake from Astles and Williams (circles) and present work (cross hatched area).

InP and on baking procedures, fall in the cross hatched area. We have chosen our growth temperatures and baking temperatures in light of these data, and presently we bake our indium at 730 °C for 20 h prior to the addition of the InP source and seed crystals. By fabricating Hall samples before and after a multiple layer run, we were able to establish the single melt repeatability results shown in table 1.

Table 1.

Prior to multiple layer growth

	μ (cm^2 V^{-1} s^{-1})	$N_D - N_A$ (cm^{-3})
RT	4386	$2 \cdot 5 \times 10^{15}$
LN	16873	$2 \cdot 06 \times 10^{15}$

After multiple layer growth

	μ (cm^2 V^{-1} s^{-1})	$N_D - N_A$ (cm^{-3})
RT	4878	$3 \cdot 38 \times 10^{15}$
LN	15078	$2 \cdot 08 \times 10^{15}$

While we have achieved liquid nitrogen mobilities up to 22 000 at $1 \cdot 2 \times 10^{16}$ cm^{-1} electron concentration with low compensation, usually the crystals have had compensation, especially at lower electron concentrations.

Paul Dean of the Royal Radar Establishment has kindly subjected some of our samples to photoluminescent studies to determine acceptor impurities. The results indicate that zinc and carbon are providing the major acceptor contributions.

In conclusion, we have inferred the diffusion coefficient of phosphorus, in an Indium-rich melt over a useful range of temperature for LPE. Using this physical parameter, we did experiments that allowed us to conclude that a step cool, moderately short, growth procedure should yield optimal results for flat doping profiles in thick device layers. This growth procedure has been employed to yield reproducibly doped crystals in the 2×10^{15} electrons/cm^3 range with tens of micrometres of controlled thickness. Photoluminescence studies have shown the major compensation to be due to zinc and carbon acceptors.

Acknowledgments

We would like to acknowledge the support of the Rome Air Development Center and the Army Research Office in Durham.

References

Astles M G and Williams E W 1972 *Electron. Lett.* **8** 2120–1
Long S I, Ballantyne J M and Eastman L F 1974 *J. Crystal Growth* **26** 13–20
Rode D L 1973 *J. Crystal Growth* **20** 13–23
Wood C E C, Wrick V L and Eastman L F 1973 *Proc. 4th Biennial Cornell Electrical Engineering, Conference* pp147–54

New structures by liquid phase epitaxy for microwave devices

F E Rosztoczy, R E Goldwasser and J Kinoshita

Varian Associates, Palo Alto, CA 94303, USA

Abstract. Further sophistication in boat design and additional improvements in growth technology make liquid phase epitaxy a suitable technique for the growth of multi-layer GaAs epitaxial films required for state-of-the-art solid state microwave devices. Uniform multi-layer epitaxial structures with individual layer thicknesses of $0 \cdot 1$ μm are reproducibly grown with good control. The doping profiles indicate abrupt carrier concentration changes at the interfaces.

Millimetre wave Gunn oscillators. 2 μm active layers with $5-12 \times 10^{15}$ carriers/cm³ are grown. Gunn devices fabricated from this type of wafer give up to 2% CW efficiency at 75 GHz. Power output is approaching 100 mW.

Low noise Gunn amplifiers. Epitaxial 4-layer structures are grown with a cathode notch profile: $n^+-n^--n^--n^+$ to improve the noise figure of Gunn amplifiers. In X-band this lowers the amplifier noise figure from 20 to 13 dB.

High efficiency avalanche devices. Epitaxial quadruple layers of $n-n^+-n-n^+$ are grown to fabricate high efficiency Lo—Hi—Lo GaAs avalanche devices. The highest observed CW efficiency is 35·6% at 10 GHz with 2·9 W.

1. Introduction

The challenge of stretching or expanding the upper limits of high efficiency GaAs Gunn device operaticj has required further sophistication and additional improvements in epitaxial growth technology.

Because the lower frequency microwave bands ($f < 18$ GHz) are becoming increasingly utilized for communications, there is more incentive to use the less-densely populated millimetre wave frequencies whenever possible. Operation of transferred electron oscillators up to 100 GHz has been predicted (Jones and Rees 1973) and 71 GHz has been reported (Ruttan 1974).

Significant improvement in transferred electron amplifier noise performance has been predicted for device structures which utilize uniform electric field profiles rather than uniform doping profiles (Thim 1971). This can be accomplished by the use of injection limiting contacts (Kroemer 1968, Kallback 1973) and cathode-notch doping profiles (Charlton and Hobson 1973).

Recent interest in the development of high efficiency GaAs IMPATT devices (Irvin 1973) has been stimulated by both theoretical calculations and experimental reports demonstrating potential efficiencies in excess of 30%. Salmer *et al* (1973) reported theoretical efficiencies for a number of potential device structures, predicting as high as 29% for a two-step field, single transit zone (Lo—Hi—Lo) structure and 35·4% for a two-step field, double transit zone (double drift Lo—Hi—Lo) structure. More recently, computer simulations (Decker 1974, Goldwasser and Rosztoczy 1974) have indicated that 42–46% efficiency for Lo–Hi–Lo GaAs IMPATT diodes is theoretically possible. Experimental observations of efficiencies in excess of 35% have been reported (Kim *et al* 1973, Goldwasser and Rosztoczy 1974).

The physical realization of these devices and their improved performance has been made possible by increasingly sophisticated epitaxial growth techniques. The purpose of this paper is to describe recent developments in liquid phase epitaxial growth technology in relation to GaAs microwave device requirements and improved performance. Progress in millimetre wave Gunn oscillators, low noise Gunn amplifiers, and high efficiency avalanche devices will be specifically discussed.

2. Experimental epitaxial growth techniques

Early liquid phase epitaxial growth methods, utilizing various schemes of tipping and dipping, were successful in growing elementary GaAs device structures for less critical applications (low frequency Gunn diodes, varactors etc). However, the more recent higher performance, higher frequency devices require significantly more control over epitaxial layer thickness and thickness uniformity than has been possible by the earlier methods. Successful control of layer thicknesses to better than $0.05\,\mu m$ dimensional tolerance are required to maintain good device yield.

As described in previous papers (Rosztoczy and Kinoshita 1974, Rosztoczy *et al* 1974), the LPE system used for this work is constructed with stainless steel gas inlet and outlet connections to the quartz furnace tube. All components are carefully leak-checked to insure a vacuum-tight system. Palladium diffused, high purity hydrogen is used as the ambient. A 2.5 inch diameter by 30 inch long tube furnace is controlled by a three-thermocouple, three-zone, high-performance controller.

A four-well graphite sliding well boat, similar to the three-well boat previously described, was designed to grow epitaxial layers on the same wafer from three different carrier concentration melts in one heat cycle. The boat is being used, for example, to grow notch profile Gunn wafers when three different carrier concentrations are required for the buffer layer, active layer, notch and the regrown contact, all grown in one operation. The boat is precision-machined for complete wipe-off of the melts from very high purity, high density graphite. These boats are cleaned through numerous boilings in different acids and organic solvents, then vacuum and hydrogen baked and further baked with gallium.

With this system, we can now grow layers with controlled, uniform reproducible thicknesses in the $0.1\,\mu m$ range.

Relatively high Hall mobilities and low degrees of compensation in the individual layers are very desirable for microwave applications. When liquid phase epitaxy is used, Sn for n-layers and Ge for p-layers are the best dopants for microwave applications (Caldwell and Rosztoczy 1972, Rosztoczy *et al* 1973, Rosztoczy and Kinoshita 1974).

3. Diode fabrication

To obtain superior microwave device performance from optimized structures, one must not only grow the precise structure, but also fabricate the device without excessive structural damage. This problem is especially acute in GaAs because of its extremely low tensile and shear properties. The high power densities required in both Gunn effect and IMPATT devices demand a near perfect heat transfer from the active region to the heat-sink. Thermocompression and ultrasonic bonds have been found to often cause severe structural damage. To reduce bonding damage, a low strain, gold plated (integrated) heat-sink technique has been developed for production of Gunn and IMPATT devices.

Figure 1 is an SEM photograph of a typical integrated heat sink (IHS) Gunn diode chip. The gold heat-sink pad is nominally $50\,\mu$m thick and $250\,\mu$m square. The diode is from 40 to $45\,\mu$m in thickness and $150\,\mu$m square. An alloyed Au–Ge–Ni substrate contact is visible on the top surface. Figure 2 shows a typical GaAs IMPATT structure. In this case, a circular mesa geometry is used, with a diode diameter of $200\,\mu$m. The plated Au heat sink, in this case, is $90\,\mu$m thick and the GaAs mesa is 40–$45\,\mu$m thick.

Figure 1. Scanning electron microphotograph of a Gunn diode chip with integral heat sink. The gold heat sink pad is nominally $50\,\mu$m thick and $250\,\mu$m square. The diode is 40–$45\,\mu$m in thickness and $150\,\mu$m square.

Figure 2. Scanning electron microphotograph of a typical GaAs IMPATT structure with integral heat sink. The diode chip is $200\,\mu$m in diameter and 40–$45\,\mu$m thick.

4. High frequency Gunn diodes

To explain the key parameters affecting millimetre wave Gunn diode performance, the doping profile shown in figure 3 will be considered. The three salient points which effect the upper technological frequency limit of Gunn diodes are:

(i) Doping profile should have sharp, abrupt interfaces
(ii) Doping density should be flat with a well defined length
(iii) Doping profile must have the exact carrier concentration and length.

To illustrate the necessary length accuracy required, it is assumed that there exists a dead space such that the active layer consists of the composite of two regions, an active region and a dead zone. Based on empirical data, we estimate that the dead space is

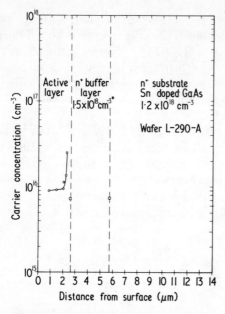

Figure 3. Doping profile of 75 GHz Gunn diode showing active layer, buffer layer and substrate regions. Asterisk; determined by separate IR measurement. Circles show data collected by depletion capacitance measurements, squares indicate thickness determined by the cleave and stain method.

about 1 μm or slightly less. The net active region of a 75 GHz diode is also approximately 1 μm. Assuming the active length random variation must be less than ±6·25% (45°) to obtain efficient oscillations, the thickness of the active region must be defined to within ±0·0625 μm, for a total length of 2 μm. Variations of the doping profile interfaces, work damage, impurities, out diffusion or any local nonuniformities at the two active layer boundaries which make the length randomness greater than 0·215 μm will prevent efficient operation at 75 GHz.

While doping profile measurements are useful tools, they are only indications. The final judgement must be the RF performance of the finished devices. Figure 4 shows CW efficiency and power output versus frequency for a 75 GHz Gunn diode. Note that at 75 GHz, the oscillator efficiency is still 2% with close to 100 mW power output. The device has an active layer length of approximately 2 μm and a carrier concentration of 9.5×10^{15} cm^{-3}. For the same diode, operation at 40 GHz was achieved with higher power and efficiency in a circuit optimized for 40 GHz operation. This is possible because of the broadband range of Gunn effect negative resistance.

The small signal impedance characteristics of Gunn diodes are valuable in the design of millimetre wave oscillators and amplifiers (De Koning *et al* 1974). These measurements also provide the necessary design feedback to improve the diode-circuit match. Figure 5 is a Smith chart plot of the negative of the diode impedance of low and high *NL* product 35 GHz devices. The data has been obtained from slotted line measurements made at a 3 mm TEM reference plane located at the tip of the diode. Note that the low (about 1×10^{12} cm^{-2}) *NL* product diode has a substantially lower negative resistance

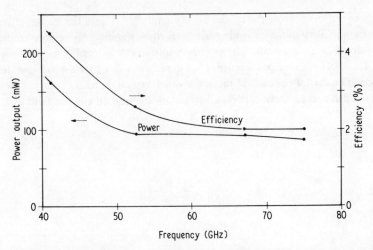

Figure 4. RF output power and efficiency of a 75 GHz Gunn diode. Measurements were made in several different circuits from 40 to 75 GHz with optimum tuning and bias for each frequency.

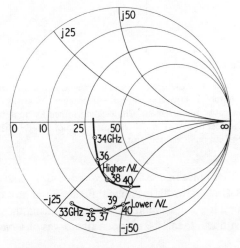

Figure 5. Smith chart plot of negative impedance of two 35 GHz Gunn diodes. One diode with a high *NL* product and one with a low *NL* product are compared. Measurements were made by slotted line techniques.

than the high *NL* (about $2 \times 10^{12} \, cm^{-2}$) product device. Impedance characterization has also been a useful technique in determining and maintaining wafer and process uniformity.

5. Low noise Gunn diodes for amplifiers

It has been shown theoretically (Thim 1971) and experimentally (Baskaran and Robson 1972) that GaAs Gunn diodes with uniform doping and good ohmic contacts exhibit noise figures in excess of 20 dB. To improve this noise figure, the electric field must be more uniform across the diode. To accomplish this field flattening, either the doping or the mobility in the cathode region has to be significantly reduced with respect to the remaining layer. Since the carrier concentration can be readily monitored, a doping notch is often preferred. Figure 6 is a doping profile of an X-band low noise Gunn diode. The carrier concentration in the notch region drops by nearly a factor of 10 and the width is accurately controlled to obtain the desired electric field distribution.

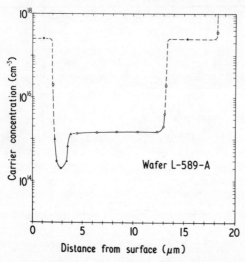

Figure 6. Composite doping profile of a low noise X-band cathode notch structure consisting of an n^+ contact, n^- notch, n active layer and n^+ buffer layer. Asterisk; determined by separate van der Pauw measurement. Circles; determined by depletion capacitance measurement. Triangles; from JAC profiler. Squares; determined by the cleave and stain method. Broken curve; estimated profile.

Utilizing this technique, Gunn amplifier noise figures have been reduced from 20–25 dB to 13–14 dB for diodes covering 3 GHz bandwidth in X-band. Figure 7 is a plot of gain and noise figure versus frequency for a low noise Gunn amplifier stage using the above diodes. Nine decibel nominal gain with less than 14·9 dB noise figure was observed from 8·9 to 11·7 GHz.

The above stage was designed using computer automated network analyser data obtained on sample diodes. Figure 8 is a Smith chart displaying the small signal impedances of a notch diode used in this work. The data is displayed as the negative of measured values to allow plotting inside the Smith chart. The device displays well behaved negative resistance from 6·5 GHz to above 12 GHz. The impedances of the devices are considerably lower than a standard flat profile device which presents a special challenge to the circuit designer. To achieve the performance shown in figure 7, two devices are operated in a series connection.

6. High efficiency Lo–Hi–Lo IMPATT diodes

Recent computer calculations by numerous workers agree that CW efficiencies of 35% or more can be expected from Lo–Hi–Lo GaAs IMPATT devices. A very narrow,

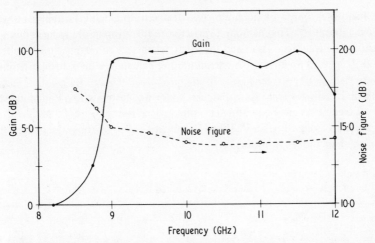

Figure 7. Plot of experimentally measured amplifier gain and noise figure as a function of frequency for a two-diode low noise amplifier circuit. Cathode notch diodes were utilized to achieve less than 14·9 dB noise figure from 8·9 to 11·7 GHz.

very highly doped spike with a very accurately designed and controlled Q and a very narrow, but well defined 'low' layer in front of it are essential to obtain maximum efficiency (Goldwasser and Rosztoczy 1974, Decker 1974). Spikes, with half widths of 100–300 Å have been grown for this purpose by vapour phase epitaxy (VPE) (Fairman and Decker 1975). With the LPE technology described in this paper, we can now grow epitaxial structures with individual layer thicknesses reproducibly controlled down to 0·1 μm.

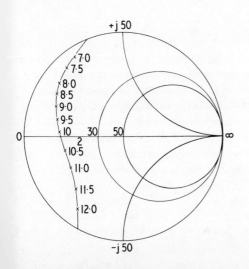

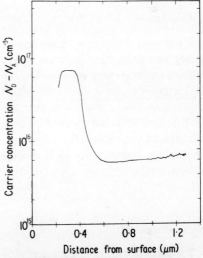

Figure 8. Smith chart plot of the negative impedance of a series resonated X-band cathode notch diode.

Figure 9. Composite doping profile of a Lo–Hi–Lo (n–n$^+$–n) GaAs IMPATT wafer. This wafer had a charge of $1 \cdot 7 – 1 \cdot 9 \times 10^{12}\,\mathrm{cm}^{-2}$ terminating the avalanche zone.

Figure 9 shows a composite doping profile of one of the high efficiency LPE wafers grown in our laboratory. The maximum carrier concentration in the spike is just below $8 \times 10^{16}\,cm^{-3}$. The first 'low' layer and the 'high' layer have thicknesses of $0.2\,\mu m$ or less. (The drift layer is $5\,\mu m$.) For this structure, we calculate a theoretical efficiency of 39%.

A group of 151 diodes from this particular wafer was tested with 51% having better than 22% efficiency. Typical operating currents varied from 150 to 400 mA with operating voltages from 40 to 56 V. Table 1 summarizes the CW operating characteristics of three of the best diodes from this wafer.

Table 1. CW operating characteristics of selected Lo–Hi–Lo GaAs IMPATT devices

Serial no.	Frequency (GHz)	Volts at 50 mA	$V_{(op)}$ (V)	$I_{(op)}$ (mA)	Output power (W)	Efficiency (%)
259	10.2	43.4	55.4	160	2.9	32.7
298	10.5	47.1	50.5	160	2.7	33.4
375	10.4	43.0	48.0	170	2.9	35.6

The observed power and efficiency is strongly dependent on input current density, however, near optimum efficiency is obtained over a wide range of drive levels. From this wafer, using only ambient air cooling, a maximum CW efficiency of 35.6% was measured at 10.4 GHz with a power output of 2.9 W.

7. Conclusion

It has been shown that improvements in GaAs liquid phase epitaxial growth technology have permitted significant advancements in solid state microwave devices. Very uniform epi-layers with close thickness tolerances and low residual background carrier concentrations have made possible operation of millimetre wave Gunn oscillators to 75 GHz, low-noise cathode notch structure Gunn amplifiers with less than 13–14 dB noise figures in X-band and high efficiency Lo–Hi–Lo IMPATT structures with over 35% efficiency. It is expected that continued optimization of the device parameters and the epitaxial growth process will result in further significant improvements in performance.

Acknowledgments

The authors wish to thank Messrs J G De Koning, R J Hamilton and Tom Ruttan for supplying some of the technical information presented in this paper and Roy Hendricks for device fabrication; a special acknowledgment to S I Long for his critical review of the manuscript and many helpful discussions.

References

Baskaran S and Robson P N 1972 *Electron. Lett.* **8** 109–10
Caldwell J F and Rosztoczy F E 1972 *Proc. 4th Int. Symp. Gallium Arsenide and Related Compounds* (London and Bristol: Institute of Physics) pp240–8
Charlton R and Hobson G S 1973 *IEEE Trans. Electron Dev.* **ED-20** 812–17
Decker D R 1974 *IEEE Trans. Electron Dev.* **ED-21** 469–79
De Koning J G, Goldwasser R E, Hamilton R J and Rosztoczy F E 1975 *IEEE Trans. Microwave Theor. Techn.* **MTT-23**
Fairman R D and Decker D R 1975 to be published
Goldwasser R E and Rosztoczy F E 1974 *Appl. Phys. Lett.* **25** 92–4
Irvin J C 1973 *Proc. 4th Cornell Elect. Eng. Conf.* 287–97
Jones D and Rees H D 1973 *J. Phys. C: Solid St. Phys.* **6** 1781–93
Kallback B 1973 *Electron. Lett.* **9** 11–12
Kim C, Steele R and Bierig R 1973 *Electron. Lett.* **9** 173–4
Kroemer H 1968 *IEEE Trans. Electron Dev.* **ED-15** 819–37
Rosztoczy F E, Caldwell J F, Kinoshita J and Omori M 1973 *Appl. Phys. Lett.* **22** 525–6
Rosztoczy F E and Kinoshita J 1974 *J. Electrochem. Soc.* **121** 439–44
Rosztoczy F E, Long S I and Kinoshita J 1974 *J. Crystal Growth*
Ruttan T G 1974 *IEEE Trans. Microwave Theor. Techn.* **MTT-22** 142–4
Salmer G, Pribetich J, Farrayre A and Kramer B 1973 *J. Appl. Phys.* **44** 314–24
Thim H W 1971 *Electron. Lett.* **7** 106–8

Multi-layer epitaxial technology for the Schottky barrier GaAs field-effect transistor

T Nozaki, M Ogawa, H Terao and H Watanabe

Central Research Laboratories, Nippon Electric Company Limited, 1753 Shimonumabe, Nakahara-ku, Kawasaki, Japan

Abstract. In order to avoid mobility degradation and the effects of deep level impurities near the GaAs epitaxial–substrate interface, the insertion of a high resistivity buffer layer between the active layer and the substrate has been examined.

Using a modified $Ga(Sn)-AsCl_3-H_2$ reaction system, a successive epitaxial growth method for obtaining a multi-layer structure which consisted of a high resistivity layer and an active layer on Cr-doped substrate is presented.

1 μm Schottky barrier gate GaAs FET have been fabricated in the multi-layer epitaxial wafer. Remarkable improvements are observed in DC and RF characteristics. f_{max} of the order of 60 GHz, *NF* of the order of 2·5 dB and *MAG* of the order of 20 dB were obtained at 6 GHz.

1. Introduction

GaAs FET fabricated using the conventional epitaxial layer, grown directly on Cr-doped substrates, often show looping and bumping in the drain current–voltage characteristics. Looping and low electron mobility near the interface between the active layer and the substrate, leading to low transconductance, have been reported to originate from the deep level impurities near the interface (Hower *et al* 1969). The large transconductance, which is related to high electron mobility, plays an important role in obtaining high performance GaAs FET.

Buffer layer insertion between the active layer and the substrate can be considered to be effective in eliminating deep level impurities and obtaining high electron mobility near the interface. Slaymaker and Turner (1973) reported that the existence of buffer layer improved not only electron mobility but the power gain and the noise figure of GaAs FET. However, the growth procedure and electrical properties of the buffer layer were not described. On the other hand, precise control of the carrier concentration as well as the layer thickness is important in controlling the pinch-off voltage from the viewpoint of FET production.

In this paper, the growth procedure of the buffer layer and the method of the precise control of the active layer carrier concentration are proposed. Multi-layer epitaxial wafers for FET application have been made by the growth of a high resistivity layer on Cr-doped substrate followed by n-active layer growth.

1 μm gate GaAs FET have been fabricated using the multi-layer epitaxial wafer. The electrical characteristics of FET have been evaluated and remarkable improvements in DC and RF performances have been confirmed.

2. Preparation and properties of epitaxial layers

2.1. *Growth system*

GaAs vapour epitaxial growth was carried out by using a Ga–AsCl$_3$–H$_2$ reaction system. Figure 1 shows a schematic diagram of the reactor system previously reported (Nozaki and Saito 1972). There were two inputs, one is a main flow of H$_2$+ AsCl$_3$ to the

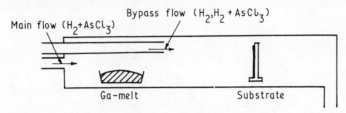

Figure 1. Schematic diagram of the reactor system.

Ga-melt and the other is a bypass flow of H$_2$ or H$_2$+AsCl$_3$ introduced in front of the substrate. Two non-bubbling type AsCl$_3$ containers, which were used to supply a main H$_2$+AsCl$_3$ flow and a bypass H$_2$+AsCl$_3$ flow, were kept constant at 2 and 10 °C, respectively. The H$_2$ flow rate was precisely controlled by using mass flow controllers. The main H$_2$ flow was kept constant at 210 ml min^{-1}. In order to obtain good uniformity in the doping level and the epitaxial layer thickness, the substrates were set nearly perpendicular to the gas flow. The Ga-melt and the deposition temperature were held typically at 850 and 750 °C, respectively. Cr-doped semi-insulating and Si-doped n$^+$-substrates of {100} orientation were used.

The carrier concentration of un-doped epitaxial layer was typically in the high 10^{14} to low 10^{15} cm^{-3} range. Tin was added to the Ga-melt for doping.

The increase of the bypass H$_2$ flow was found to increase the carrier concentration, while the growth rate was not changed. Rough adjustment of the carrier concentration to the value desired for FET was made by the amount of tin added in the Ga-melt and precisely done by controlling the bypass H$_2$ flow rate. The bypass H$_2$+AsCl$_3$ flow influenced not only the growth rate but the carrier concentration. Growth rate and carrier concentration decreased with increasing the bypass H$_2$+AsCl$_3$ flow rate. This carrier concentration change was large and a high resistivity epitaxial layer was obtained at a higher H$_2$+AsCl$_3$ flow rate.

The bypass H$_2$+AsCl$_3$ flow was also used to gas-etch the substrates before the growth of a buffer layer.

2.2. *Carrier concentration dependence on bypass H$_2$ flow rate*

The influence of the bypass H$_2$ flow rate on the carrier concentration of the epitaxial layer was examined. The results are shown in figure 2. The carrier concentrations of three groups, which were obtained by different Ga-melts contained different amounts of tin, increased in the same manner, with increasing bypass H$_2$ flow rate. The desired carrier concentration of active layer, which was needed for GaAs FET was easily obtained merely by changing the bypass H$_2$ flow rate. H$_2$ flow effect on carrier concen-

48   *T Nozaki, M Ogawa, H Terao and H Watanabe*

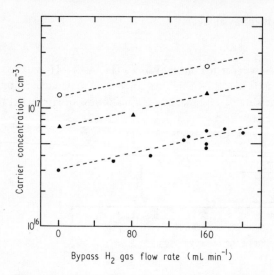

Figure 2. Carrier concentration dependence on bypass H_2 flow rate.

tration is useful for precise control of the active layer carrier concentration. The carrier concentration dependence on the bypass H_2 flow could be explained by the change of the tin incorporation reaction into the epitaxial layer, as will be discussed later. The surface of the epitaxial layer remained mirror-like up to a 400 ml min^{-1} bypass H_2 flow rate but it became cloudy with further increase.

2.3. Carrier concentration dependence on bypass $H_2 + AsCl_3$ flow rate

GaAs epitaxial growth was carried out on Si-doped n$^+$-substrate by introducing the bypass $H_2 + AsCl_3$ flow followed by the n-layer growth by stopping the bypass $AsCl_3$ gas supply. In order to measure the change of the carrier concentration, the bypass $AsCl_3$ gas supply was varied by changing the bypass H_2 flow rate with a constant $AsCl_3$ concentration. Figure 3(a) shows the representative carrier concentration profiles of the epitaxial layers. Carrier concentration profiles were measured by a conventional $C-V$ method. It is clear that the high resistivity buffer layers were grown by introducing the bypass $H_2 + AsCl_3$ flow.

The carrier concentration dependence on the bypass $H_2 + AsCl_3$ flow rate can be estimated by plotting the minimum carrier concentration in the carrier concentration profiles versus the bypass $H_2 + AsCl_3$ flow rate. The results are shown in figure 3(b). The carrier concentration was drastically reduced by increasing the bypass $H_2 + AsCl_3$ flow rate, as shown by the dotted line, from about 3×10^{16} to about 1×10^{12} cm^{-3}, while the growth rate was changed from 0·25 to 0·08 μm min^{-1}.

2.4. Multi-layer epitaxial technology for GaAs FET

Figure 4 shows the growth diagram for obtaining multi-layer epitaxial wafers. As described before, the main $H_2 + AsCl_3$ flow rate was kept constant, 210 ml min^{-1}, during

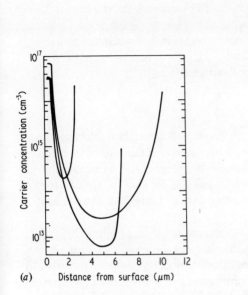

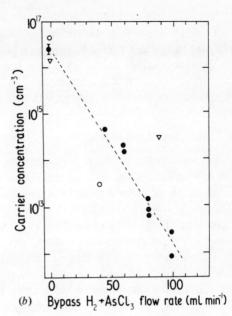

(a) Distance from surface (μm)

(b) Bypass $H_2 + AsCl_3$ flow rate (mL min^{-1})

Figure 3. (a) Carrier concentration profile of a high resistivity layer on n$^+$-substrate. (b) Carrier concentration dependence on bypass $H_2 + AsCl_3$ flow rate.

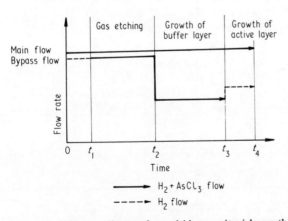

Figure 4. Flow rate diagram for multi-layer epitaxial growth.

the epitaxial growth. After the Ga-melt and the substrate temperature reached the operating temperature, the $AsCl_3$ gas was mixed to the main H_2 flow ($t = 0$). The steady state in Ga-melt was obtained in about 5 minutes, then the bypass $H_2 + AsCl_3$ flow was introduced and gas etching was initiated ($t = t_1$). After about 4 μm of the substrates was etched off, the bypass $H_2 + AsCl_3$ flow rate was reduced to a desired value for making a high resistivity buffer layer ($t = t_2$). The active layer was deposited successively by stopping the bypass $AsCl_3$ supply ($t = t_3$), while it was possible to control the bypass H_2 flow rate to obtain a desired carrier concentration.

The main $H_2 + AsCl_3$ flow rate was kept constant at $210\,\mathrm{ml\,min^{-1}}$. The bypass $H_2 + AsCl_3$ flow rate was $50-100\,\mathrm{ml\,min^{-1}}$ for the buffer layer growth. The bypass flow rate was $160\,\mathrm{ml\,min^{-1}}$ for the active layer growth.

The time schedule for the multi-layer epitaxial growth is as follows. Initial source saturation time is 5 min, gas etching time is 20 min, the growth time of buffer layer is 40–90 min and the active layer growth time is about 2 min.

2.5. Mobility distribution

In order to estimate the mobility distribution in epitaxial layers, Hall effect measurements were carried out. Two wafers, one with a buffer layer and one without a buffer layer, were used. After bridge shape type Hall elements were prepared by mesa etching, Al-Schottky barrier contact was made on the surface of Hall element, in addition to the electrodes. Changing the reverse bias to the Schottky barrier, Hall effect measurements were carried out at room temperature.

The results are shown in figure 5, with the carrier concentration profiles. The electron mobility gradually decreased towards the interface for a wafer without a buffer layer, while, for a wafer with a buffer layer, the electron mobility gradually increased to the value of the buffer layer. This mobility improvement yields good FET performance characteristics in microwave frequency operation.

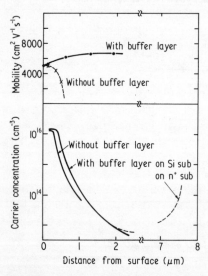

Figure 5. Electron mobility distribution and carrier concentration profile in an epitaxial multi-layer.

3. Device application

Figure 6(a) shows a Schottky barrier gate field effect transistor, with a gate length of $1\,\mu\mathrm{m}$, fabricated by the conventional photo-contact masking technique using a multi-layer epitaxial wafer. The carrier concentration profile is shown in figure 6(b). The

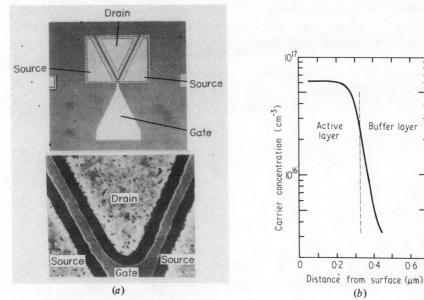

(a) (b)

Figure 6. Structure of Schottky barrier gate field effect transistor; (a) optical microscope image and scanning electron microscope image, (b) carrier concentration profile of an active layer and a buffer layer.

n-type active layer was $0.3 \, \mu m$ thick with the carrier concentration of $6.4 \times 10^{16} \, cm^{-3}$. The buffer layer was about $10 \, \mu m$ thick with the carrier concentration of below $1 \times 10^{13} \, cm^{-3}$. A Schottky gate made from aluminium had two branches which were $1 \, \mu m$ long, $150 \, \mu m$ wide and $0.6 \, \mu m$ thick. Source and drain contacts were separated from the gate by $1 \, \mu m$. They were obtained by forming Au–Ge alloy films on the substrate, followed by a platinum film. To reduce the metallization resistance, additional gold films were plated on these contacts. In order to determine the effect of the buffer layer, FET without a buffer layer were also fabricated on a wafer having the same carrier concentration and thickness.

Figure 7 shows $I-V$ curves of FET with and without a buffer layer. Looping and bumping that were present in the drain current of an FET without a buffer layer were not observed in the drain current of an FET with a buffer layer. Moreover, a higher transconductance ($gm = 24 \, m\Omega^{-1}$) was obtained in the latter than the transconductance ($gm = 18 \, m\Omega^{-1}$) in the former.

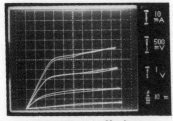

FET without buffer layer

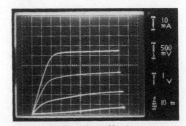

FET with buffer layer

Figure 7. $I-V$ characteristics of FET with and without a buffer layer.

Looping and low electron mobility near the interface between the active layer and Cr-doped substrate, leading to low transconductance, have been reported to originate from the deep level impurities near the interface. Therefore, these improvements were obtained by introducing a buffer layer which eliminated the deep level impurities.

Figure 8 shows the unilateral power gain calculated from S-parameters measured from 2 to 12 GHz under bias conditions of I_d = 60 mA and V_d = 3·5 V. The extrapolated f_{max} was 60 GHz for an FET with a buffer layer, and 40 GHz for an FET without a buffer layer.

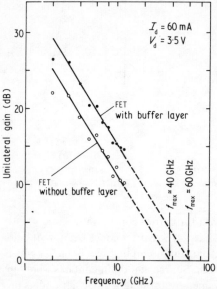

Figure 8. Unilateral power gain against frequency for FET with and without a buffer layer.

The noise figure and power gain results measured directly at 6 GHz are shown in table 1. With the buffer layer, a remarkable improvement was achieved. The optimum noise figure was about 2·5 dB under bias conditions of I_d = 10 mA and V_d = 2 V. The maximum available gain was 20 dB under bias conditions of I_d = 60 mA and V_d = 3·5 V.

FET with a buffer layer thickness from 1 to 14 μm were also fabricated and it was confirmed that no looping and no bumping occurred.

Table 1. RF characteristics of FET with and without a buffer layer

	I_d = 10 mA V_d = 2 V		I_d = 60 mA V_d = 3·5 V
	NF_{op} (dB)	G_{av} (dB)	G_{mag} (dB)
FET without buffer layer	4·0	8	14
FET with buffer layer	2·5	11	20

4. Discussion

It was found that the carrier concentration in the epitaxial layer was changed by the bypass H_2 or $H_2 + AsCl_3$ flow rate. Tin doping to epitaxial GaAs obtained by a Ga(Sn)–AsCl$_3$–H$_2$ system has been reported by Wolfe *et al* (1968). They obtained the result that the carrier concentration of the epitaxial layer depends upon its growth rate, which can be explained by a time-dependent tin adsorption at the vapour–solid interface. Present results mentioned before, however, can not be understood by this model.

In order to understand the observed results, an equilibrium model, which is similar as reported by DiLorenzo and Moore (1971) in the case of silicon incorporation reaction, was applied.

Following chemical reactions were considered,

$$2GaCl + \tfrac{1}{2}As_4 + H_2 \rightleftharpoons 2GaAs + 2HCl$$

and

$$SnCl_2 + H_2 \rightleftharpoons Sn \text{ (in GaAs)} + 2HCl,$$

which are the growth reaction and the tin doping reaction, respectively.

The equilibrium state of the vapour phase near the growing surface is changed according to the bypass H_2 or $H_2 + AsCl_3$ flow rate. The equilibrium HCl pressure near the crystal surface becomes lower by introducing the bypass H_2 flow, resulting in a higher tin activity. On the other hand, the bypass $H_2 + AsCl_3$ flow gives a higher equilibrium HCl pressure and results in a lower tin activity. This model can qualitatively explain the observed carrier concentration change by the bypass flow of H_2 or $H_2 + AsCl_3$.

The Hall mobility ($6600 \, cm^2 \, V^{-1} \, s^{-1}$) of the buffer layer was low as compared with the mobility of a high purity layer having an electron concentration of about $10^{12} \, cm^{-3}$. It was supposed that the buffer layer was compensated in some degree. Further experiments on the properties of the high resistivity layer will be needed.

GaAs FET fabricated using the successively grown multilayer wafers showed no looping and no bumping. This suggests that deep level impurities, which may affect these anomalously, were reduced by using the buffer layer.

5. Conclusion

Using the carrier concentration change caused by the bypass H_2 and $H_2 + AsCl_3$ flow rate, a multi-layer epitaxial wafer, which consisted of an active layer, a high resistivity buffer layer and the substrate, was obtained. The insertion of a buffer layer between an active layer and the substrate was found to be effective in obtaining high performance GaAs FET.

Looping and bumping, which was observed in FET without a buffer layer, disappeared in FET with a buffer layer. The high frequency performances were also improved. f_{max} of the order of 60 GHz and *NF* of the order of 2·5 dB and *MAG* of the order of 20 dB were obtained at 6 GHz.

Acknowledgments

The authors would like to express their appreciation to S Asanabe, Y Seki and N Kawamura for their encouragement and for stimulating discussions.

References

DiLorenzo J V and Moore G E Jr 1971 *J. Electrochem. Soc.* **118** 1823
Hower P L, Hooper W W, Tremere D A, Lehrer W and Bittmann C A 1969 *Proc. 3rd Int. Symp. Gallium Arsenide and Related Compounds* (London and Bristol: Institute of Physics) p187
Nozaki T and Saito T 1972 *Japan. J. Appl. Phys.* **11** 110
Slaymaker N A and Turner J A 1973 *Proc. 1st European Microwave Conf.* A.5.1
Wolfe C M, Stillman G E and Lindley W T 1969 *Proc. 3rd Int. Symp. Gallium Arsenide and Related Compounds* (London and Bristol: Institute of Physics) p43

Schottky barrier diodes fabricated on epitaxial GaAs using electron beam lithography

G T Wrixon†
Bell Telephone Laboratories, Holmdel, NJ, USA

and R F W Pease
Bell Telephone Laboratories, Murray Hill, NJ, USA

Abstract. Schottky barrier diodes, for use at millimetre wavelengths, shaped as crossed stripes of width 0·25 μm and 0·4 μm have been fabricated on epitaxial GaAs using electron beam lithography. The electron lithographic apparatus consisted of a commercial scanning electron microscope together with a simple digital pattern generator. The electron resist was polymethylmethacrylate. The junctions were formed by electroplating gold through the cross shaped cuts in an oxide film on the GaAs. This shaping of the diode has resulted in a 30% reduction in spreading resistance over that of photolithographically formed circular diodes with approximately the same junction area and capacitance. Initial RF measurements at 175 GHz, though subject to some uncertainty at this frequency, indicate the cross shaped diodes to have similar noise properties, and a reduced conversion loss over that of equal area circular diodes.

1. Introduction

Current interest in extending the techniques of coherent detection to the 1 mm wavelength region and beyond has focused attention on the problems of producing high quality Schottky-barrier diodes for use at very high frequencies. These diodes, characterized by their low noise, are now commonly used as the non-linear element in millimetre wave mixers. The factors limiting their performance are the barrier capacitance (C_0) and the spreading resistance (R_s).

It is an extremely complex problem to relate mixer performance to R_s and C_0 quantitatively. It has been shown by Torrey and Whitmer (1948) however that for small signals

$$L_0 \propto \frac{\alpha}{2} \frac{1}{1 + \omega^2 C_0^2 R_j R_s} \tag{1}$$

where L_0 is the conversion loss of a diode, $\alpha = q/\eta kT$, where η is the exponential or quality factor of the diode and R_j is the nonlinear diode resistance.

A cutoff frequency at which L_0 equals half of its maximum value can be defined by:

$$f_c = \frac{1}{2\pi C_0 (R_j R_s)^{1/2}}. \tag{2}$$

† Present address: University College, Cork, Ireland.

Equations (1) and (2) show how the *high frequency* performance of a Schottky barrier diode can be limited by the presence of the parasitics R_s and C_0. Other debilitating effects caused by these parasitics such as the thermal noise associated with R_s, which is present at all frequencies, also exist however. Clearly it is important to find ways of reducing the values of C_0 and R_s in order to extend the range of usefulness of Schottky barrier diodes to higher frequencies and to improve, if possible, their performance as mixers at all frequencies.

2. Photolithography

Schottky barrier diodes are normally made using a mask consisting of an array of circular dots, each of which defines a single diode. If r is the radius of such a diode then $C_0 \sim r^2$ and $R_s \sim 1/r^n$ where n lies between 1 and 2 depending on the thickness of the epi-layer and the diameter of the diode, see for example Sato *et al* (1972).

It is seen from equation (2) that the detection sensitivity high frequency cutoff f_c can be increased by decreasing the diode radius, that is to say

$$f_c \propto 1/r^m \qquad \text{where } 1 < m < 1\cdot5.$$

In practice, however, one cannot keep decreasing r indefinitely in order to obtain diodes with higher and higher cutoff frequencies. In the first place, diffraction effects impose a limit of about $1\,\mu\text{m}$ on the diameter of photolithographically produced diodes. In the second place, when using epitaxial material, even before this limit of $1\,\mu\text{m}$ is reached the spreading resistance has already reached an unacceptably high value.†

For example, diodes of diameter $1\,\mu\text{m}$, having $C_0 = 0\cdot003$ pF and $R_s = 20\,\Omega$ exhibited a slightly inferior noise figure when used by Schneider and Wrixon (1974) in a 230 GHz mixer, than diodes of $2\,\mu\text{m}$ diameter with $C_0 = 0\cdot012$ pF and $R_s = 10\,\Omega$. This, in spite of the fact that the $1\,\mu\text{m}$ diodes have a cutoff frequency $f_c = 1677$ GHz,‡ while the $2\,\mu\text{m}$ diodes have a cutoff frequency $f_c = 593$ GHz. That is, a reduction in diode size while resulting in an increased cutoff frequency gave a degraded R F performance. This degradation is thought to be due to two factors: (*a*) the higher value of spreading resistance which combined with the smaller diode area leads to excessive ohmic heating at the junction resulting in increased diode noise and (*b*) the fact that the performance of the 230 GHz mixer was limited by the low level of local oscillator at the mixer. Thus, doubling the spreading resistance doubles the amount of local oscillator power dissipated in it, leaving that much less available for the mixing process, and consequently increasing the conversion loss. Currently, because of a lack of suitable generators, all diode mixers operating above about 160 GHz do so with less than the optimum amount of local oscillator power. In addition, at 160 GHz a $2\,\mu\text{m}$ diameter diode with a cutoff frequency $f_c = 593$ GHz has already added 0·3 dB to its minimum conversion loss. Clearly a diode with a higher cutoff frequency is needed; however, if this is achieved by reducing the diameter of the diode, any resulting decrease in conversion loss will be offset by an

† Of course non-epitaxial (ie, bulk) GaAs could be used to circumvent this problem. This however, would cause an increase in junction capacitance as well as a reduction in the reverse breakdown voltage. This latter effect could also lead to an increase in diode noise due to reverse current flow during the negative half of the local oscillator cycle.

‡ Diode biased so that $R_j = 50\,\Omega$.

increase caused by a reduced level of local oscillator power. It is thus desirable to prevent this concomitant increase in spreading resistance as the cutoff frequency is increased.

3. Non-circular diodes

The spreading resistance of a diode, in addition to being directly related to material constants, is also affected by any current 'bunching' which might occur due to the geometrical profile of the junction. Thus, if it were possible to reduce this bunching, for instance by increasing the ratio of the diode perimeter to its area, a reduction in spreading resistance should also be obtained. We have attempted to implement this procedure using electron beam lithography to fabricate diodes in the form of crossed stripes. These diodes have been made in two sizes, $2 \times 0.25 \mu m$ and $4 \times 0.40 \mu m$ and are thus approximately equal in area to the 1 and $2 \mu m$ diameter diodes mentioned above.

4. Electron lithography

For this experiment epitaxial GaAs supplied by the Monsanto Corporation was used. This was the same material used by Schneider and Wrixon (1974) to make the $2 \mu m$ circular diodes, and its characteristics are given in table 1. The diode processing was

Table 1. Characteristics of the epi-layer material from which the cross diodes and the $2 \mu m$ diameter circular diodes were fabricated.

Supplier	Monsanto Corporation
Orientation	[100]
Conductivity	n^+
Dopant	Te
Bulk resistivity	$4.5 \times 10^{-4} \Omega\, cm$
Epi-dopant	S
Epi-carrier density	$3.56 \times 10^{17}\, cm^{-3}$
Epi-resistivity	$0.0045 \Omega\, cm$
Epi-mobility	$3840\, cm^2\, V^{-1}\, s^{-1}$
Epi-thickness	$0.24 \mu m$

carried out as follows. An SiO_2 passivating layer was R F sputtered onto the surface of the semiconductor to a thickness of about 2500 Å. The wafer was then lapped to a thickness of $150 \mu m$ and a plated gold, tin–nickel layer was alloyed into the back to give the ohmic back contact. The electron resist used was polymethylmethacrylate – a relatively insensitive positive electron resist described by Haller *et al* (1968). The resist was exposed using a commercial scanning electron microscope (Cambridge Scientific Instruments, Sterescan IIA) with 20 kV electrons at a dose of $3 \times 10^{-5}\, cm^{-2}$. The beam of the SEM was controlled by a simple digital pattern generator and the resulting format was that of arrays of orthogonally crossed stripes $0.25 \times 2 \mu m$ spaced at about $10 \mu m$ and $0.4 \times 4 \mu m$ spaced at about $15 \mu m$. Following resist development, the SiO_2 was etched through, the resist stripped, the GaAs surface cleaned and the diodes gold plated.

Figure 1 is an electron micrograph of the $0.25 \times 2.0\,\mu m$ diode array after developing the resist and figures 2 and 3 are photographs of the same array before and after gold plating.

The DC parameters, which were measured for the diodes before the material was cut into chips are shown in table 2. It is seen that for both the 1 and $2\,\mu m$ circular diodes, making an approximately equal area cross structure resulted in diodes with about the same capacitance but with a 30% reduction in the average value of spreading resistance together with no significant degradation of the diode quality factor†.

Figure 1. Electron micrograph of the $0.25 \times 2\,\mu m$ cross diodes after development of the electron sensitive resist.

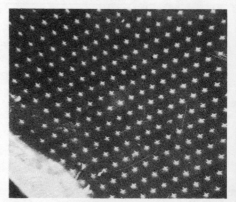

Figure 2. Electron micrograph of the $0.25 \times 2\,\mu m$ cross diodes after etching of the SiO_2 passivating layer and removal of the electron sensitive resist.

Figure 3. Electron micrograph of the $0.25 \times 2\,\mu m$ cross diodes after gold plating.

Table 2. DC characteristics of circular and cross shaped Schottky barrier diodes. Values quoted for R_s represent average values of many measurements on each diode type; for each type range of value is within 10% of the mean. The cutoff frequency values refer to diodes biased so that $R_j = 50\,\Omega$.

Diode	Spreading resistance R_s (Ω)	Junction capacitance C_0 (pF)	Diode quality factor η	Cutoff frequency $f_c = \dfrac{1}{2\pi C_0 (R_j R_s)^{1/2}}$ (GHz)
1 μm diameter circular	20	0·003	1·15–1·25	1677
Cross shaped 2 × 0·25 μm	14	0·0027	1·2–1·25	2228
2 μm diameter circular	10	0·0120	1·1–1·18	593
Cross shaped 4 × 0·4 μm	7	0·0123	1·15–1·2	692

circular diodes were similarly mounted. The RF measurements were made at 175 GHz using a klystron, coupled through a high Q cavity, as the local oscillator. The first IF amplifier was a 0·8 dB noise-figure parametric amplifier operating at 1·4 GHz. The conversion loss L and the noise temperature ratio t were measured for the diodes using a method similar to that described by Weinreb and Kerr (1973).

The RF results for the two most satisfactorily mounted diodes of each group are shown in table 3. It is seen that while the noise contribution of each diode was approximately the same the conversion loss of the cross diode was 0·5 dB less than that of its equal area circular counterpart.

Table 3. DSB conversion loss and noise temperature ratio for 4 × 0·4 μm cross diode and 2 μm diameter circular diode operating as mixers at 175 GHz.

Sharpless wafer No.	DSB conversion loss (dB)	Noise temperature ratio
BC-5 (4·0 × 0·4 μm cross diode)	8·3	0·95
50-25 (2 μm diameter circular diode)	8·8	0·96

decreased LO level has added more than 4 dB to the conversion loss. At this decreased LO level the dependence of conversion loss on LO power is extremely critical (see for example Weinreb and Kerr 1973). In the present experiment therefore the additional 0·4 dB of conversion loss possessed by the circular diode may be due to LO power loss in its larger series resistance although the relatively high uncertainty of the conversion loss measurements at 175 GHz, about 0·2 to 0·3 dB, must also be a factor.

No noticeable decrease in the diode noise was detected with the cross diode. At the low LO levels which were used however it is possible that no significant heating of the junction took place.

7. Conclusions

It has been shown that the spreading resistance of a small Schottky barrier diode of given area (capacitance) can be reduced by increasing its perimeter/area ratio. In the present case, a reduction in R_s of 30% was achieved for diodes with approximately the same area and capacitance by changing from a circular to a cross shape. This results in a higher cutoff frequency which at short enough millimetre wavelengths can result in a decreased conversion loss. More importantly perhaps, degraded mixer performance caused by the unavailability of strong LO sources at frequencies above about 160 GHz can be somewhat alleviated by using a cross diode with an inherently lower spreading resistance.

In practice the DC parameters have usually been a measure of the best RF performance that can be attained when all other aspects of making the diode mixer are optimized. Thus in spite of the uncertainty in the RF measurements, it is not unexpected that they show conversion loss performance exceeding that of the best of the equivalent circular shaped diodes.

Acknowledgments

We would like to thank M V Schneider and R A Linke for many helpful suggestions and R D Standley for making the electron micrographs. Professor P Thaddeus of Columbia University kindly loaned us the 175 GHz klystron used in the RF tests.

References

Haller I, Hatzakis M and Strinivason R 1968 *IBM J. Res. Dev.* **12** 251
Sato Y, Ida M, Uchida M and Shimada K 1972 *Electronics and Communications in Japan* **55**-C, No. 1
Schneider M V and Wrixon G T 1974 *Proc. Int. Microwave Cong. Atlanta*
Torrey H C and Whitmer C A 1948 *Crystal Rectifiers* (New York: McGraw-Hill)
Weinreb S and Kerr A R 1973 *IEEE J. Solid St. Circuits* **SC-8** 58

Performance and characterization of X-band GaAs Read-type IMPATT diodes

F Hasegawa, Y Aono and Y Kaneko†

Central Research Laboratories, †Semiconductor Division, Nippon Electric Company Ltd, Kawasaki, Japan

Abstract. The high efficiency, high power operation of X-band GaAs Read-type IMPATT diodes has been investigated by estimating the bias field distribution in the active region, and by comparing the diode parameters with those of conventional GaAs IMPATT diodes.

Diodes having lower bias fields in the drift region, that is, lower operating bias voltages, gave higher efficiencies. The high efficiency operation at low bias voltage is inferred to be due to the low threshold field for electron velocity saturation in GaAs. It was found that the junction area of GaAs Read diodes could be made much larger than conventional GaAs IMPATT diodes without deterioration of the efficiency. CW output powers of about 10 W with efficiencies greater than 20% were obtained at 9·5 GHz. The bias voltages of these diodes were about two thirds, the diode areas were about 2−3 times, the efficiencies were over 1·5 times, and the output powers were about 3−4 times those of the best conventional GaAs IMPATT diodes made using the same technology.

1. Introduction

Recent experimental work has shown that GaAs Read-type IMPATT diodes can give extraordinary high efficiencies and high powers (Salmer *et al* 1973, Kim *et al* 1973, Wisseman *et al* 1973, Irvin *et al* 1973). To date, efficiencies as high as 35% (Kim *et al* 1973, Goldwasser and Rosztoczy 1974) and CW output powers as high as 5−9 W for C−X bands (Kim *et al* 1973, Irvin *et al* 1973, Wisseman *et al* 1974) have been reported. However, it has not yet been clarified experimentally why the GaAs is so promising, although there have been several theoretical publications (Schroeder and Haddad 1973, Huang 1973, Su and Sze 1973).

The purpose of this paper is to present experimental results which might give some answers to the above question and to present details of the performance that has been obtained in our laboratory.

Section 2 provides a brief explanation of the fabrication process. Variation of the efficiency with bias current density characteristics for diodes with different breakdown voltages is presented in §3.1. Section 3.2 shows that the area of GaAs Read diodes can be increased larger without deterioration of the efficiency, as compared with conventional GaAs IMPATT diodes. The performance and the diode parameters will be compared with those of conventional GaAs IMPATT diodes in §3.3 and the dynamics of the GaAs Read diode will be discussed in §3.4.

2. Fabrication and measurements

High–low doping profile (n^+–n–n^{2+}) epitaxial wafers were grown by the $Ga-AsCl_3-H_2$ vapour transport method. The substrates used for the growth were tellurium or silicon doped (low 10^{18} cm^{-3}) and cut $2°$ off $\langle 100 \rangle$ orientation. A buffer layer several micrometres thick ($n \sim 1 \times 10^{17}$ cm^{-3}), a $3 \mu m$ thick drift region ($n \sim 2 \times 10^{15}$ cm^{-3}) and an n^+ surface region less than $1 \mu m$ ($n \sim 4 \times 10^{16}$ cm^{-3}) were successively grown by using sulphur as a dopant. Examples of doping profiles can be seen in figure 1(a) and figure 3(a). The n^+–n transition region was generally less than $0.5 \mu m$.

After etching the surface of the epitaxial layer slightly, platinum and gold were evaporated successively with platinum as the Schottky metal, onto which gold was plated to a thickness of about $150 \mu m$ to form a plated heat sink. The substrate was thinned down to a total thickness of about $60 \mu m$, and Au–Ge–Ni ohmic contacts were made. The semiconductor wafer was then selectively etched off to make mesas on the plated heat sink. The diode diameter and spacing were about 180 and $500 \mu m$, respectively. The plated heat sink pellet with four GaAs diode mesas was mounted in a package with a solder, and the mesas were connected in parallel with gold ribbons.

The doping profile and the field distribution of each diode was calculated from capacitance–voltage measurements by determining the diode area with a microscope. The doping profiles agreed reasonably with those measured on the wafer with the profile plotter.

The diodes were measured in an X-band waveguide test cavity with a matching disc or hat. By using a low-pass filter, it was confirmed that none of the measured output power was due to higher harmonics.

3. Experimental results and discussions

3.1. Efficiency and field distribution

Figures 1(a) and (b) show, respectively, the field distributions at a bias voltage near the breakdown and the efficiency against bias current density, measured in two diodes of different breakdown voltage. In both diodes, the donor density at the drift region is low enough to be swept out at the breakdown voltage. In contrast to the efficiency against bias current density relation of a conventional GaAs IMPATT diode shown by the dotted line in figure 1(b), the efficiencies of these diodes increase rapidly and reach a maximum at relatively low bias current densities. Further increase of the efficiency in the high bias current density region suggests some change of the dynamics of operation, since the bias voltage saturates and sometimes decreases slightly at this point. It might be due to a large field modulation by the space charge effect as shown later.

An experimental fact which should be noted is that the maximum efficiency of a diode with a lower breakdown voltage is higher than that of a diode with a higher breakdown voltage. This is summarized in figure 2, where the efficiencies of a number of diodes at the same bias current density are plotted against operating bias voltage. The diodes have different breakdown voltages owing to slight difference in the surface n^+ layer thickness. The physical explanation for this dependence has been discussed in a previous paper (Hasegawa and Aono 1974).

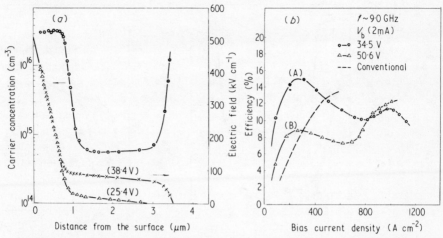

Figure 1. (*a*) Doping profile and field distributions at bias voltages near the breakdown; (*b*) efficiency against bias current density characteristics for GaAs Read diodes whose donor density in the drift region is low enough to be swept out at the breakdown voltage. The broken curve plots efficiency against bias current density relation of a representative conventional GaAs IMPATT diode.

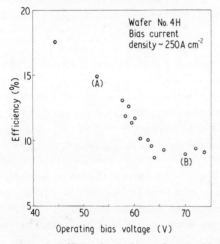

Figure 2. Efficiencies at almost the same bias current densities plotted against operating bias voltages.

Figures 3(*a*) and (*b*) show similar characteristics of Read diodes from another epitaxial wafer whose doping density in the drift region is higher than that of the diodes in figure 1. The field distributions at the breakdown voltage shown in figure 3(*a*) indicate that a diode with a low breakdown voltage has a wide unswept region. Because of this wide unswept region, the efficiency of diode A does not increase very much even if the bias current density is increased to more than 1000 A cm^{-2}. For diode B, whose unswept region is narrower than that of diode A, the efficiency increases steeply at a certain bias current density where the depletion region sweeps away the whole of the low doped region.

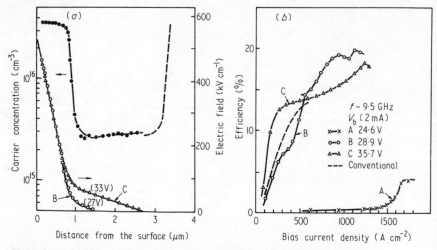

Figure 3. (*a*) Doping profile and field distributions at the breakdown voltages; (*b*) efficiency against bias current density characteristics for GaAs Read diodes with different unswept regions at the breakdown voltages.

For diode C which is slightly 'before punch through' at breakdown, the efficiency increases rapidly at low bias current density, but the maximum efficiency is lower than that of diode B with lower breakdown voltage. This means that even for 'before punch through' diodes, higher efficiency can be obtained for a diode with a lower breakdown voltage, unless the unswept region at the breakdown is too thick. In a 'before punch through' diode, the depletion region extends deeper towards the n^+ substrate when the bias voltage is increased from the breakdown to the operating point, because of the field strength increase due to the temperature rise and to the carrier space charge effect. The highest efficiency is expected to be obtained for a diode which is a slightly 'over punch through' at the operating bias voltage, and in that condition, higher bias current density will give higher efficiency.

An important feature of GaAs Read diodes is that we can choose the bias current density, therefore the input power density, where the highest efficiency is obtained, by changing the thickness of the high doping region so that the diode is a little 'over punch through' at that bias current density. This is one of the reasons why high power and high efficiency operation is possible for GaAs Read diodes.

3.2. *cw output power and diode area*

The area of Read diodes could be made much larger than that of conventional GaAs IMPATT diodes without deterioration of efficiency. This was demonstrated by changing the diode area, by changing the number of mesas connected in parallel. The area of each mesa was almost constant and was about $2 \cdot 4 \times 10^{-4} \, cm^2$.

Figure 4(*a*) shows output–input power characteristics of one, two, three and four mesa diodes. As is shown in the figure, the efficiency decreases as the number of mesas increases, and the output power does not increase much with increase of the diode area, although thermally limited maximum input power has been doubled by the connection

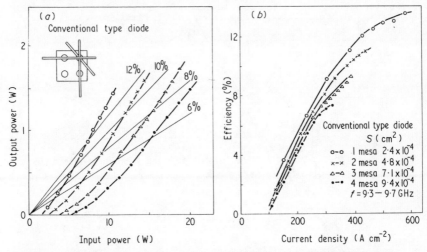

Figure 4. (*a*) Dependence of output power and (*b*) efficiency on the diode area for conventional GaAs IMPATT diodes. The diode area was increased by increasing the number of mesas, each mesa having the same area (about $2\cdot4 \times 10^{-4} \, cm^2$).

of four mesas. (The thermal resistance of the four mesa diode is about a half of that of a single mesa diode, while the area of the former is four times that of the latter.) As shown in figure 4(*b*), the efficiency is an increasing function of the current density. On the other hand, the maximum current density is limited thermally and decreases as the number of mesas increases.

In Read diodes, however, the current density dependence of the efficiency is much less. Therefore, as shown in figure 5, the efficiency does not decrease so much with the increase of number of mesas, even if the thermally limited maximum bias current density decreases to a half.

The negative conductance per unit area of the GaAs Read diode is considerably larger than that of the conventional diode, and negative Q of the former is lower (Takayama 1974). This fact must be one of the reasons why the power combining between mesas is more successful in Read diodes.

As described before, for GaAs Read diodes, we can choose the bias current density that gives the maximum efficiency, by changing the breakdown voltage or the field distribution at the breakdown voltage. Even if the efficiency–bias current density relation of a diode is like the ones shown in figure 3(*b*), and even if the thermally limited maximum bias current density is low, the diode can be designed to give the maximum efficiency at that maximum bias current.

3.3. Performance and diode parameters

CW output powers of about 10 W with more than 20% efficiency were obtained at around 9·5 GHz with four mesa type GaAs Read diodes. Diode doping profiles and field distributions at the breakdown voltage were similar to those shown in figure 3(*a*).

Figure 6 shows representative input–output characteristics of the diodes with breakdown voltages of about 25 V. These diodes have an unswept region of more than 2 μm

66   *F Hasegawa, Y Aono and Y Kaneko*

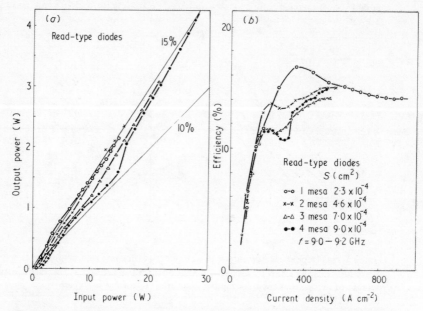

Figure 5. (*a*) Dependence of output power and (*b*) efficiency on the diode area for GaAs Read-type IMPATT diodes. The diode area was increased in the same way as in figure 4.

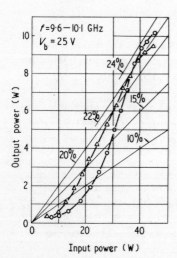

Figure 6. Representative input–output characteristics of high-power GaAs Read diodes.

at breakdown, so that the efficiencies at lower input power density are low. As the input power increases and the operating bias voltage increases, the diode becomes 'slightly over punch through', and the maximum efficiency is obtained. A maximum output power of 10·2 W (9·6 GHz, 22·1%) and a maximum efficiency of 22·9% (9·25 W, 9·52 GHz) were obtained.

Oscillation characteristics and diode parameters of the high power Read diodes made from two different epitaxial wafers are summarized in table 1, and compared

Table 1. Performances and parameters of high-power GaAs Read diodes and conventional IMPATT diodes.

Diode	V_b (V)	S (cm²)	V_d (V)	I_{op} (mA)	P_{out} (W)	η (%)	f (GHz)
A-1†	25·5	$8\cdot1 \times 10^{-4}$	52·9	850	9·50	21·1	9·73
9	25·2	$9\cdot6 \times 10^{-4}$	51·4	890	10·1	22·1	9·60
9	25·2	$9\cdot6 \times 10^{-4}$	50·5	800	9·25	22·9	9·52
17	24·6	$10\cdot5 \times 10^{-4}$	49·8	925	10·2	22·1	9·60
20	24·8	$7\cdot9 \times 10^{-4}$	51·5	920	9·60	20·3	9·70
B-13†	27·8	$7\cdot7 \times 10^{-4}$	52·1	800	8·50	20·4	9·60
15	30·9	$7\cdot4 \times 10^{-4}$	56·3	850	9·20	19·2	9·26
23	28·8	$6\cdot7 \times 10^{-4}$	54·1	820	8·90	20·1	9·04
26	29·7	$7\cdot0 \times 10^{-4}$	55·0	800	8·80	20·0	9·26
28	31·0	$6\cdot3 \times 10^{-4}$	56·0	860	9·25	19·2	9·46
Conventional‡	54·5	$3\cdot4 \times 10^{-4}$	73·0	260	2·40	12·5	9·50

† Read diodes from wafers A and B. ‡ Conventional GaAs IMPATT diode with the best performance. V_b, breakdown voltage; S, diode area; V_d, operating bias voltage; I_{op}, bias current; P_{out}, CW output power; η, efficiency; f, frequency.

with those of the best four mesa-type conventional GaAs IMPATT diodes made using the same technology. CW output powers of about 10 W were obtained for the diodes from wafer A, whose high and low doping densities are $4\cdot2 \times 10^{16}$ and $2\cdot2 \times 10^{15}$ cm^{-3}, respectively. The high and low doping densities of wafer B are $3\cdot5 \times 10^{16}$ and $2\cdot8 \times 10^{15}$ cm^{-3}, respectively. The diode group with the lower breakdown voltages (wafer A) gives larger diode areas, lower operating bias voltages and higher powers.

The remarkable features of the Read diodes compared with conventional IMPATT diodes are as follows: (i) the breakdown voltages are about a half and the operating bias voltages are about two thirds, (ii) the optimum diode areas are about 2–3 times, (iii) the efficiencies are more than 1·5 times that of conventional diodes. Thermal resistances of the Read and the conventional diodes were about 7 °C W^{-1} and about 13 °C W^{-1}, respectively. As a result of these differences, CW output powers of three or four times those of conventional diodes have been obtained.

3.4. Dynamic field distribution

In order to see the dynamics of the high efficiency operation of GaAs Read diodes, a bias field distribution and a dynamic field distribution were estimated for diode No. A-9 in table 1. The field distribution at breakdown voltage V_b (25·2 V) and the one corresponding to the operating bias voltage V_d (50·5 V) are shown in figure 7. (In the latter distribution, the carrier space charge effect is not incorporated.)

At phase $3\pi/2$ when the terminal voltage takes a minimum, the 'averaged' bias field distribution is lowered as shown by the broken line in figure 7. The dynamic field distribution at this instant must be the one drawn by the dotted line in figure 7, which differs from the 'average' one due to the space charge of the drifting carrier bunch.

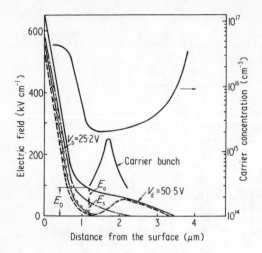

Figure 7. Estimated bias and dynamic field distributions for a high efficiency GaAs Read diode.

The RF field amplitude E_a or RF voltage amplitude V_a is restricted by the fact that the velocity of electrons in the drift region must not fall below the saturated velocity. The field E_t at the trailing edge of the carrier pulse is expressed as $E_t = E_0 - (E_a + E_s)$, where E_0 is the bias field at the position in question, E_a the average RF field amplitude and E_s the field drop due to space charge. E_t should be larger than 3 kV cm^{-1} (threshold field of Gunn effect) at phase $3\pi/2$. E_s is expressed as $E_s = I/2\epsilon\epsilon_0 f$ at phase $3\pi/2$ (Read 1958, Nishida 1968) where I is the bias current density, f the frequency.

For diode No. 9, the maximum efficiency of 22·9% was obtained at $I = 830$ A cm^{-2}, $V_d = 50·5$ V and $f = 9·5$ GHz. The field drop E_s due to the space charge effect is calculated to be 44 kV cm^{-1}. If we assume the effective width of the space charge to be 1 μm and bias field E_0 at the trailing edge of the carrier pulse is about 100 kV cm^{-1} (see figure 7), then $E_a \leqslant E_0 - (E_s + 3 \text{ kV cm}^{-1}) = 53$ kV cm^{-1}, $V_a \leqslant 18$ V and $V_a/V_d \leqslant 0·36$.

The conversion efficiency, η, can be expressed as follows:

$$\eta = \frac{1}{2} \frac{V_a}{V_d} \frac{I_a}{I_d} \cos \psi \tag{1}$$

where V_a and I_a are voltage and current amplitudes at the fundamental frequency, V_d and I_d bias voltage and current and ψ the phase angle between V_a and I_a. Since (I_a/I_d) is usually less than $4/\pi$ (Read 1958), the theoretically estimated maximum efficiency of diode No. A-9 is about 23%, which is very close to the experimental value. But, the theoretical efficiency is considered to be too low, because in the actual operation the phase factor $\cos \psi$ must be less than unity (the RF current and voltage are not completely in phase) and there must be some power losses in diode series resistance and in the cavity circuit. Even if the bias field is assumed to be 110 kV cm^{-1}, the theoretical efficiency should be less than 27%.

The above fact suggests that (I_a/I_d) or (V_a/V_d) in the GaAs Read diodes are not restricted as mentioned above or as in Si Read diodes. Probably the ratio (I_a/I_d) can be larger than $4/\pi$ for GaAs Read diodes because of acceleration of the electron bunch due to the negative differential mobility in GaAs, as suggested by Shroeder and Haddads' (1973) computer calculation or by the simple analysis of Culshaw and Giblin (1974).

If a Si Read diode has a similar bias field distribution to the one in figure 7, the efficiency should be less than 15%, because travelling electrons lose the velocity at the field below 20 kV cm^{-1}. The (V_a/V_d) ratio might be improved if the field in the drift region is increased, but the phase relation must deteriorate due to the increase of RF field amplitude as discussed in a previous paper (Hasegawa and Aono 1974), so that efficiency as high as that of GaAs Read diodes is not to be expected for Si Read diodes.

4. Summary

GaAs Schottky barrier high–low doped Read diodes for X-band have been investigated experimentally, with particular emphasis on the relationship between the oscillation efficiency and the field distribution in the active semiconductor region.

It has been shown that a high efficiency is obtained in a diode whose field distribution at the operating bias point is a little 'over punch through', or the field distribution at the breakdown voltage is an almost 'just punch through' or a little 'before punch through' type, depending on the bias current density of the operating point. This relation is quite consistent with the experimental observation that a diode with a lower breakdown voltage gave a higher efficiency (Hasegawa and Aono 1974).

One of the advantageous features of the Read diode is that the diode area can be made considerably larger than that of a conventional (uniformly doped) GaAs IMPATT diode. Efficient power combination between four mesas on a Au-plated heat sink has been successfully accomplished. Because of this large area capability and with a well designed field distribution, cw output powers of 10 W with efficiencies of over 20% have been achieved at 9·5 GHz. The bias voltages of these diodes were about two thirds, the diode areas were about 2–3 times, the efficiencies were more than 1·5 times, and the output powers were about 3–4 times that of the best conventional GaAs IMPATT diodes made using the same technology.

Investigation of the dynamic field distribution has suggested that the relatively low threshold field of electron velocity saturation in GaAs and probably the acceleration effect due to the negative differential mobility in GaAs, are responsible for the high efficiency of GaAs Read-type IMPATT diodes.

Acknowledgments

The authors would like to thank Dr K Sekido for his valuable suggestions and discussions, Dr Y Okuto and Mr N Muneta for their contributions to the realization of the high-power diodes, and Dr K Ayaki and Dr M Uenohara for their encouragement.

References

Culshaw B and Giblin R A 1974 *Electron. Lett.* **10** 285–6
Goldwasser R E and Rosztoczy F E 1974 *Appl. Phys. Lett.* **25** 92–4
Hasegawa F and Aono Y 1974 *Proc. IEEE Lett.* **62** 631–3
Huang H C 1973 *IEEE Trans. Electron Dev.* **ED-20** 541–3
Irvin J C, Luther L C and Coleman D J 1973 *Device Research Conf., Boulder, Colorado, IEEE Trans. Electron Dev.* **ED-21** 1178

Kim C K, Matthei W G and Steele R 1973 *Proc. 4th Biennial Cornell Electrical Engineering Conf.*
     (Ithaca: Cornell UP)
Nishida K 1968 *Japan. J. Appl. Phys.* **7** 1484–90
Read W T 1958 *Bell Syst. Tech. J.* **47** 401–46
Salmer G, Pribetich J, Farrayre A and Kramer B 1973 *J. Appl. Phys.* **44** 314–24
Schroeder W E and Haddad G I 1973 *Proc. IEEE* **61** 153–82
Su S and Sze S 1973 *IEEE Trans. Electron Dev.* **ED-20** 482–6
Takayama Y 1974 *Technical Digest Int. Electron Devices Mtg, Washington DC* pp130–3
Wisseman W R *et al* 1973 *Device Research Conf., Boulder, Colorado*
—— 1974 *IEEE Trans. Electron Dev.* **ED-21** 317–23

High efficiency avalanche diodes from 10 to 15 GHz

A Farrayre, B Kramer and A Mircea

Laboratoires d'Electronique et de Physique Appliquée, 3, Avenue Descartes,
94450 Limeil-Brevannes, France

Abstract. Efficiencies up to 26% have been obtained at LEP with GaAs IMPATT diodes from 10 to 15 GHz.

The doping profile is of the two-step type with a high N_{D1}/N_{D2} ratio (of the order of 25). Characterization of the doping profile is made with a high precision before starting the fabrication of diodes. The thickness of the n⁺ superficial layer which is a critical parameter is — if necessary — reduced to the optimum value and controlled with a precision of 100 Å. Ionic beam etching was previously used; then an anodic etching technique has been developed to tailor the doping profile. Thus the vapour-grown epitaxial wafers can be made thicker to obtain higher yields. Six out of seven wafers have given efficiencies exceeding 20% with a good homogeneity in the results. As a typical result on such a slice at 12 GHz, 85% of the diodes have an efficiency greater than 16% and 65% greater than 20%.

To avoid the problems associated with Pt diffusion Mo and Al have been tried for the Schottky barrier. These metals do not significantly diffuse in GaAs: the measured diffusion depth is less than 100 Å after 1800 hours at 300 K.

Recently, high efficiencies with GaAs IMPATT diodes have been reported (Goldwasser and Rosztoczy 1974, Kim *et al* 1974). There are two major problems in the design of these devices:

(i) the charge density in the peak with the Lo–Hi–Lo profile or in the n⁺ superficial layer with the Hi–Lo profile must be controlled with high precision;
(ii) the platinum layer diffuses in GaAs.

With the Lo–Hi–Lo profile the charge density in the peak must be controlled during the epitaxial process. With the Hi–Lo profile an eventual excess in the thickness of the n⁺ layer can be corrected by appropriate etching. Thus the epitaxial wafers can be made thicker than required and higher yields obtained. For this reason we have chosen the Hi–Lo profile.

The optimum thickness of the n⁺ layer for different doping levels is given on the figure 1. An electric field of $100 \, \text{kV cm}^{-1}$ at breakdown threshold and at the entrance of the Lo region was found experimentally to be the optimum value. This condition is given on curve A. On the same curve boundaries (curves B and C) are defined which delimits a zone where efficiencies exceeding 15% have been obtained. On curve B the value of the electric field at the entrance of the Lo zone is zero and on curve C about $150 \, \text{kV cm}^{-1}$. The tolerance in the thickness is 300 Å at $2 \times 10^{17} \text{cm}^{-3}$ and $0.3 \, \mu\text{m}$ at $2 \times 10^{16} \text{cm}^{-3}$. As the maximum efficiency increases with the ratio $N_{D \, Hi}/N_{D \, Lo}$, a doping level between 8×10^{16} and 2×10^{17} is required in practice as the best compromise.

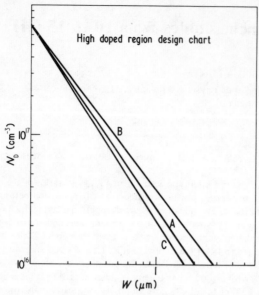

Figure 1. Hi–Lo profile: high-doped region design chart.

To tailor the profile we have tried two techniques.

(*a*) First, after epitaxy, the doping profile was measured with high precision by the $C(V)$ method on evaporated planar diodes. Since the total thickness of the epitaxial layer is greater than the depletion width at breakdown, steps are made on a small sample of the wafers. The height of these steps is measured by a Talysurf with a precision of 100 Å. Then the thickness of the n^+ layer is reduced to the optimum value by ionic beam etching. The apparatus consists of a thermionically sustained and magnetically confined argon arc plasma. An accelerating voltage of 300 V and a 45° incidence angle were found to be the appropriate conditions to minimize the damage at the surface of the sample and to maintain a sufficiently high sputtering yield. The etch rate was $3.2\ \text{Å s}^{-1}$ under these conditions. After etching surface charges appear on the sample which contribute to an important error in the doping profile measurement (figure 2). This negative charge density is of the order of $0.1 \times 10^{12}\ \text{cm}^{-2}$.

By this technique the two best results obtained are 26% at 12 GHz with 1500 mW output power and 24·4% at 15 GHz with 1550 mW. Figure 3 shows the homogeneity in the results on a typical slice at 12 GHz. Eighty-five per cent of the diodes have an efficiency greater than 16% and 65% greater than 20%. On these slices the donor concentration was between 1·5 and $2 \times 10^{17}\ \text{cm}^{-3}$ for the Hi region, and 5 to $6 \times 10^{15}\ \text{cm}^{-3}$ for the Lo region. We observe a net dependence of the efficiency on the breakdown voltage. The optimum is about 28–30 V (figure 4).

(*b*) We have now developed an anodic oxidation technique (Logan *et al* 1973, Rode *et al* 1974) to adjust the doping profile. The basic element of this method is that the anodization bath† behaves as a Schottky barrier and the GaAs sample as a reverse

---

† An adequate composition for this bath has been given to us by B C Easton of Mullard Research Laboratories.

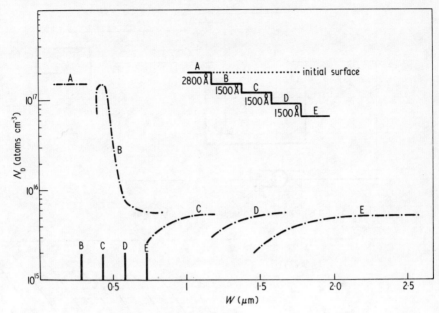

Figure 2. Surface effects after ionic beam etching.

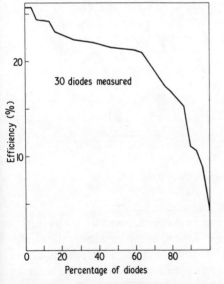

Figure 3. Cumulative diagram: percentage of diodes against efficiency.

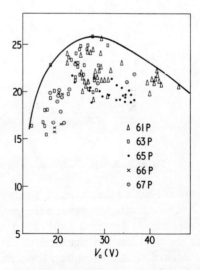

Figure 4. Influence of the breakdown voltage on efficiency.

biased avalanche diode. When the anodization cell passes a constant current we observe the change of the voltage with time (figure 5). In figure 6 a typical record can be seen.

(i) We see a sudden increase up to a threshold value. We have verified on a series of flat profiles that this is the value of the avalanche breakdown voltage.

(ii) Then the voltage increases linearly in time with the formation of the oxide film.

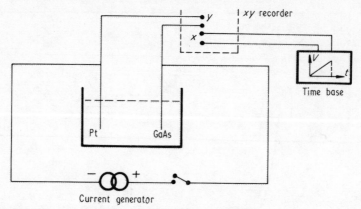

Figure 5. Anodic oxidation: diagram of the apparatus.

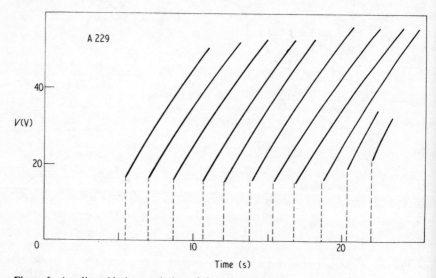

Figure 6. Anodic oxidation: variation of the voltage in time.

After dissolving the oxide, the process can be repeated as desired.

A Hi–Lo profile behaves as a flat profile as long as the thickness of the n^+ layer is greater than the depletion width. Then, when the depleted region extends in the Lo zone the breakdown voltage increases. Thus we can stop the anodization process when the required breakdown voltage is reached. A higher precision can be obtained because the breakdown voltage is very sensitive to the thickness of the Hi region. Then precise measurements of the thickness are not needed.

We have seen that the n^+ layer thickness is critical, thus it becomes important to avoid an eventual translation of the barrier with time.

It is now known that the platinum barrier diffuses in GaAs (Rode 1974, Sinha and Poate 1973, MacFinn *et al* unpublished†). This diffusion tends to reduce the thickness

† H Y MacFinn, P Hong, W T Findley, R A Murphy, E B Owens and A J Strauss; Paper presented at the 1973 DI Conference, Las Vegas, Nevada.

of the n^+ layer and to modify the performance of the diodes. We have observed at 300 °C a diffusion rate of the same order as that published by Coleman *et al* (1974) with a small difference at small diffusion times. This is probably due to a difference in the initial diffusion during the Pt sputtering, caused by the superficial temperature rise of the sample (figures 7 and 8).

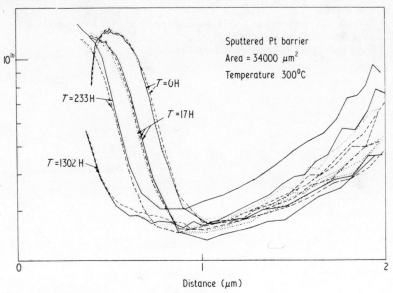

Figure 7. Diffusion of Pt in a GaAs Hi–Lo profile.

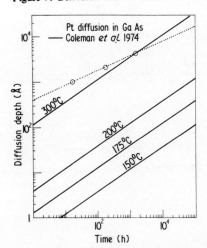

Figure 8. Diffusion rate of Pt.

To replace the Pt by a metallurgically nonreactive contact, we have tested the behaviour of Mo and Al with GaAs. Dots of these metals have been sputtered on a Hi–Lo profile and the thickness of the Hi region was measured after different times. After 1800 hours at 300 °C no diffusion has been observed with either of these two metals. In these measurements the precision reached was 100 Å.

The new diodes are currently being assessed.

References

Coleman D J Jr, Wisseman W R and Shaw D W 1974 *Appl. Phys. Lett.* **24** 335–7
Goldwasser R E and Rosztoczy F E 1974 *Appl. Phys. Lett.* **25** 92–4
Kim C K, Matthei W G and Steele R 1974 *Cornell Conf. Proc. 1973* (Ithaca: Cornell UP) pp299–305
Logan R A, Schwartz B and Sundburg W J 1973 *J. Electrochem. Soc.* **120** 1385–90
MacFinn H Y, Hong P, Findley W T, Murphy R A, Owens E B and Strauss A J 1973 *DI Conference, Las Vegas, Nevada*
Rode D, Lorenzo J Di and Schwartz B 1974 *Workshop on Compound Semiconductors for Microwave Devices, Philadelphia 1974*
Sinha A K and Poate J M 1973 *Appl. Phys. Lett.* **23** 666–8

New design criteria of Gunn diode contacts

T Sebestyen†, H Hartnagel and L Herron
Department of Electrical and Electronic Engineering, University of Newcastle upon Tyne,
Newcastle upon Tyne NE1 7RU, UK

Abstract. Low breakdown, hole injection and other phenomena connected with
evaporated and alloyed metal anode contacts of Gunn diodes are explained by generation
of hot holes in the anode metal due to an energy step at the semiconductor–metal interface and domain quenching. The processes in the highly doped intermediate semiconductor
anode and cathode layers are examined. It is shown that not only the concentration but
also the length of the highly doped semiconductor layers can play a dominant role in
determining the characteristics of a Gunn diode. By changing the length of the highly
doped intermediate semiconductor layer of given concentration the contact can be
altered from an ohmic type to a Schottky barrier or tunnelling reactive cathode contact.
Requirements and order of magnitude values are given for the carrier concentration and
length of the highly doped intermediate semiconductor anode and cathode layers and
compared with the estimated values of some practical devices. Energy band models are
given for the Schottky barrier and tunnelling type current limiting reactive cathode
contacts and experimental findings connected with the alloying of the evaporated metal
contacts are explained by these models. Throughout the paper an X-band GaAs Gunn
diode is used as a numerical example for showing the order of magnitude values estimated.

1. Introduction

During its early days the Gunn diode was said to be a rather simple symmetrical
metal–semiconductor–metal structure in which the metal–semiconductor contacts are
low loss 'ohmic' contacts with negligible length and structure. The main features of
Gunn diode characteristics were said to be determined by the homogeneous semiconductor active layer length l and carrier concentration n or rather their products nl.
It was stated that the more homogeneous the doping profile in the active semiconductor
part and the more linear and ohmic the I–V characteristics of the Gunn diode at low
voltage are, the more efficient is the Gunn diode operation. More recently it turned
out that a current limiting 'reactive' cathode contact with rather high and non-ohmic
contact resistance and with about 0·15 to 0·25 eV energy barrier is advantageous in
achieving higher efficiency and breakdown voltages (Yu *et al* 1971, Colliver *et al* 1972).
It was expected that a continuous change can be made from a reactive cathode contact
to a low resistance ohmic contact by changing the doping concentration in the cathode
(Colliver *et al* 1972). This paper will discuss that besides the doping concentration n_c
the length l_c of the cathode or rather their product $n_c l_c$ can be dominant in determining
Gunn diode properties. We consider only ideal contacts, where any of the often observed,
deep interface–crystallographic damage is neglected.

† On leave from Research Institute for Technical Physics of the Hungarian Academy of Sciences,
H-1047 Budapest, Foti ut 56, Hungary.

On the other hand Gunn diodes are often experimentally found to have electrical asymmetry with respect to polarity of bias voltage, and a low breakdown voltage was obtained if the evaporated and alloyed metal contact was applied as the anode. This phenomenon was usually explained by a non-uniform conductivity profile in the active n-layer with a short low conductivity region near the anode of low breakdown voltage (Hasty *et al* 1968, Harris *et al* 1969). We shall show that a small length l_a^+ and concentration n_a^+ of the anode region or rather their small product can cause a low breakdown voltage even if the conductivity is homogeneous in the active layer. We introduce as new design parameters the anode and the cathode layer lengths and concentration or rather their products.

2. Design of anode contacts

2.1. Thermal asymmetry in Gunn diodes

In sandwich or mesa type Gunn diodes the thick n^+ substrate layer needed during processing for mechanical support of the thin active n-layer is generally employed as an anode layer and heat sinking is provided through a thin cathode contact layer prepared usually by alloying a metal layer deposited by vacuum evaporation. The main reason for this is the low anode breakdown voltage found if the evaporated metal layer is biased as the anode. Near the anode end of the active layer, however, generally more heat is inherently generated. The reasons of the asymmetric heat generation are:

(i) the time average of the electric field is higher towards the anode since during their transit towards the anode, the electrons are transferred from the central valley to satellite valleys (domain growth), where their mobility is reduced and thus the resistivity increased;
(ii) possible contact resistance difference of the backward biased cathode and forward biased anode contact;
(iii) the positive feedback of the temperature gradient since the higher the temperature the lower the mobility both in the central and the satellite valleys.

Additionally there are other reasons of thermal asymmetry, namely heat sinking through one contact only, and inhomogeneities in cross section, doping, etc, caused by technological factors.

For a Gunn diode the maximum length l^+ permitted for the highly doped contact layer through which heat sinking takes place is determined by the acceptable temperature drop T^+ across it. If this is, say, 10% of the temperature drop T along the active layer, one can estimate l^+ to a first approximation as follows (Narayan and Paczkowski 1972):

$$0 \cdot 1 = T^+/T = 2K \frac{l^+}{lK^+}$$

where K and K^+ are the thermal conductivity of the n-GaAs and the n^+-GaAs, respectively. For a usual X-band Gunn diode we would obtain then

$$l_{max}^+ = l/25 = 4000 \text{ Å}.$$

If the carrier concentration of this contact layer is about $2 \times 10^{18} \text{ cm}^{-3}$, then $n^+l^+ = 8 \times 10^{13} \text{ cm}^{-2}$.

There exists an experimental observation of hole—electron recombination radiation, too, originating at the anode and decreasing towards the cathode. This phenomenon was proposed to be the most important reason of material migration and thermal breakdown through recombination heat (Jeppsson and Marklund 1967, Ullrich 1971). There was, however, insufficient reason given for hole injection. We shall explain the hole injection phenomenon in the next section.

2.2. Schottky barrier anode contact made directly on the active layer

A Schottky barrier type anode contact attached directly to the heat sink would be ideal from a thermal point of view since l^+ would be very small. The electrical contact resistance would also be comparatively low at the high bias current densities needed for a Gunn diode since the built-in potential ϕ_{Bn} at an n-GaAs—metal Schottky barrier junction is in the range of $0{\cdot}7{-}1$ V. Therefore about 10% of a bias voltage of 7 to 10 V of a usual X-band Gunn diode is sufficient to remove the energy barrier. The contact resistance R_c at high bias current densities J can be estimated from

$$R_c = \phi_{Bn}/J$$

and is in the order of $10^{-4}\,\Omega\,\mathrm{cm}^2$.

The active semiconductor layer, however, injects very hot electrons into the metal because of the high step of about 1 eV for GaAs in the band structure and the lack of domain extinction in an intermediate layer. The hot electrons injected into the anode metal lose their energies in a very short time (about 10^{-13} s) and distance (about 200 Å) by impacting on phonons, impurities and cold electrons of the anode metal (Sze 1969). They excite electrons into higher energy states from well below the Fermi level bringing about hot holes which are able to overcome the relatively low hole barrier height (about 0·5 eV in GaAs) as shown in figure 1. The holes fed back into the active layer have

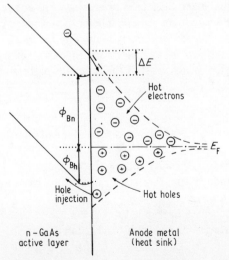

Figure 1. Schematic energy band diagram of a strongly forward biased Schottky barrier anode contact on an n-GaAs active layer.

long drift and diffusion lengths there (Ryan and Eberhardt 1972) and travel back to the cathode causing recombination heat and radiation (Jeppsson and Marklund 1967, Ullrich 1971) enhanced conductivity and material migration caused partly by drag of the relatively heavy holes on atoms, low breakdown voltages, etc. If any domain is created in the active layer then their quenching in the anode metal means a strong surge of hot electrons which causes a surge of hot holes and their injection into the active layer making the oscillation noisy and incoherent. The breakdown is caused by enhanced conductivity of the active layer due to minority carriers and the voltage generator type bias circuit which does not limit the current. This breakdown at the energy step resembles a trap-assisted breakdown phenomenon. The analysis of this structure shows that a Schottky diode anode contact can not be applied directly on the active layer of a Gunn diode.

2.3. Criteria for a sufficiently highly doped anode layer

The injection of holes into the active layer decreases if in a long enough highly doped semiconductor anode layer

(i) the domains are attenuated or extinguished before reaching the semiconductor—metal interface
(ii) the recombination of holes increases
(iii) the effective ϕ_{Bn} is decreased due to tunnelling and
(iv) the electric field decreases.

All of these requirements can be satisfied by an intermediate n^+ semiconductor anode layer being sufficiently long and highly doped.

If the heat sink is applied on the cathode side there is no serious limit for the length and the doping of the intermediate anode layer and the above requirements can be satisfied easily (eg, using the highly doped and thick substrate layer as the anode). If we want to apply the heat sink at the anode side where, inherently, more heat is generated, the length of the highly doped layer is seriously limited by the small temperature drop requirements as was shown above. The whole length of the highly doped and sufficiently long anode layer has to consist of the following parts (shown schematically in figure 2):

(i) Length l_1 of the depletion in the highly doped anode layer caused by either
the diffusion of electrons due to their concentration gradient at the n^+–n high—low contact or
the surface charge depth on the anode electrode opposing the cathode or
the surface charge depth on the anode as opposing the accumulation layer of the domain during the quenching.
Both the first and the second effects give a negligible value as compared with the third effect. Assuming full depletion we get for a usual X-band Gunn diode:

$$l_1 \simeq l_d n/n^+ \simeq 60 \text{ Å}$$

assuming that the domain depletion length $l_d \simeq 6\,\mu\text{m}$ and $n/n^+ \simeq 10^{-3}$.

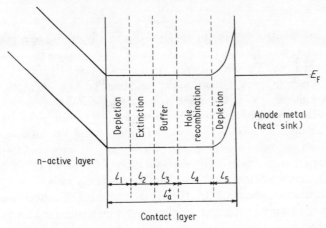

Figure 2. Constituent layers of the anode contact able to avoid any hole diffusion into the n-GaAs active layer.

(ii) Length l_2 for domain extinction which is roughly determined by the dielectric relaxation time of the highly doped layer. Its order of magnitude can be estimated as

$$l_2 \simeq \ln/n^+ \simeq 100 \, \text{Å}$$

for our example.

(iii) Length l_3 of an additional buffer layer to incorporate a safety factor.

(iv) Length l_4 for electron–hole recombination to avoid any hole injection into the high-field areas. The hole diffusion length is about $2 \, \mu$m at 300 K in the n-GaAs if $n < 10^{18} \, \text{cm}^{-3}$, but above $n \simeq 2 \times 10^{18} \, \text{cm}^{-3}$ its value decreases very rapidly and at $n \simeq 7 \times 10^{18} \, \text{cm}^{-3}$ its value is about 100 Å for bulk GaAs (Hwang 1971). For good epitaxial layers the decrease is not so fast (Casey *et al* 1973).

(v) Length l_5 of the depletion layer at the n^+-GaAs–metal interface which is about 300 Å at 300 K for $n = 10^{18} \, \text{cm}^{-3}$ and zero voltage drop. This value decreases towards zero at high forward bias current densities.

The requirements for a usual X-band GaAs Gunn diode is therefore:

$$l_1 + l_2 + l_3 + l_4 + l_5 = l_a^+ < l_{\text{max}}^+ \simeq 4000 \, \text{Å}.$$

This can only be satisfied if the carrier concentration n^+ of the highly doped anode layer is raised into the $10^{19} \, \text{cm}^{-3}$ range. This is mainly because of the high diffusion length of holes in the n^+-GaAs. The high doping concentration also helps to satisfy the above-mentioned requirements on ϕ_{Bn} and electric field.

3. Design of cathode contacts

Holes eventually created at cathode contacts on an n-type semiconductor layer are removed from the semiconductor because of the current flow direction. The cathode contact is backward biased; therefore, the depletion layer length increases with higher bias and can penetrate into the active layer.

3.1. *Ohmic cathode contacts*

A highly doped intermediate semiconductor layer is generally needed to achieve low loss 'ohmic' contacts with specific contact resistance not depending strongly on temperature and applied bias. The ohmicity of a contact, of course, depends on the ratio of contact resistance to series resistances. For an X-band GaAs Gunn diode the specific contact resistance $R_a = \rho l$ of the active layer is about $10^{-3} \, \Omega \, \text{cm}^2$; therefore, the specific cathode contact resistance should be in the range of $10^{-4} \, \Omega \, \text{cm}^2$ or below. Such a low value can be obtained for n-GaAs and zero bias voltage if the concentration of the ionized donors beneath the metal contact is about $2 \times 10^{19} \, \text{cm}^{-3}$ (Chang *et al* 1971). At such high concentration the transport of electrons takes place mainly through the tunnelling process and the specific contact resistance depends only weakly on temperature. Its thermal coefficient is about $1\% \, ^\circ\text{C}^{-1}$ (Chang *et al* 1971). The critical tunnelling length X for this energy barrier profile and specific contact resistance can be estimated to be in the range of 50 Å.

The other important requirement from our viewpoint is that the highly doped intermediate layer thickness l_c^+ must be larger than the depletion layer length W^+ at applied bias. Then the depletion layer does not penetrate into the active layer as shown in figure 3. The situation is more complicated if there are deep donor levels in the highly

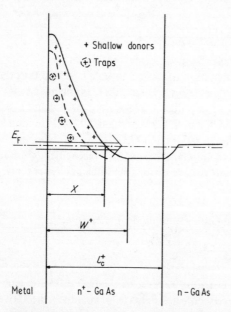

Figure 3. Schematic energy band diagram of an ohmic cathode contact. W^+ and X are the depletion layer width and the effective tunnelling length for this energy barrier profile, respectively.

doped layer since in this case they are ionized near the surface of the semiconductor layer, contribute to the space charge and make alterations both in the profile and the width of the energy barrier as schematically shown by a dotted line in figure 3. From the viewpoint of the low contact resistance their overall effect, however, can be advantageous. The requirements for this highly doped ohmic cathode layer, if heat sinking is employed through it, can be written as follows:

$$n_c^+ \gtrsim 2 \times 10^{19} \, \text{cm}^{-3}$$

$$50 \, \text{Å} \simeq X < W^+ < l_c^+ < l_{max}^+ \simeq 4000 \, \text{Å}$$

and

$$n_c^+ l_c^+ \gtrsim 2 \times 10^{13} \, \text{cm}^{-2}$$

where the numerical values are related to our example of an X-band GaAs Gunn diode.

3.2. Reactive cathode contacts

If the depletion layer length W^+ is larger already at zero or low bias voltage than the thickness l_c^+ of the highly doped intermediate semiconductor contact layer then the depletion layer penetrates into the active layer and its width increases very rapidly at higher bias voltages and the current is limited. In the first type of the current limiting cathode, the current limitation was attributed to thermionic emission over an energy barrier with height from 0·15 to 0·25 eV (Colliver *et al* 1972). Figure 4 shows schematically the possible band structure of this Schottky barrier type cathode and typical low voltage $I-V$ characteristics obtained (Colliver *et al* 1972). In this case the n$^+$ layer length l_c^+ is smaller than the critical tunnelling length X; therefore, the voltage drop across it is also very small and the effective barrier height depends weakly on the applied bias.

The same structure made from Si ought to operate as a majority type BARRITT diode since it would show a negative differential resistance due to transit time effect and the saturation type velocity–field characteristics of Si. In the case of materials with inherent negative differential resistance due to the transferred-electron effect a better BARRITT

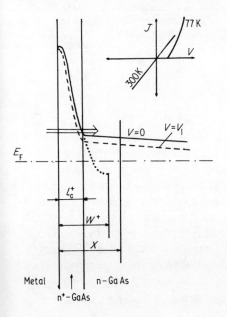

Figure 4. Schematic diagram of the energy band model for a Schottky barrier type current limiting reactive cathode contact for which the typical $I-V$ characteristics (obtained by Colliver *et al* 1972) are also shown. The depletion layer length would be W^+ if the highly doped layer were thick enough. X is the critical effective tunnelling length which means that the highly doped intermediate semiconductor layer can be considered to be transparent for electrons if its width is smaller than X.

diode operation could be obtained because of the more advantageous velocity–field characteristics. This structure, if considered now as a Gunn diode, compensates for the electron relaxation and dynamical effects by shifting the average electric field distribution towards the anode (Yu *et al* 1971, Colliver *et al* 1972, Bosch and Thim 1974).

The order of magnitude values for our numerical example can be estimated as follows:

$$l_c^+ \lesssim X \simeq 50 \,\text{Å}$$

$$n_c^+ \gtrsim 2 \times 10^{19} \,\text{cm}^{-3} \qquad l_c \lesssim 50 \,\text{Å} \qquad \text{and} \quad n_c^+ l_c^+ \simeq 10^{13} \,\text{cm}^{-2}.$$

In the second type of current limiting cathode contact the current limitation was attributed to a tunnelling process since at larger reverse voltages the current was large and showed only a weak temperature dependence (Colliver *et al* 1972). For this case we assume a band structure model as schematically shown in figure 5 together with the low voltage *I–V* characteristics. The important feature of this model is that, contrary

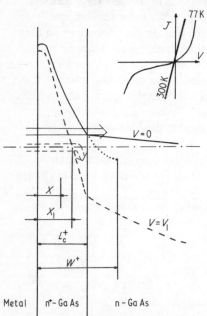

Figure 5. Schematic diagram of the energy band model for a tunnelling type current limiting reactive cathode contact for which the typical *I–V* characteristics (obtained by Colliver *et al* 1972) are also shown. X and W^+ are the critical effective tunnelling and depletion layer lengths in the same sense as explained in figure 4. X_1 is the effective tunnelling length at bias voltage V_1.

to the model of the first type of reactive contact shown in figure 4 the critical effective tunnelling length X is smaller than the thickness l_c of the highly doped intermediate semiconductor layer. The consequence of this is that the voltage drop across l_c is higher and the barrier profile and the effective tunnelling length X_1 depend very much on the bias voltage. At higher biases the effective tunnelling length and barrier height decrease and instead of field-assisted thermionic emission a pure field emission transport process takes place. This type of energy barrier profile also helps to accelerate the electrons since there are strong DC built-in electric fields behind the barrier. This could cause the lack of a 'dead zone' (Bosch and Thim 1974) and a higher efficiency. This model is again a good example to emphasize that not only the concentration but also the thickness of the highly doped intermediate semiconductor layer can be important.

The order of magnitude values for our numerical example are as follows:

$$50 \, \text{Å} \simeq X < l_c^+ \simeq 100 \, \text{Å} < W^+ \simeq 150 \, \text{Å}$$

and if

$$l_c^+ \simeq 100 \, \text{Å} \qquad n_c^+ \simeq 7 \times 10^{18} \, \text{cm}^{-3} \qquad \text{then} \qquad n_c^+ l_c^+ \simeq 7 \times 10^{12} \, \text{cm}^{-2}.$$

The current limiting cathode is also advantageous regarding heat sinking through the cathode since in this case the maximum of the heat generation density is shifted towards the cathode electrode.

4. Technological aspects

The structure of the usual contacts are influenced by the technologies employed. The values of some parameters are shown in table 1 for X-band GaAs Gunn diodes.

Table 1. Estimated order of magnitude values for some highly doped intermediate semiconductor contact layers obtained by usual Gunn diode technologies

Contact layer	Order of magnitude values		
	$n^+ \, (\text{cm}^{-3})$	$l^+ \, (\mu\text{m})$	$n^+ l^+ \, (\text{cm}^{-2})$
Thick substrate	$1-3 \times 10^{18}$	$100-400$	$10^{17}-10^{16}$
Thin substrate (integrated heat sink)	$1-3 \times 10^{18}$	$20-30$	$10^{16}-10^{15}$
Epitaxial contact layer	$0 \cdot 5-2 \times 10^{18}$	$1-5$	$10^{15}-10^{14}$
In evaporated and alloyed contacts (estimated)	$10^{17}-10^{19}$	$10^{-3}-10^{-2}$	$10^{13}-10^{11}$

It is surprising how few data are available in the literature for the evaporated and alloyed metal contacts of great importance. We have estimated them solely from the liquid and solid solubility data supposing that the diffusion rate of donors into GaAs is negligible at comparatively low temperatures. One of the main difficulties of Gunn diode technology is that there is no doping element giving a very high donor concentration at the alloying temperatures. In fact, every donor-type element becomes self-compensating if its concentration is raised into the range of 10^{18} to $10^{19} \, \text{cm}^{-3}$; hence this seems to be an upper limit of alloyed contacts. The thickness of the evaporated metals is between $0 \cdot 1$ and $1 \, \mu\text{m}$ and they can dissolve only a few per cent of GaAs at the usual alloying temperatures; therefore the thickness of the highly doped regrown layer might be in the range $10-100 \, \text{Å}$. The ideal heat sinking ohmic anode contact with a thickness of at least $1000 \, \text{Å}$ and with a concentration of about $10^{19} \, \text{cm}^{-3}$ (that is, with $n^+ l^+$ product about $10^{14} \, \text{cm}^{-2}$) cannot easily be obtained with alloyed metals.

Table 2 shows possible combinations of the highly doped contact layers in various types of Gunn diodes, and the nature of the junction between the contact metal and the highly doped semiconductor layer.

Table 2. Some combinations of the highly doped intermediate semiconductor ohmic and reactive layers of Gunn diodes. Signs +, − and ? mean desirable, undesirable and questionable combinations, respectively. O = ohmic, R = reactive contact

				Contact layer			
				Epitaxial		Metal	
Nature of metal−semiconductor contact		$\rightarrow$		O	R	O	R
		$\downarrow$	Case	a	b	c	d
Substrate layer	Anode	O	1	+	−	+	+
		R	2	+	−	+	+
	Cathode	O	3	+ ?	?	?	−
		R	4	−	−	−	−

In cases 1 and 2 the highly doped anode layer is a relatively thick substrate; therefore the heat sinking should take place through the much thinner cathode layer. The reactive cathode contacts (cases b and d) are advantageous from heat sinking aspects in these cases since the peak of the heat generation density is shifted towards the cathode if a current limiting cathode contact is employed.

Cases 1a and 1c are usual combinations. In cases 1b and 2b, the reactive cathode contacts could hardly be realized by epitaxially deposited contact layers since very thin epitaxial contact layers would be needed for reactive contact layers. Case 1d has been realized as an anomalous InP Gunn diode (Colliver *et al* 1972). Cases 2a, 2c and 2d have also been realized at the Plessey Company (Brookbanks *et al* 1974) and seem to be promising. Gunn diodes corresponding to these cases were prepared by integrated heat sink technology. The cathodes were epitaxially deposited ohmic or evaporated and alloyed reactive or ohmic metal contacts. On the anode side the evaporated metal layer was not alloyed; that is, a reactive (Schottky barrier) anode contact was employed behind a *thick* (substrate) n^+ layer. In this thick n^+ anode layer the domains are easily quenched and the holes eventually fed-back, are recombined.

In cases 3a−d the cathode is the thick substrate layer hence the heat sinking should take place through the thin anode layer. In these cases Gunn diodes with ohmic contacts would be desirable since the peak of the heat generation density in the active layer is shifted towards the anode. The main problem in these cases arises from the difficulty in avoiding the feed-back of holes and low breakdown voltages.

Cases 4a−d are not practical since for reactive cathode contacts very thin n^+ layers are needed.

4.1. Interpretation of contact alloying phenomena

The energy band models introduced to interpret the cathode contacts are suitable to explain the phenomena observed in connection with metal contact alloying. It is well established experimentally (Paola 1970, Heime *et al* 1974, Salow and Grobe 1968) that the specific contact resistance increases again after going through a minimum as

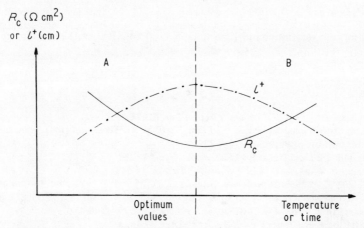

Figure 6. Specific contact resistance and regrown layer length versus temperature or time.

shown in figure 6, if either the maximum alloying temperature or the alloying time is increased from low values. In the range A of figure 6 the quantity of the dissolved GaAs is too little to give a sufficiently thick regrown layer (dissolution-limited range). In the range B the quantity of the dissolved GaAs is enough but because of the high As pressure, the high diffusion rate of the dissolved As in the evaporated metal system and the high surface—volume ratio of the metal contact, the As loss is too high to have sufficient As in the metal system during the regrowth part of the temperature cycle (As evaporation-loss range).

The unduly high As loss can be avoided by a very short non-equilibrium process (as is usually done) or by applying a sufficiently high As pressure during alloying from a separate As source and by employing a heat cycle similar to that of liquid phase epitaxy (as is done in a new alloying method called Thin Phase Epitaxy, Sebestyen *et al* 1974). In this latter case the limit in dissolution can also be avoided by evaporating Ga, too, into the metal system.

5. Summary

The considerations given in this paper indicate that not only the concentration but also the length of the highly doped semiconductor layers between the contact metals and the active semiconductor layer can play a dominant role in determining the characteristics of a Gunn diode. At a given concentration of the highly doped intermediate layer the energy barrier profile at the cathode contact can be controlled by controlling the length of this layer. By changing the barrier profile the contact can be altered from ohmic to a thermionic or tunnelling type current limiting reactive cathode contact. On the anode side of a Gunn diode the length of the highly doped intermediate semiconductor layer determines also the nature of the contact which can be an ohmic and non-injecting or Schottky barrier type injecting anode. The experimentally found alloying behaviour of Gunn diode contacts is also consistent with the band structure models given in this paper.

Acknowledgments

The authors are very much indebted to Dr K Gray of RRE for the stimulating discussions on this topic.

References

Bosch R and Thim H W 1974 *IEEE Trans. Electron Dev.* **ED-21** 16
Brookbanks M, Griffith I and White P M 1974 *Proc. Conf. Metal–Semiconductor Contacts* (London and Bristol: Institute of Physics) pp116–22
Casey H C Jr, Miller B I and Pinkas E 1973 *J. Appl. Phys.* **44** 1281
Chang C Y, Fang Y K and Sze S M 1971 *Solid St. Electron.* **14** 541
Colliver D J, Gray K W, Jones D, Rees H D, Gibbons G and White P M 1972 *Proc. 4th Int. Symp. Gallium Arsenide and Related Compounds* (London and Bristol: Institute of Physics) p286
Harris J S, Nannichi Y, Pearson G L and Day G F 1969 *J. Appl. Phys.* **40** 4575
Hasty T E, Stratton R and Jones E L 1968 *J. Appl. Phys.* **39** 4623
Heime K, König U, Kohn E and Wortmann A 1974 *Solid St. Electron.* **17** 835
Hwang C J 1971 *J. Appl. Phys.* **42** 4408
Jeppsson B and Marklund I 1967 *Electron. Lett.* **3** 213
Narayan S Y and Paczkowski J P 1972 *RCA Rev.* **33** 752
Paola C R 1970 *Solid St. Electron.* **13** 1189
Ryan R D and Eberhardt J E 1972 *Solid St. Electron.* **15** 865
Salow H and Grobe E 1968 *Z. Angew. Phys.* **25** 137
Sebestyen T, Hartnagel H and Herron L H 1974 *Electron. Lett.* **10** 372
Sze S M 1969 *Physics of Semiconductor Devices* (New York: Wiley Interscience) p599
Ullrich D 1971 *Electron. Lett.* **7** 193
Yu S P, Tantraporn W and Young J D 1971 *IEEE Trans. Electron Dev.* **ED-18** 88

Indium phosphide CW transferred electron amplifiers

R Corlett, I Griffith and J J Purcell

Allen Clark Research Centre, The Plessey Company Limited, Caswell, Towcester,
Northants, UK

Abstract. The results of measurements of noise, gain, bandwidth and gain compression
of CW indium phosphide transferred electron amplifiers are presented. J-band noise
figures below 12 dB have been obtained with devices from a number of slices, with a
lowest result of less than 9 dB at 14·5 GHz.

1. Introduction

Solid state devices such as avalanche diodes (IMPATT) and transferred electron
devices, have been successfully used as reflection amplifiers in X-band.

Noise figures published to date range from:

	30 to 40 dB for	Si IMPATT
	15 to 30 dB	GaAs IMPATT
and	15 to 20 dB	GaAs TEA

with a lowest figure of 10·5 dB reported by Rabier and Spitalnik (1973).

The impedance-field method of Shockley *et al* (1960) predicts a minimum thermal
noise figure, in the limit of negligible space-charge growth, of

$$\text{Noise figure} \;\rightarrow\; 1 + (q/kT)\,(D/|\mu|).$$

Amplifiers produced from materials having minimum high field values of $D/|\mu|$ should
therefore afford the possibility of realizing lowest noise figures.

Although diffusion information is limited, predictions of Hammer and Vinter
(1973) indicate that indium phosphide amplifiers offer an improved noise performance,
relative to gallium arsenide.

In this paper, the results of J-band CW noise figure measurements are reported. Noise
figures below 12 dB have been obtained from a large number of devices processed from
11 slices of indium phosphide (Corlett *et al* 1974).

2. Device technology

The slices are produced by vapour epitaxial growth of n-type InP onto Sn doped
substrates. An n^+–n–n^+ structure is generally used, having S doped contact and buffer
layers grown with carrier densities of greater than $10^{17}\,\text{cm}^{-3}$. The active layer are 7 to
14 μm in width, with carrier concentrations below $10^{15}\,\text{cm}^{-3}$. A typical profile is shown
in figure 1.

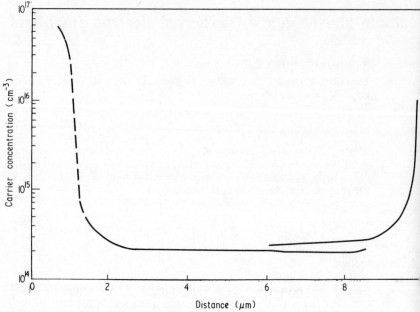

Figure 1. Typical doping density profile.

Devices are processed using the integral heat sink (IHS) technique, with an evaporated Ti–Pd–Au back contact, and annealed Au–In–Ge front contacts. The layer is plated with a 50 to 75 μm thick gold heat-sink and mesa etched into devices of approximately 75 μm diameter.

3. Noise and gain measurements

Microwave assessment of the packaged diodes is performed using a 7 mm, 50 Ω, coaxial air-line, with a single matching slug.

Noise figure measurements are obtained using the Y-factor method (Mumford and Scheibe 1968). A schematic diagram of the system used for both noise and gain measurement is shown in figure 2. Gain is measured by replacing the amplifier by a short-circuit, and the noise figure is measured by comparison with a calibrated noise tube. The results of a noise measure determination across the frequency band 12 to 17 GHz is shown in figure 3. At each frequency, the bias and circuit tuning have been adjusted to minimize the noise, and to maintain gain in excess of 10 dB.

In table 1 are shown data obtained from 11 slices of InP. More than 50% of the slices processed produced diodes with noise figures below 12 dB, measured at 14·5 GHz. Also shown in the table are results of pulsed noise figure measurements made at RRE, using the equipment described by Braddock and Gray (1973). It is evident, that the noise figure is not strongly temperature dependent, and that the design of the active layer profile is not critical. The lowest cw noise figure obtained is 8·5 dB at 14·5 GHz. This result is somewhat anomalous, however, as it was obtained from a slice with an I–V characteristic which exhibited a pronounced hysterisis, and annealing effects. By

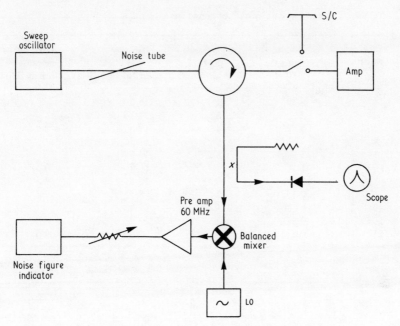

Figure 2. Schematic diagram of noise and gain measurement system.

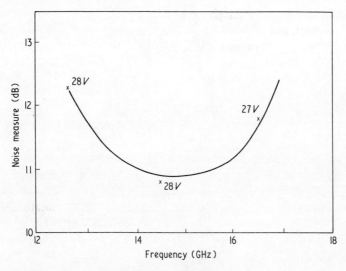

Figure 3. CW noise measure determination in J-band.

quenching the device in liquid nitrogen, the current change could be halted, thus enabling gain and noise measurements to be made.

In figure 4, is shown the amplifier performance in J-band, over a temperature range of 25 °C to 100 °C. By adjustment of the bias voltage, the gain can be maintained essentially constant.

Table 1. CW and pulse noise data obtained from 11 InP slices

Slice No.	Doping (cm^{-3}) $\times 10^{-14}$	Thickness (μm)	IHS noise measure (dB at 14·5 GHz)	Layer-up noise measure (dB at 10·5 GHz)
1	6−9	11	16·5	
2	2	8	11·15	
3	5	14	12·5	8·4
4	2	9	11·35	11
5	5	8·5	14·8	
6	1·5	8	10·7	10·5
7	3·5	10·5	13·8	14·6
8	0·15	12·5	8·55	9·6
9	4−20	11	12·8	
10	0·2	12	8·9	9·0
11	0·8	13	9·8	9·6

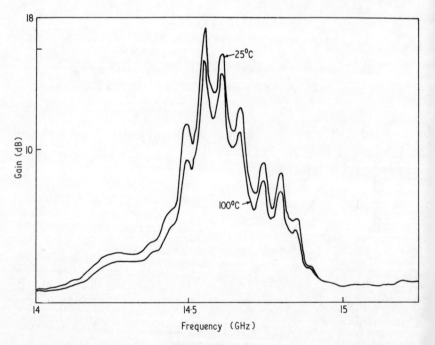

Figure 4. Amplifier performance with ambient temperatures between 25 and 100 °C.

Gain saturation curves are shown in figure 5. Maximum gain falls to about 10 dB at a 1 mW drive level. At lower bias, the gain curve is practically flat until saturation at about −10 dBm. AM/PM conversion with saturation is below 1° dB^{-1}.

Slotted-line impedance measurements have shown instantaneous negative resistance bandwidths of about 4 GHz. Adjustment of the bias voltage enables gain to be realized over octave bandwidths.

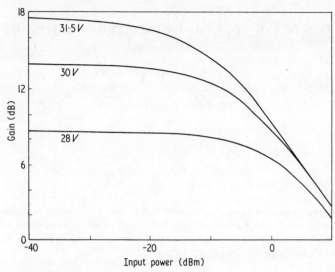

Figure 5. Gain saturation performance.

4. Conclusions

In conclusion, it has been shown that indium phosphide can be reproducibly processed to produce cw amplifiers for J-band applications, having noise figures below 12 dB.

Measurements of gain, noise temperature stability and saturation indicate their potential for many systems applications.

References

Braddock P W, Gray K W and Hodges R D 1973 *Proc. European Microwave Conf. Brussels 1973*
Corlett R M, Griffith I and Purcell J J 1974 *Electron. Lett.* **10** 307–5
Hammer C and Vinter B 1973 *Electron. Lett.* **9** 1
Mumford W W and Scheibe E H 1968 *Noise Performance Factors in Communication Systems* (Dedham: Horizon House)
Rabier A and Spitalnik R 1973 *Proc. European Microwave Conf. Brussels 1973*
Shockley W, Copeland J A and James R P 1960 *Quantum Theory of Atoms, Molecules and the Solid State* (New York: Academic)

Characteristics of Ga$_x$In$_{1-x}$Sb transferred-electron devices

J Michel, P Esquirol, A Joullié and E Groubert

Centre d'Etude d'Electronique des Solides associé au CNRS, Laboratoire de Physique Appliquée, Université des Sciences et Techniques du Languedoc, 34060 Montpellier, France

Abstract. It is now well known that Ga$_x$In$_{1-x}$Sb mixed crystals are suitable systems for transferred-electron devices. Indium rich monocrystals grown by travelling solvent method (TSM) are presently studied.

In this paper we define electrical characteristics of Ga$_x$In$_{1-x}$Sb bulk material transferred-electron oscillators (TEO) and present current—voltage characteristics obtained by a subnanosecond time domain reflectometry (TDR) technique. Preliminary data on microwave performance of TEO are also reported.

1. Introduction

Well known theoretical aspects of transferred-electron effects in III—V compounds point out the interest of various materials such as InP, InAsP, InGaAs, GaInSb, for transferred-electron devices (Hilsum and Rees 1970a,b, McGroddy *et al* 1969, McGroddy 1970).

Considering the necessity of high electron mobility for good microwave performance and convenient band structure to ensure easy electron transfer with great peak-to-valley ratio it appeared in agreement with the band structure engineering method (Hilsum 1971) that GaInSb alloy was a very interesting material.

McGroddy *et al* (1969) gave the first experimental confirmation using polycrystalline material but with recent progress in crystal growth it was possible to obtain more significant results as well in the indium-rich composition range (Joullié *et al* 1974), as in the gallium-rich range (Hojo and Kuru 1974).

A more complete material information necessitates knowledge of the velocity—field characteristic; for technical reasons we only present the current—voltage graph obtained by a subnanosecond reflectometry technique. In spite of its restricted possibilities, this method gives valuable comparative data and permits an interesting comparison with commercial GaAs devices.

According to these studies we state fundamental conditions to obtain current instabilities and give a valuable estimate of energy gap variation as a function of the alloy parameter x.

As a confirmation of the practical interest of this material we show room temperature coherent microwave oscillations.

2. Device preparation

2.1. Crystal growth

The Ga$_x$In$_{1-x}$Sb crystals are obtained by a travelling solvent method. The growth procedure consists of setting high thermal gradient (about 100 °C cm^{-1}) along a sandwich formed by a seed (indium antimonide), a solvent (pure indium) and a charge (polycrystalline Ga$_x$In$_{1-x}$Sb obtained by a vertical Bridgman method). The charge is regularly dissolved, and then crystallizes on the InSb seed. The runs are operated at low temperatures (about 400 °C) under vacuum (10^{-6} Torr), and without any contact (the charge is floating over the solvent). So impurity incorporation during the growth is negligible and the crystal purity depends on the purity of the sandwich components. n-type (111) oriented Ga$_x$In$_{1-x}$Sb crystals were obtained in the concentration range $0 < x < 0.5$.

2.2. Diode preparation

Wafers are mechanically lapped to about 100 or 150 μm in thickness, surface dislocations are then removed by chemical etching with HCl and bromine—methanol. Contacts are deposited by classical vacuum evaporation technique. Metallic alloys employed for GaAs seem satisfactory for GaInSb. We sometimes use Au, Ge, Ni and generally Au, Sn with alloying in H$_2$ atmosphere at 200 °C.

Wafers are cleaved in small pellets, cross dimensions are usually in the range 100 to 400 μm.

Diodes are pinched between the two conic electrodes of a special package easy to demount. This arrangement is useful to study a great number of diodes but does not have good power dissipation so we report only results obtained in the pulsed mode. The package may be inserted in a coaxial air line for reflectometry measurements or in a coaxial tunable cavity (about 1 to 8 GHz) for current instability studies.

3. Experimental procedure

Current instabilities in coaxial cavity and current—voltage characteristics are both studied.

3.1. Cavity measurement

To detect the current instabilities we use a coaxial tunable cavity covering the range 1 to 8 GHz. This very broad band is not consistent with a very high quality factor, moreover the demountable package does not have a good high frequency response and a practical result is that a lower band 0.8 to about 3 GHz is enhanced. Current signal, very noisy instabilities or coherent oscillations, are sent to a spectrum analyser or sampling oscilloscope.

3.2. Current—voltage characteristic measurement

It is well known that the velocity—field characteristic is necessary to understand electron transfer mechanism and forecast material performances. Satisfactory methods developed for GaAs are not applicable to GaInSb alloys because it is not yet possible to obtain large homogeneous samples or donor density below $10^{15}\,\mathrm{cm}^{-3}$.

Presently we have only obtained current—voltage measurements using a subnanosecond time domain reflectometry technique (Jantsch and Heinrich (1970)). This method may be compared to the ultra-short pulse techniques proposed by Gunn and Elliott (1966) and Elliott (1968) to obtain the static characteristics by measurements before significant buildup of instabilities. It also presents the same defect that is to give too low a value of the negative differential resistance, electric field redistribution being nonuniform in subnanosecond time.

A diagram of the experimental arrangement is given in figure 1 with typical oscillogram records. The sample is mounted at the end of a coaxial transmission air line just before a short circuit. A voltage probe is inserted in the line so that incident and reflected pulses may be observed on a sampling scope.

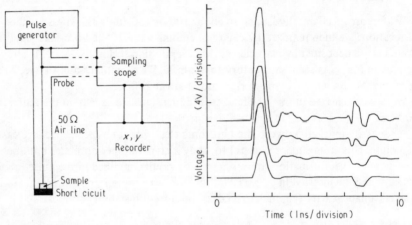

Figure 1. Diagram of the experimental arrangement for reflectometry measurements and typical waveform records. Lower trace with voltage under threshold, upper traces with voltage increasing above threshold and growth of a separate positive spike due to negative resistance effect.

V_i and V_r being the algebraic amplitudes of incident and reflected pulse, the classical reflectometry relations give:

Sample resistance $\qquad R = Z\,\dfrac{V_i + V_r}{V_i - V_r}$

Voltage across sample $\quad V = V_i + V_r$

Current $\qquad\qquad\quad I = \dfrac{V_i - V_r}{Z}.$

The accuracy of the method is best when R and Z are not very different, 3% may be expected if $0\cdot3 < R/Z < 3$ and 6% between $0\cdot1$ and 10.

The main advantage of this method is the very good impedance matching of the coaxial arrangement since the sample is directly mounted against the terminal short circuit, no current measuring series resistor being needed. As sample resistance is compared with the line impedance, we use a 50 Ω air line to have a stable reference. With a mercury wetted relay pulse generator, high frequency voltage pick-off and sampling scope, a fast response may be expected; at present a 0·2 ns rise time is obtained. This is greater than the typical nucleation time of a domain, but analysis of the phenomenon within the rise time gives interesting information although current displacement effects cannot be avoided.

Figure 1 shows variation of the reflected pulse with increasing voltage. At the beginning of the pulse within 0·1 ns one can see the appearance and variation of a separate positive peak. Plotting the corresponding calculated values on the $I-V$ characteristic we find the negative differential resistance part of the curve.

In the future time response improvements might be expected to ensure better transient phenomenon analysis. Also using longer pulses it would be possible to apply the velocity–field calculation technique proposed by Bastida *et al* (1971) if a clear transit domain mode appears.

4. Experimental results

4.1. Current instabilities related to band structure

Numerous Ga$_x$In$_{1-x}$Sb crystals in the concentration range $0 < x < 0.5$ were studied. In a previous paper (Joullié *et al* 1974) we have defined the principal material characteristics to obtain current instability namely: composition $x > 0.32$, donor concentration $n(300 \text{ K}) < 10^{16} \text{ cm}^{-3}$ and Hall mobility $\mu(300 \text{ K}) > 20\,000 \text{ cm}^2 \text{V}^{-1} \text{s}^{-1}$. Results are summarized in figure 2 which gives donor concentration against composition for the various crystals studied, the black points represent samples giving Gunn effect.

The threshold field for current oscillation is variable from 400 to 800 V cm^{-1} but uncertainties in sample length, contact technology and metal diffusion depth can explain these differences. The average value of about 600 V cm^{-1} is in good agreement with other experimental results (Hojo and Kuru 1974, McGroddy *et al* 1969) and the theoretical value from the calculated velocity–field characteristic (T Ikoma and K Sakai

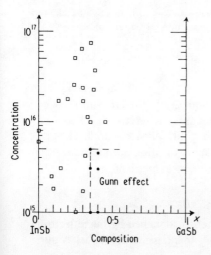

Figure 2. Carrier concentration plotted against alloy composition for various crystals studied. Solid points represent samples which exhibit the Gunn effect.

1974, private communication). Assuming in the first instance a simple electron transfer between the two Γ and L minima, it may be admitted that Gunn oscillations occur if the following approximate relation is satisfied:

$$Eg > \Delta E_{\Gamma L} > 4kT. \tag{1}$$

According to our experimental results in the composition range $0.35 < x < 0.50$ and those of Hojo and Kuru for $0.8 < x < 1$ it appears that room temperature Gunn effect oscillations may occur in the extended range:

$$0.35 < x < 0.9. \tag{2}$$

By comparison of relations (1) and (2) we can make reasonable assumptions on conduction band structure

$$\Delta E_{\Gamma L} \simeq Eg \qquad \text{for} \qquad x = 0.35 \tag{3}$$

$$\Delta E_{\Gamma L} \simeq 4kT \qquad \text{for} \qquad x = 0.9. \tag{4}$$

In relation to known data concerning the conduction band of GaSb, InSb and GaInSb (Auvergne *et al* 1974, Coderre and Woolley 1969, Kudman and Seidel 1967) these results allow us to draw an approximate curve of $\Delta E_{\Gamma L}$ or E_L against composition x (figure 3).

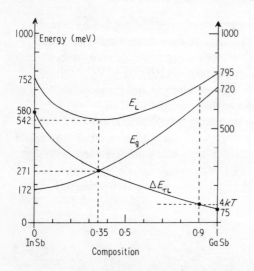

Figure 3. Band structure as a function of composition for $Ga_x In_{1-x} Sb$ alloys.

We have to note an important contradiction between these results and previous data relative to variation of energy level E_L with alloy composition (Coderre and Woolley 1969, Kudman and Seidel 1967). The E_L marked minimum is defined by reliable experimental results showing that current instabilities appear only when $x > 0.35$. Then it must be assumed that either our curve is a valuable approximation or its basic relation $Eg > \Delta E_{\Gamma L}$ is not consistent with an actual electron transfer mechanism.

4.2. Current–voltage characteristics related to device efficiency

Figure 4 shows a typical current–voltage characteristic obtained by the reflectometry method. The part OA of the curve corresponds to a classical ohmic behaviour.

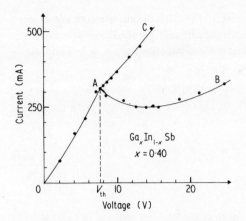

Figure 4. Typical current–voltage characteristic for a Ga$_x$In$_{1-x}$Sb sample.

When the voltage reaches the threshold value V_{th} (point A) the curve splits in two and the AB section displays the negative differential resistance calculated from the initial anomalous spike in the reflected pulse. Section AC from the normal part of the reflected pulse corresponds to the overall resistance of the device when a domain is grown.

At the present state of the work we cannot deduce any valuable velocity–field characteristic but important relative data may be given by comparison of different diodes under the same experimental conditions.

In figure 5 we have a comparative display of I–V curves from various samples. Normalized coordinates are used for an easy comparison avoiding the threshold current and voltage variations.

To verify this reflectometry method we have also plotted some characteristics of low power GaAs diodes, a typical curve is shown as A in figure 5. No strong discrepancy with other results is observed and that is a confirmation of the importance of both the reflectometry method and the new GaInSb material.

Generally we have found a good agreement with other current instability measurements; in particular a comparable threshold field value for a given diode. Three sorts of characteristics were observed:

(i) normal I–V curves with slow saturation effects in samples without current instabilities

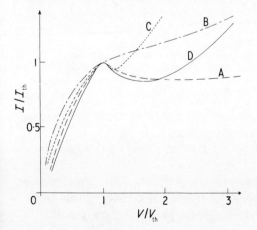

Figure 5. Comparative current–voltage characteristics: curve A: low power GaAs diode; curve B: GaInSb diode with noise; curves C, D: GaInSb diodes giving coherent microwave oscillations.

(ii) anomalous characteristics with curve splitting and important saturation or slight negative resistance in samples giving only noisy instabilities (figure 5, curve B)
(iii) anomalous characteristics with more or less marked negative resistance in samples giving coherent microwave oscillation (curves C and D of figure 5).

The general curve shape with pronounced negative differential resistance followed by a positive differential resistance above the valley field seems in agreement with the theoretical calculation made by Hilsum and Rees (1970a,b) with a three-level model.

4.3. Room temperature microwave oscillation

Coherent room temperature microwave oscillations were observed with several high quality crystals (Groubert *et al* 1974).

As an example we present results obtained with $Ga_{0.4}In_{0.6}Sb$ samples from TSM crystal B8-3. The diode parameters are:

carrier concentration: $n\,(77\,K) = 2.9 \times 10^{15}\,cm^{-3}$
$n\,(300\,K) = 3.7 \times 10^{15}\,cm^{-3}$
Hall mobility: $\mu\,(77\,K) = 35\,000\,cm^2\,V^{-1}\,s^{-1}$
$\mu\,(300\,K) = 34\,500\,cm^2\,V^{-1}\,s^{-1}$
Threshold field: $E_{th} = 400\,V\,cm^{-1}$
Sample length: $L = 110\,\mu m.$

The diode mounted in the coaxial cavity operates in a pulsed mode (pulse duration about 200 ns, repetition rate 0.5 to 1 kHz). The signal from current pick-off is applied to a 50 Ω load and displayed on a sampling oscilloscope. Figure 6 shows the corresponding 1 GHz waveform.

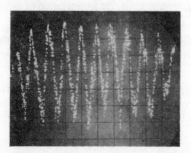

Figure 6. Current waveform through 50 Ω load for a $110\,\mu m$ length sample. Alloy composition $x = 0.4$, Bias field: 500 V cm^{-1}. Horizontal scale: 1 ns/division. Vertical scale: 4 mA/division.

In this example we have to assume that the diode operates in a domain mode because the frequency is constant over the whole cavity tuning range. Electron velocity deduced from transit frequency and sample length is $v = 1.1 \times 10^7\,cm\,s^{-1}$ not very different from the threshold calculated value $v_{th} = \mu E_{th} = 1.38 \times 10^7\,cm\,s^{-1}$.

This result seems in contradiction with previous interpretation of the current–voltage characteristic but other phenomena such as contact effects may enhance domain mode. It is also possible that the electron transfer mode may depend on alloy composition.

5. Conclusion

In the present state of the work we cannot draw definite conclusions but the observation of room temperature coherent microwave oscillation emphasizes the practical interest of Ga$_x$In$_{1-x}$Sb. On the other hand we have obtained some interesting information concerning the band structure of the alloy defining the concentration range suitable for the Gunn effect.

In the future we hope that improved experiments with high quality crystals over an extended composition range will permit us to obtain valuable information on the electron transfer mechanism.

Acknowledgments

This work was supported by DRME under agreement No. 701168.

References

Auvergne D, Camassel J, Mathieu H and Joullié A 1974 *J. Phys. Chem. Solids* **35** 133
Bastida E M, Fabri G, Svelto V and Vaghi F 1971 *Appl. Phys. Lett.* **18** 28–31
Coderre M and Woolley J C 1969 *Can. J. Phys.* **47** 2553
Elliott B J 1968 *IEEE Trans. Instrum. Meas.* **IM-17** 330–2
Groubert E, Michel J, Esquirol P and Joullié A 1974 *Japan. J. Appl. Phys.* to be published
Gunn J B and Elliott B J 1966 *Phys. Lett.* **22** 369–71
Hilsum C 1971 *Proc. European Conf. Solid St. Dev.* (London and Bristol: Institute of Physics) pp77–85
Hilsum C and Rees H C 1970a *Electron. Lett.* **9** 277–9
—— 1970b *Proc. Int. Conf. Semiconductors, Cambridge, Mass.* (USAEC: Oak Ridge, Tenn.) pp45–51
Hojo A and Kuru I 1974 *Electron. Lett.* **10** 61–2
Jantsch W and Heinrich H 1970 *Rev. Sci. Instrum.* **41** 228–30
Joullié A, Esquirol P and Bougnot G 1974 *Mater. Res. Bull.* **9** 241–50
Kudman I and Seidel T E 1967 *J. Appl. Phys.* **38** 4379
McGroddy J C 1970 *Proc. Int. Conf. Semiconductors, Cambridge, Mass.* (USAEC: Oak Ridge, Tenn.) pp31–40
McGroddy J C, Lorentz M R and Plaskett T S 1969 *Solid St. Commun.* **7** 901–3

A comparison of $In_xGa_{1-x}As$:Ge and GaAs:Si prepared by liquid phase epitaxy

M Ettenberg and H F Lockwood

RCA Laboratories, Princeton, New Jersey 08540, USA

Abstract. The electrical and optical properties of amphoterically doped p–n junctions of $In_xGa_{1-x}As$:Ge and GaAs:Si prepared by liquid phase epitaxy exhibit significant similarities. The liquid phase epitaxial growth of $In_xGa_{1-x}As$:Ge is described, including thermodynamic analysis of growth conditions. Some important properties of Ge-doped $In_xGa_{1-x}As$ light-emitting diodes emitting at $1·06\,\mu m$ ($x \simeq 0·09$) will be described and in particular, an analysis of the quantum efficiency will be given. By comparing GaAs:Si with $In_xGa_{1-x}As$:Ge we show that the dislocation density plays a dominant role in determining the quantum efficiency of the latter. This is done by introducing dislocations into GaAs:Si comparable in density to that induced in $In_xGa_{1-x}As$ by the lattice mismatch with the GaAs substrate. Electroluminescent spectra and minority carrier lifetime data suggest that dislocations act as nonradiative recombination centres.

1. Introduction

Of available light emitting diodes (LED) those fabricated from Si-compensated GaAs (GaAs:Si) have the highest quantum efficiency (Ladany 1971). These diodes emit in the wavelength range $0·92-1·0\mu m$ and are grown in a single step by liquid phase epitaxy. This same technology has been used to extend the spectral range to $0·75\,\mu m$ (Kressel *et al* 1969) by the addition of Al to the GaAs to form higher bandgap (AlGa)As alloys. It would also be useful to extend this range to longer wavelengths for potential applications in conjunction with Nd:YAG lasers at $1·06\,\mu m$ and at longer wavelengths where optical fibres have very low losses.

Recently it was shown (Takahashi *et al* 1971) that Ge acts amphoterically in (InGa)As alloys just as Si does in (AlGa)As alloys. While Ge-compensated (InGa)As diodes have been fabricated (Kurihara *et al* 1973) they have had disappointingly low quantum efficiencies of only $0·001-0·01\%$.

In this paper we show that (InGa)As:Ge diodes emitting at about $1·06\,\mu m$ can be made with efficiencies that are 20 times higher (0·2%) than previously reported. However, this efficiency is still not comparable to GaAs:Si diodes which are commercially available with efficiencies exceeding a few percent. We also demonstrate, by analogy with GaAs:Si, that dislocations arising from the lattice mismatch between the GaAs substrate and the (InGa)As epitaxial layer can account for the low efficiency of the Ge-compensated (InGa)As LED. A similar dislocation density intentionally introduced into GaAs:Si LED results in a ten-fold decrease in efficiency.

2. Growth

The Ge-doped (InGa)As epitaxial layers were grown on (111)B GaAs substrates (Si-doped, n-type $2 \times 10^{18}\,cm^{-3}$) by liquid phase epitaxy using the thin solution slider boat previously described (Lockwood and Ettenberg 1972). The starting solutions consisted of 2 g of In containing from 200 to 500 mg of Ga and from 20 to 600 mg of Ge. A GaAs polycrystalline source wafer above the solution was used for saturation with As at 900 °C in an ambient of Ag–Pd diffused H_2. The GaAs polycrystalline source was maintained in contact with the Ga–In solution for 45 min at 900 °C prior to the introduction of the substrate wafer for growth. The furnace was then cooled at a rate of 1 °C min^{-1} to 820 °C where the substrate was wiped clean of solution in the slider boat and the quartz tube containing the boat was then withdrawn from the furnace. In general, it was difficult to wipe the 2 cm^2 area of the layer entirely clean of solution; only those areas of the grown layer (about half the area) that wiped clean were studied. The reason for the lack of complete wiping is not clear since GaAs and (AlGa)As layers can easily be wiped completely clean of solution.

The grown layers were from 40–100 μm thick with the p–n junction appearing in etched cross section (Olsen and Ettenberg 1975) between 25 and 50 μm from the top surface. The composition of the grown layers was determined by x-ray lattice parameter measurements to be from 6 to 14 molar percent InAs ($In_{0.06}Ga_{0.94}As–In_{0.14}Ga_{0.86}As$). The absence of spreading in the back reflection lines of x-ray powder patterns of material taken from the entire thickness of the layer indicated that the grading of In in the layer was not more than 0·5%.

The use of GaAs source wafers to saturate an In–Ga–As solution presents an interesting situation since at the so-called saturation temperature the solution is probably not saturated with respect to an $In_xGa_{1-x}As$ solid. In thermodynamic terms this means that an In–Ga–As solution equilibrated with a GaAs solid has a lower As activity (a_{As}) than an (InGa)As solid. The activity of As can be expressed as

$$a_{As} = \frac{P_{As}}{P_{As}^0} \tag{1}$$

where P_{As} is the partial pressure of As in the system and P_{As}^0 is the pressure over pure As at the same temperature. Since the In–Ga–As solution equilibrated with a GaAs source has a lower partial pressure of As than the (InGa)As solid that can form, significant cooling is necessary before the precipitation point (liquidus) is reached. At the liquidus the partial pressures of As in the In–Ga–As liquid solution and precipitating (InGa)As solid are, of course, equal.

It is interesting to note that the situation is quite different in the Al–Ga–As system when GaAs is used to saturate an Al–Ga solution. In this case, the GaAs has a higher As activity than the resulting (AlGa)As solid, so ideally, precipitation of (AlGa)As should continue indefinitely. What happens in practice is that when the As concentration reaches the saturation point, a thin coherent layer of (AlGa)As is probably formed on the surface of the GaAs thus establishing equilibrium and preventing further dissolution.

The thermodynamic analysis of the growth of (InGa)As using a GaAs source is different from that usually employed for ternary systems, although the basic equations are the same. First the activity of As in the GaAs source is calculated, then with the initial In and Ga concentrations in the solution the final In, Ga and As concentrations in equilibrium with the GaAs source are calculated. These solution concentrations are then used to calculate the liquidus temperature and solidus composition. Because of this difference in procedure the entire set of equations employed are reproduced here. The thermodynamic parameters are available from the work of Antypas (1970), and are reproduced in table 1. Adequate fit of the theory to the limited experimental data was

Table 1. Thermodynamic parameters for In—Ga—As system (Antypas 1970)

	In—As	Ga—As	In—Ga
α	$-9 \cdot 16\,T + 4300$	$-3 \cdot 7\,T$	1000
ΔS_f	14·52	16·64	
T_f (K)	1215	1509	

β (InAs—GaAs) = 2000

also achieved with data from Stringfellow and Greene (1969). The equations describing the ternary system were developed by Ilegems and Pearson (1968). To solve for the solidus composition and liquidus temperature, the activity of As in the GaAs source is first calculated at the so-called saturation temperature T_s. This is the temperature at which the GaAs source and In—Ga solution are brought into equilibrium. From equations (2—4), the activity of As in a Ga—As liquid solution in equilibrium with a GaAs solid is calculated. It is assumed in this thermodynamic analysis that the activity of As in the GaAs source is fixed on the Ga-rich side of the compound and not enough time is allowed for diffusion in the solid and a change in the stoichiometric composition of the GaAs source. Only the As is thus equilibrated between source and solution.

$$1 = 4X'_{As}(1 - X'_{As})\,\gamma_{As}\gamma_{Ga}\exp\left(\frac{\Delta S_f^{GaAs}(T_f^{GaAs} - T_S)}{RT_S}\right)\exp\left(\frac{-\alpha_{Ga-As}}{2RT_S}\right) \tag{2}$$

$$\gamma_{As} = \exp\left(\frac{\alpha_{Ga-As}(1 - X'_{As})^2}{RT_S}\right) \tag{3}$$

$$\gamma_{Ga} = \exp\left(\frac{\alpha_{Ga-As}(X'_{As})^2}{RT_S}\right) \tag{4}$$

$$a_{As} = \gamma_{As}(X'_{As}). \tag{5}$$

X'_{As} is the equilibrium atomic fraction of As in a saturated Ga solution at the temperature T_s. ΔS_f and T_f are the entropy and temperature of fusion, respectively; α is the interaction parameter, γ the activity coefficient, and R is the gas constant.

At the saturation temperature T_s, a_{As} in solid GaAs will equal a_{As} in the liquid

In—Ga solution. From the initial In and Ga atomic fractions, the concentrations of In, Ga, and As in equilibrium with the GaAs source may be determined from

$$a_{As} = X_{As} \exp \left[\frac{\alpha_{Ga-As}(X_{Ga})^2 + \alpha_{In-As}(X_{In})^2}{RT_s} \right.$$
$$\left. + \frac{(\alpha_{Ga-As} + \alpha_{In-As} - \alpha_{In-Ga})X_{In}X_{Ga}}{RT_s} \right] \tag{6}$$

where the X are the equilibrium atomic fractions at T_S. X_{Ga} in equation (6) is not simply the initial concentration of Ga in the solution but must be modified because for each As atom entering the solution a Ga atom enters from the source.

Given the composition of the In—Ga—As solution, the liquidus temperature, T_G, and the composition of the precipitating solid Y may be calculated from equations (7—15) which are taken from Ilegems and Pearson (1968).

$$\gamma_{As} = \exp \left[\frac{\alpha_{Ga-As}(X_{Ga})^2 + \alpha_{In-As}(X_{In})^2 + (\alpha_{Ga-As} + \alpha_{In-As} - \alpha_{In-Ga})X_{In}X_{Ga}}{RT_G} \right] \tag{7}$$

$$\gamma_{Ga} = \exp \left[\frac{\alpha_{Ga-As}(X_{As})^2 + \alpha_{In-Ga}(X_{In})^2 + (\alpha_{In-Ga} + \alpha_{Ga-As} - \alpha_{In-As})X_{In}X_{As}}{RT_G} \right] \tag{8}$$

$$\gamma_{In} = \exp \left[\frac{\alpha_{Ga-In}(X_{Ga})^2 + \alpha_{In-As}(X_{As})^2 + (\alpha_{In-Ga} + \alpha_{In-As} - \alpha_{Ga-As})X_{As}X_{Ga}}{RT_G} \right] \tag{9}$$

$$\gamma_{In}^{SL} = \gamma_{As}^{SL} = \exp \left(\frac{\alpha_{In-As}}{4RT_G} \right) \tag{10}$$

$$\gamma_{Ga}^{SL} = \gamma_{As}^{SL} = \exp \left(\frac{\alpha_{Ga-As}}{4RT_G} \right) \tag{11}$$

$$\gamma_{GaAs} = \exp \left[\frac{\beta(Y^2)}{RT_G} \right] \tag{12}$$

$$\gamma_{InAs} = \left[\frac{\beta(1-Y)^2}{RT_G} \right] \tag{13}$$

$$\gamma_{GaAs}(1-Y) = \frac{4\gamma_{As}\gamma_{Ga}}{\gamma_{Ga}^{SL}\gamma_{As}^{SL}} X_{Ga}X_{As} \exp \left[\frac{\Delta S_f^{GaAs}}{RT_G} (T_f^{GaAs} - T_G) \right] \tag{14}$$

$$\gamma_{InAs}Y = \frac{4\gamma_{As}\gamma_{In}}{\gamma_{In}^{SL}\gamma_{As}^{SL}} X_{In}X_{As} \exp \left[\frac{\Delta S_f^{InAs}}{RT_G} (T_f^{InAs} - T_G) \right]. \tag{15}$$

The superscript SL refers to stoichiometric liquid, so that γ_{Ga}^{SL} is the activity of Ga in a Ga–As stoichiometric liquid, and β is the interaction coefficient between GaAs and InAs in the solid. With the atomic fractions X_{As}, X_{Ga} and X_{In} calculated from equations (2–6), equations (7–15) were solved by a simple iterative method on a computer for the liquidus temperature T_G and the InAs in the solid Y.

For atomic fractions of Ga (X_{Ga}) in the initial In–Ga solution ranging from about 10^{-2} to $1 \cdot 0$, Y and T_G were calculated from equations (2–15) for 700, 800 and 900 °C saturation temperatures. These results are shown in figures 1 and 2. Figure 1 shows the calculated mole percent InAs as a function of the initial Ga concentration and includes experimental data points from this work at 900 °C and from prior work (K Takahashi *et al* 1971 and M Kurihara 1973) at 700 and 800 °C.

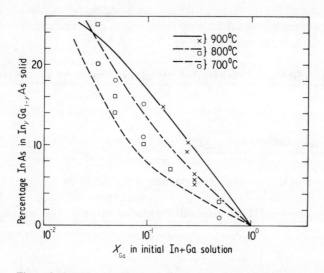

Figure 1. In$_y$Ga$_{1-y}$As solid composition as a function of starting In+Ga solution composition for saturation temperatures of 700, 800, and 900 °C. Lines represent calculated results, points at 900 °C are experimental results from this study, points at 800 and 700 °C are experimental data of Takahashi (1971) and Kurihara (1973).

While in some cases there is good fit to the theory, in general there is significant deviation. The random scatter of the experimental points suggests that most of the difference between the theoretical curves and experimental points is probably due to departure from equilibrium during growth due to such causes as loss of As. The theory, thus, is useful only as a first order approximation. It should be noted that the experimental points at 900 °C do not include the Ge dopant which in some cases comprised a substantial part of the solution. This was not the case for the data taken from the other references at 700 and 800 °C, where the dopant contributed less than 5 atom% to the solution.

The degree of cooling, $(T_S - T_G)$, necessary to initiate growth as a function of the starting solution composition is plotted in figure 2 for saturation temperatures of 700, 800 and 900 °C. A limited number of experiments were made to determine $T_S - T_G$;

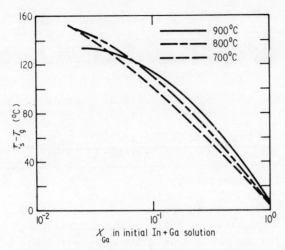

Figure 2. Calculated temperature drop $(T_S - T_G)$ necessary to initiate growth as a function of initial In+Ga solution composition for saturation temperatures T_S of 700, 800 and 900 °C.

these experiments were complicated by the difficulty in wiping the wafer clean of solution. The results indicate that the values given in figure 2 are overestimated by 40–60 °C.

One of the intentions of this detailed thermodynamic analysis of the growth is to point out the peculiar situation that may result during the LPE growth of mixed III–V pseudo-binary epitaxial layers when the compound with the lower vapour pressure or activity of the group V element (invariably the compound with the higher melting point) is employed to saturate the group III solution, viz, GaP as a source in the growth of (InGa)P or AlSb for the growth of (GaAl)Sb.

3. Electroluminescent device fabrication and characterization

As indicated earlier, the goal of this study was to determine whether Ge-compensated (InGa)As diodes could be made to emit at about $1 \cdot 06 \mu m$ with efficiencies comparable to those obtained with Si-compensated (AlGa)As devices. Those layers fabricated into LED devices were processed by lapping from the substrate side of the wafer to about $100 \mu m$ thickness and applying a Sn–Ni–Au contact to the n-side. Diodes were then saw cut $250 \mu m$ square, and an Au wire was bonded to the p-side. Finally the diodes were solder mounted to a TO-5 header.

The spontaneous emission spectra of the diodes were measured with a Spex spectrometer and an S-1 photomultiplier (RCA-7102) with the diodes driven at 25 mA (40 A cm^{-2}). The spectral outputs were corrected for the detector response.

Table 2 shows the peak of the incoherent emission, the spectral half width (Δ) and the composition of (InGa)As layer as determined from lattice parameter measurements. The efficiencies of the best diodes from each of the wafers as measured in an integrating sphere with a calibrated Si photodiode are also tabulated.

Table 2. Results of (In, Ga)As : Ge study

Run No.	Wt Ga† (mg)	Wt Ge† (mg)	Solid composition y In$_y$Ga$_{1-y}$As	(Peak of EL 300 K) (μm)	Δ-FWHM 300 K (μm)	η_{ex} %
412	400	240	0·089	1·065	0·075	0·2
424	400	20	0·103	1·000	0·050	0·01
428	500	120	0·055	1·025	0·070	0·05
501	500	600	0·064	1·080	0·100	0·1
711	500	600	0·57	1·080	0·100	0·1

† Weight in initial solution containing 2 g In

The 300 and 77 K emission spectra of one of the diodes which emit near 1·06 μm at 300 K is shown in figure 3. The spectral output is rather broad and well below the bandgap as it is for GaAs : Si. For example, Zn-doped diodes emitting at about 1·06 μm contain 16% InAs (Nuese and Enstrom 1972) while Ge-doped (InGa)As can have only 6% InAs for the same peak emission wavelength. Table 2 shows that the room tempera-

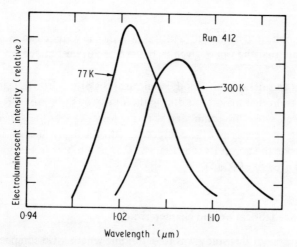

Figure 3. Electroluminescent spectra at 77 and 300 K at 25 mA (40 A cm^{-2}) drive current of In$_{0·09}$Ga$_{0·91}$As : Ge diode emitting near 1·06 μm.

ture efficiencies were generally in the range of about 0·1% even with changes in the InAs and Ge concentrations. The major exception is run 424 made with 20 mg of Ge, where the efficiency is only 0·01%. This lower value is approximately the efficiency obtained in the study of In$_x$Ga$_{1-x}$As : Ge by Kurihara *et al* (1973). There are two differences between our results and that study. First, the starting saturation temperatures were different, 900 rather than 800 °C in the prior study, but more importantly, the Ge-doping concentrations in the Kurihara study (1973) were comparable to that of run 424. In our study the Ge concentrations were an order of magnitude higher; this heavy compensation depresses the emission energy further below the bandgap energy, thus reducing the internal losses and increasing the external efficiency.

Nonetheless, these efficiency values are still rather low compared to (AlGa)As : Si. One probable cause for this is the high dislocation density in the (InGa)As caused by the lattice mismatch between the GaAs substrate and the epitaxial layer. Recently, it was shown in GaAs : Ge that if the average dislocation spacing is less than the minority carrier diffusion length there is a substantial decrease in the luminescent efficiency (Ettenberg 1974).

To reveal the dislocation density in the (InGa)As : Ge layer, the surface of one of the wafers (run 711) was etched in A—B etch (Abrahams and Buiocchi 1965). The etch pits at the (111)B surface are shown in the photomicrograph, figure 4. The dislocation density from the pit count is $4 \times 10^7 \, \mathrm{cm}^{-2}$, giving an average spacing of about $2 \, \mu\mathrm{m}$. Etching of the surface of some of the other runs gave similar densities.

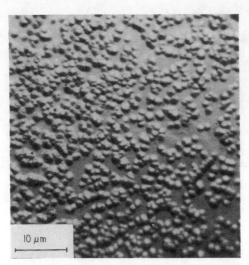

Figure 4. As-grown surface of $In_{0.09}Ga_{0.91}As$: Ge layer after etching to reveal dislocation etch pits, dislocation density is about $4 \times 10^7 \, \mathrm{cm}^{-2}$.

Because of the relatively low free carrier concentration near the p—n junction it is anticipated that the effective diffusion length of minority carriers in this region is considerably greater than $2 \, \mu\mathrm{m}$; for example, the diffusion length in GaAs : Ge at $1 \times 10^{17} \, \mathrm{cm}^{-3}$ is $25 \, \mu\mathrm{m}$ (Ettenberg *et al* 1973). It is thus thought that the relatively low quantum efficiency is due to the presence of a large density of dislocations acting as nonradiative recombination centres. In a recent study of Zn-doped (InGa)As (Ettenberg *et al* 1975) dislocations were also found to be responsible for low LED efficiencies.

4. Comparison to GaAs : Si LED

To establish that the high dislocation density is responsible for the low efficiencies, Si-doped GaAs amphoteric diodes with controlled dislocation densities were studied.

A Si-compensated p—n junction was grown on a (111)B face of an n-type (Si-doped $2 \times 10^{18} \, \mathrm{cm}^{-3}$) GaAs substrate. The thin solution technique was employed with a 1 g Ga solution containing 10 mg of Si, and the layers were grown from 900 °C at 1 °C min^{-1} to

750 °C where the melt was wiped clean from the surface. Before growth, half of the surface of the substrate was damage-free polished, and half of the surface was mechanically damaged by roughening with abrasive. The surface of the grown layer after treatment with the A−B dislocation revealing etch is shown in figure 5. The part of the layer grown on the polished half of the wafer had a dislocation density below 4×10^3 and that grown on the roughened half of $5 \times 10^7 \, \text{cm}^{-2}$. In this manner the effect of dislocation density on identically grown Si-compensated device layers could be evaluated.

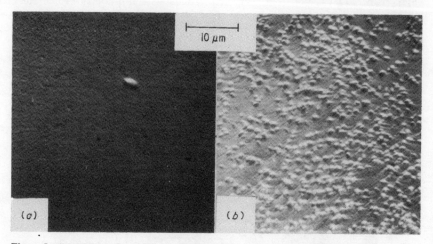

Figure 5. As-grown surface of GaAs: Si layer after etching to reveal dislocation etch pits (*a*) layer grown on polished substrate, dislocation density is below $4 \times 10^3 \, \text{cm}^{-2}$; (*b*) layer grown on substrate with surface damaged (roughened), dislocation density is about $5 \times 10^7 \, \text{cm}^{-2}$.

Diodes were fabricated from each half of this material in a manner identical to that employed for the (InGa)As: Ge. The electroluminescence spectra of the high and low dislocation diodes are shown in figure 6. The quantum efficiency of uncoated diodes from the high and low dislocation halves was 0·1% and 1·0%, respectively. This large difference in efficiency is directly attributed to the difference in dislocation density.

The difference in the electroluminescence spectra can also be attributed to the effect of the dislocations. In amphoterically doped diodes, the region near the p−n junction is highly compensated forming deep bandtail states, and thus, the emission arising from regions close to the p−n junction is lower in energy than that from regions somewhat removed. Photoluminescence scans perpendicular to the p−n junction (Mettler and Pawlik 1972) show that within about 10 μm of the p−n junction the peak shift in the spectrum is as much as 200 Å. If the dislocation spacing is about 2 μm we would thus expect that the carriers do not penetrate very far into the bulk of the material due to recombination at the dislocations. Thus, the sample with the higher dislocations should exhibit a diminished spectral output at shorter wavelengths. This effect is clearly seen in the electroluminescence of diodes from both halves of the wafer as shown in figure 6.

The effect of dislocations is also reflected in the minority carrier lifetime. Figure 7 shows the electroluminescent decay for diodes from the high and low dislocation density portions of the wafer. The faster light decay in the sample with the higher dislocation

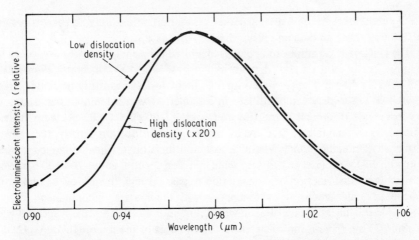

Figure 6. Room temperature electroluminescent spectra of GaAs: Si diodes from half of wafer containing high and low dislocation densities, corresponding to the roughened and polished substrates.

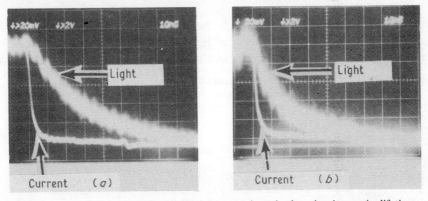

Figure 7. Electroluminescent decay, which is approximately the minority carrier lifetime for (a) GaAs: Si diode with low dislocation density, and (b) GaAs: Si diode with high dislocation density.

density is indicative of a shorter lifetime and thus lower internal quantum efficiency. This result again shows the effect of dislocation in limiting the diffusion length or lifetime and thus the radiative efficiency.

5. Conclusions

The growth of (InGa)As liquid phase epitaxial layers on GaAs substrates using a GaAs source has been studied and analysed thermodynamically. The interesting feature of this system is that the melt may not be saturated at the initial so-called saturation temperature so that the solution must be cooled considerably to initiate growth (as opposed to the (AlGa)As system where the solution is theoretically 'oversaturated' by the source).

Using published thermodynamic parameters, we find a first-order agreement between the limited experimental data and phase diagram calculations.

The (InGa)As:Ge amphoteric diodes exhibited efficiencies a factor of twenty higher than previously reported, but they are still at least an order of magnitude lower than (AlGa)As:Si diodes. The spectral output of these devices can easily be made to peak at about $1.06\,\mu m$ so they can be utilized in conjunction with Nd:YAG lasers and in the low loss region of optical fibres. However, even if (InGa)As:Ge diodes were made with efficiencies comparable to Si:GaAs by reducing the dislocation density, they have certain characteristics which would limit their utility. First, when employed in conjunction with Nd:YAG lasers, their very wide emission spectra, about 70–100 nm, would allow only a small fraction of the radiation to pass through the narrow bandpass filters (10 nm) normally employed in conjunction with these lasers. In comparison, Zn-doped (InGa)As emitting at about $1.06\,\mu m$ with 1% quantum efficiency has a spectral width of about 35 nm (Nuese and Enstrom 1972). In terms of the use of (InGa)As:Ge diodes in optical communications systems, their broad spectrum presents a potential problem of dispersion in the fibre. Furthermore, the electroluminescence response time is similar to that found in Si:GaAs and is thus very long. A limited number of measurements of the electroluminescent response time made in this study show rise and fall times (10–90%) on the order of 50–100 ns, which limits the information rate these diodes are capable of transmitting. In contrast, Zn-doped (InGa)As devices exhibit less than 20 ns response times (Nuese and Enstrom 1972). Thus, (InGa)As:Ge LED even with increased efficiencies do not necessarily appear attractive in comparison to Zn-doped (InGa)As for most $1.06\,\mu m$ applications and to (AlGa)As:Si diodes as high efficiency emitters.

References

Abrahams M S and Buiocchi C J 1965 *J. Appl. Phys.* **36** 2855

Antypas G A 1970 *J. Electrochem. Soc.* **117** 1393

Ettenberg M 1974 *J. Appl. Phys.* **45** 901

Ettenberg M, Kressel H and Gilbert S L 1973 *J. Appl. Phys.* **44** 827

Ettenberg M, Nuese C J, Gannon J J, Appert J and Enstrom R E 1975 *J. Electron. Mater.* to be published

Ilegems M and Pearson G L 1968 *Proc. 2nd Int. Symp. Gallium Arsenide and Related Compounds* (London and Bristol: Institute of Physics) p3

Kressel H, Hawrylo F Z and Almeleh N 1969 *J. Appl. Phys.* **40** 2248

Kurihara M, Moriizumi T and Takahashi K 1973 *Solid St. Electron.* **16** 763

Ladany I 1971 *J. Appl. Phys.* **42** 654

Lockwood H F and Ettenberg M 1972 *J. Crystal Growth* **15** 81

Mettler K and Pawlik D 1972 *Siemens Forsch. U. Entwickl. Ber. Bd.* **1** 274

Nuese C J and Enstrom R E 1972 *IEEE Trans. Electron Dev.* **ED-19** 1067

Olsen G H and Ettenberg M 1975 *J. Appl. Phys.* to be published

Stringfellow G B and Greene P E 1969 *J. Phys. Chem. Solids* **30** 1779

Takahashi K, Moriizumi T and Shirose S 1972 *J. Electrochem. Soc.* **118** 1639

Growth and properties of $Ga_yAl_{1-y}As-Ga_xIn_{1-x}P$ heterostructure electroluminescent diodes

H Beneking, N Grote, P Mischel and G Schul

Institut für Halbleitertechnik, Rheinisch-Westfälische Technische Hochschule, Aachen, W Germany

Abstract. $Ga_yAl_{1-y}As-Ga_xIn_{1-x}P$ heterostructures have been grown on GaAs substrates by liquid phase epitaxy from Ga-rich and In-rich solutions respectively. Melt compositions were adjusted to achieve precise matching between the GaAs ($Ga_yAl_{1-y}As$) and $Ga_xIn_{1-x}P$ lattice constants.

Thus, heterostructures of excellent crystalline perfection can be produced. Detailed information on parameters affecting growth of the layers is given.

We have directed special attention to the following sequence of layers: GaAs-substrate (n-type), $Ga_{0.51}In_{0.49}P$ (active layer), $Ga_{0.32}Al_{0.68}As$ (p-type), GaAs (p+-type). Devices fabricated from this heterostructure exhibit efficient electroluminescence.

1. Introduction

Increasing interest has been shown in recent years in the gallium indium phosphide ($Ga_xIn_{1-x}P$) alloy system, because it offers the possibility of direct recombination luminescence up to photon energies of approximately 2·2 eV at 300 K (Hilsum and Porteous 1968, Lorenz *et al* 1968, Pitt *et al* 1974). It is one of the most promising materials for the preparation of luminescent and laser diodes throughout the near infrared and visible spectrum almost to the green. In this respect it offers considerable advantage compared with the more commonly used $Ga_yAl_{1-y}As$ and $GaAs_xP_{1-x}$ alloy systems: its progress, however, has been hampered by the difficulty of its preparation.

It is essential to grow $Ga_xIn_{1-x}P$ crystals of locally very uniform composition, because otherwise the large lattice mismatch between InP (5·8697 Å) and GaP (5·4510 Å) will severely affect the crystalline quality. Epitaxial layers of $Ga_xIn_{1-x}P$ have been grown from the vapour phase (Tietjen *et al* 1970) and by liquid phase epitaxy (Hakki 1971). In particular $Ga_{0.51}In_{0.49}P$ (direct energy gap approximately 1·9 eV at 300 K) matches the lattice parameter of GaAs (Stringfellow 1972) and can be deposited on these readily available substrates.

Injection electroluminescence of $Ga_xIn_{1-x}P$ p–n junctions has been reported (Onton and Lorenz 1970). In order to achieve efficient electroluminescence and laser emission, it seems desirable to use the principle of close-confinement structures, which is well known from the fabrication of $Ga_yAl_{1-y}As$ diodes and has proved to be highly successful (Alferov *et al* 1968, Beneking *et al* 1972). For this purpose it is necessary to grow a material with an energy gap larger than the energy gap of $Ga_xIn_{1-x}P$ onto this $Ga_xIn_{1-x}P$ layer.

The present paper reports a method of growing high quality GaAs–Ga$_{0.51}$In$_{0.49}$P–Ga$_y$Al$_{1-y}$As–GaAs heterostructures by liquid phase epitaxy. In this case Ga$_y$Al$_{1-y}$As with $E_g \simeq 2$ eV is used as confinement. It is possible to obtain such structures with excellent crystalline quality, because the lattice parameter of Ga$_y$Al$_{1-y}$As closely matches the lattice constants of GaAs as well as of Ga$_{0.51}$In$_{0.49}$P.

2. Preparation of the heterostructures

The heterostructures were prepared in an apparatus which is shown in figure 1. A liquid phase epitaxial growth process is performed in a boat of high-purity graphite, which is contained in a silica tube under a flow of Pd-purified hydrogen.

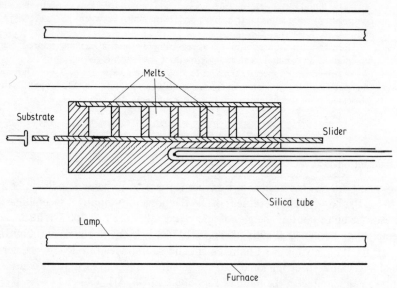

Figure 1. Apparatus for LPE growth of heterostructures.

Heating is accomplished by a set of 12 quartz infrared lamps, the light of which is focused onto the graphite boat. Due to its small thermal capacity and the high available power (30 kW) this set-up allows for fast heating or cooling and precise control of the desired cooling cycle. The substrate is carried by a graphite slider and is positioned successively under different melts which are contained in the graphite boat. Thus, a series of thin epitaxial layers with desired properties can be grown. The first step in growing GaAs–Ga$_x$In$_{1-x}$P–Ga$_y$Al$_{1-y}$As–GaAs heterostructures was to study the conditions under which high-quality Ga$_x$In$_{1-x}$P layers may be deposited on GaAs substrates. Ga$_x$In$_{1-x}$P layers have been grown from In-rich melts saturated with Ga and InP at 800 °C. In and Ga of 99.9999% purity and undoped polycrystalline InP were used as starting materials. The composition of the Ga$_x$In$_{1-x}$P layer was determined by the Ga : In ratio (m_{Ga}/m_{In}) of the melt. By varying the ratio m_{Ga}/m_{In} we have deposited Ga$_x$In$_{1-x}$P with compositions $0.37 \leqslant x \leqslant 0.57$. Deposition was carried

out on polished and etched (111) Ga, (111) As and (100) surfaces of Te-doped GaAs
wafers ($n \simeq 10^{18}$ cm^{-3}). The average growth velocity was approximately 0·5 μm per
minute.

We have investigated the influence of m_{Ga}/m_{In} on the composition and crystalline
perfection of the layers. As a result, figure 2 shows the dependence of the $Ga_x In_{1-x}$P-
layer composition on m_{Ga}/m_{In}. The composition has been determined by measuring
the lattice constant a_0 using an x-ray diffractometer. As the InP–GaP alloy system
obeys Vegard's law (Heritage *et al* 1970) x can be calculated from a_0. Another method
for the determination of x was photoluminescence measurements. The results obtained
by both methods were found to agree within ±3%.

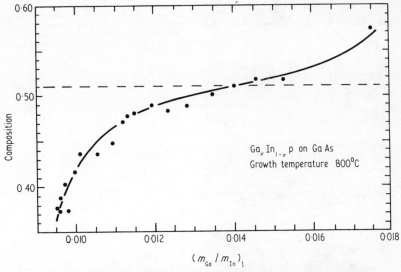

Figure 2. Composition x of $Ga_x In_{1-x}$P as a function of m_{Ga}/m_{In} of the melt.

The curve in figure 2 shows a strong deviation from linearity. Linearity in this
temperature and composition range would be expected from theoretical calculations
using the regular solution model (Isozumi *et al* 1973). This anomalous behaviour was
observed in a similar way by Stringfellow (1972) and explained by the excess energy
due to lattice parameter mismatch which perturbs the solid composition from the
chemical equilibrium composition towards the composition which minimizes the
mismatch between substrate and epitaxial layer.

Very good epitaxial $Ga_x In_{1-x}$P layers have been grown with a composition of
$x = 0·51$. A SEM photograph of the cleaved and stained cross section of such a layer is
shown in figure 3. Dislocation densities of those layers were found to be 6000–
10 000 cm^{-2} and did not exceed the dislocation density of their respective substrates.

Deviation from this optimum composition will result in layers of inferior quality,
as can be seen in figure 4 from the cross section of a $Ga_{0·48} In_{0·52}$P layer. Crystals with
stronger deviation from $x = 0·51$ were highly imperfect with a large number of dislocations and melt inclusions. Highest quality of $Ga_{0·51} In_{0·49}$P layers was achieved by

Figure 3. $Ga_{0.51}In_{0.49}P$ layer.

Figure 4. $Ga_{0.48}In_{0.52}P$ layer.

deposition on (111) As substrate surfaces. Similar, but slightly inferior results, were obtained on (111) Ga surfaces, whereas layers on (100) oriented substrates were of very poor crystalline quality. For that reason, the layers used for the preparation of $Ga_xIn_{1-x}P$–$Ga_yAl_{1-y}As$ heterostructures described in the following were always deposited on (111) As oriented GaAs substrates.

Favourably, GaAs–$Ga_{0.51}In_{0.49}P$–$Ga_yAl_{1-y}As$–GaAs heterostructures for electro-luminescent devices are grown in a single run. First a $Ga_{0.51}In_{0.49}P$ active layer is grown as described above. After having achieved the desired layer thickness, the substrate is brought into contact with a second melt from which a confining $Ga_yAl_{1-y}As$ layer is grown. Afterwards the substrate is moved to a third melt, from which the GaAs top layer is grown. This final layer considerably facilitates the fabrication of ohmic contacts. Figure 5 shows the cleaved edge of a heterostructure described above. Surfaces are mirrorlike, as can be seen from figure 6. The photograph has been taken immediately after removal of the wafer from the furnace. Only a few small droplets

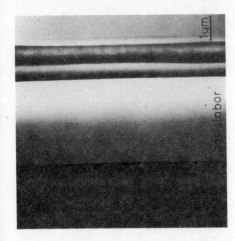

Figure 5.
$GaAs - Ga_{0.51} In_{0.49} P - Ga_{0.32} Al_{0.68} As - GaAs$
heterostructure.

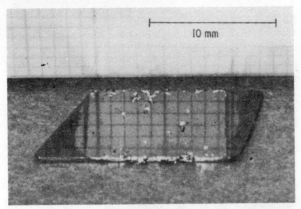

Figure 6. Surface of heterostructure wafer after removal from epitaxy apparatus;
$Ga_{0.51} In_{0.49} P - Ga_{0.32} Al_{0.68} As - GaAs$.

of gallium melt have remained on the surface. Diodes have been prepared by n-type doping of the $Ga_{0.51} In_{0.49} P$ layer with tellurium and p-type doping of the subsequent $Ga_y Al_{1-y} As$ and GaAs layers with zinc. Dopants were added to the respective melts. p–n junctions in the $Ga_{0.51} In_{0.49} P$ layers were generated by the inward diffusion of zinc from the adjacent $Ga_y Al_{1-y} As$ layer during the growth process. The position of the p–n junction can be controlled by variation of several process parameters (ie, the Te concentration of the $Ga_{0.51} In_{0.49} P$ layer, the Zn concentration of the $Ga_y Al_{1-y} As$ layer, and the time taken for the $Ga_y Al_{1-y} As$ and GaAs layer growth). Figure 7 shows the energy band diagram of the obtained structures.

The positions of the p–n junctions have been determined by a scanning electron microscope in the induced current mode.

Figure 8 shows the current–voltage characteristics of a diode, the p–n junction of which is positioned $0.8 \mu m$ below the $Ga_y Al_{1-y} As$ layer boundary. The $Ga_{0.51} In_{0.49} P$ layer thickness is about $3.3 \mu m$ in this device. The tellurium doping level of this layer was selected to obtain a free electron concentration of $n \simeq 1.2 \times 10^{18}$ cm^{-3}. Efficient

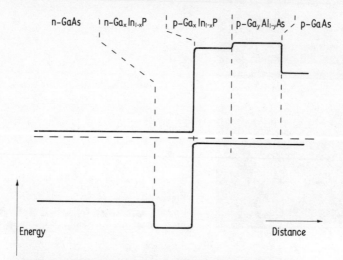

$$\text{n–GaAs} \quad \text{n–Ga}_x\text{In}_{1-x}\text{P} \quad \text{p–Ga}_x\text{In}_{1-x}\text{P} \quad \text{p–Ga}_y\text{Al}_{1-y}\text{As} \quad \text{p–GaAs}$$

Figure 7. Energy band diagram of heterostructure p–n diode.

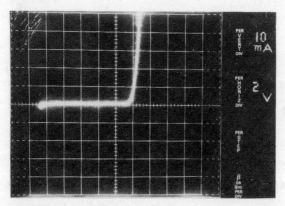

Figure 8. Current–voltage characteristics of heterostructure p–n diode;
$Ga_{0.51}In_{0.49}P–Ga_{0.32}Al_{0.68}As–GaAs$.

electroluminescence was obtained from those diodes. The measured external quantum efficiencies were 0·1% at 300 K and 1·5% at 77 K. The electroluminescence spectra of such a diode at 300 and 77 K are shown in figure 9.

These spectra were measured at a current density of 10 A cm^{-2}. The peak emission wavelength is 680 nm at 300 K and 655 nm at 77 K, which is about 30 nm larger than the wavelength corresponding to the energy gap at the p–n junction. The spectral half width is 30 nm at 300 K and 15 nm at 77 K. With increasing forward bias the emission shifts towards shorter wavelengths, a behaviour that may be explained by diagonal tunnelling or band-filling mechanisms.

3. Conclusions

We have demonstrated the feasibility of growing GaAs–$Ga_{0.51}In_{0.49}P$–$Ga_yAl_{1-y}As$–GaAs heterostructures of good crystalline quality, with p–n junctions generated by

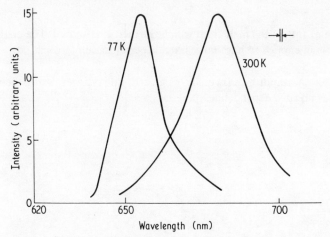

Figure 9. Electroluminescence spectra of heterostructure p–n diode at 300 and 77 K.

doping with tellurium and zinc. Diodes fabricated from these heterostructures have been demonstrated to be efficient light emitters.

Acknowledgments

The authors are grateful to W Alt for expert technical assistance and to H E Krüger, E Menzel and L Balk for x-ray diffractometer and scanning electron microscope evaluation. We thank Dr M Blätte, Dr Jacob and Dr Kuhn-Kuhnenfeld, and Wacker Chemitronic GmbH for the donation of high-quality GaAs substrates. This work was supported by the Deutsche Forschungsgemeinschaft/Sonderforschungsbereich 56, Festkörperelektronik.

References

Alferov Zh I, Andreev V M, Korol'kov V I, Portnoi E L and Tret'yakov D N 1968 *Fiz. Tek. Poluprov.* **2** 1545–7
Beneking H, Mischel P and Schul G 1972 *Electron. Lett.* **8** 16–17
Hakki B W 1971 *J. Appl. Phys.* **42** 4981–95
Heritage R J, Porteous P and Sheppard B J 1970 *J. Mater. Sci.* **5** 709–10
Hilsum C and Porteous P 1968 *Proc. 9th Int. Conf. Physics of Semiconductors* **2** 1214
Isozumi S, Komatsu Y, Kotani T and Ryuzan O 1973 *Japan. J. Appl. Phys.* **12** 306–7
Lorenz M R, Reuter W, Dumke W P, Chicotka R J, Pettit G D and Woodall J M 1968 *Appl. Phys. Lett.* **13** 421–3
Onton A and Lorenz M R 1970 *Proc. 3rd Int. Symp. Gallium Arsenide and Related Compounds* (London and Bristol: Institute of Physics) pp222–30
Pitt G D, Vyas M K R and Mabbitt A W 1974 *Solid St. Commun.* **14** 621–5
Stringfellow G B 1972 *J. Appl. Phys.* **43** 3455–60
Tietjen J J, Enstrom R E and Richman D 1970 *RCA Rev.* **31** 635–46

Discussion

M Rodot (CNRS, Meudon)

What amount of supersaturation do you use to grow GaInP? Is this amount much dependent on the Ga: In ratio?

Dr Beneking

There is a piece of InP on the melt from which the melt is saturated. The cooling rate ($2°$ min^{-1}) is small enough to have no marked supersaturation (growth rate $\sim 0\cdot 5\,\mu$m min^{-1}).

The second question cannot be answered; the growth rate changes. Using the method described we notice no such dependence.

The preparation and properties of bulk indium phosphide crystals and of indium phosphide light emitting diodes operating near 1·05 μm wavelength

K J Bachmann, E Buehler, J L Shay† and D L Malm

Bell Laboratories, Murray Hill, New Jersey 07974, USA

Abstract. The preparation of InP LED by liquid phase epitaxy in a sliding boat arrangement is described. Diodes emitting near 1·05 μm wavelength with 3% external efficiency (30% internal efficiency) at room temperature are obtained. Also, a critical comparison is made between the properties of bulk InP crystals grown by the gradient freeze technique, zone melting, the horizontal Bridgman method and liquid encapsulated Czochralski pulling. The lowest dislocation density $N_d \simeq 10^3\,\mathrm{cm}^{-2}$ and highest mobility of electrons $\mu_e = 4510\,\mathrm{cm}^2\,\mathrm{V}^{-1}\,\mathrm{s}^{-1}$ and holes $\mu_h = 120\,\mathrm{cm}^2\,\mathrm{V}^{-1}\,\mathrm{s}^{-1}$ at room temperature are observed in LEC-pulled InP crystals. The residual impurity levels in InP crystals are revealed by spark source mass spectrometric analyses. Thermodynamic arguments and analytical results indicate that the two major impurities Si and O are introduced during synthesis and crystal growth from the silica envelope via the reaction

$$10\,SiO_2(s) + P_4(g) \rightarrow 10\,SiO(g) + P_4O_{10}(g).$$

The coefficient of thermal expansion of InP between room temperature and 800 °C is $\alpha = 5\cdot3 \times 10^{-6}\,°C^{-1}$.

1. Introduction

Advances in the development of low loss fibre optical waveguides in the past two years have brought optical communications closer to reality than previously anticipated. These advances constitute presently an incentive for materials research directed towards the invention of new light sources, detectors and integrated optical components which match the characteristics of the present state-of-the-art optical fibres. In this paper we describe the results of an effort to develop a new InP light emitting diode (LED) operating near 1·05 μm wavelength. Figure 1 shows a plot of total loss against wavelength λ for a germania-doped BTL fibre, which is typical of the loss spectra generally observed for optical fibres with minimum loss less than 4 dB km^{-1} (French *et al* 1974, MacChesney *et al* 1974). Also, we have indicated in figure 1 the range of a number of light sources and detectors which are available in the most useful regions of the spectrum. Two windows exist in the total loss curve, one near 0·9 μm and the other between 1·05 and 1·1 μm, being separated by the OH-absorption peak at 0·95 μm. $Ga_xAl_{1-x}As$ LED and CW laser diodes at room temperature are suitable light sources near 0·9 μm and the technology related to these devices is more advanced than for corresponding sources near 1·05 μm. Although ultimate bandwidth may be expected from a combination of

† Holmdel, New Jersey 07733, USA

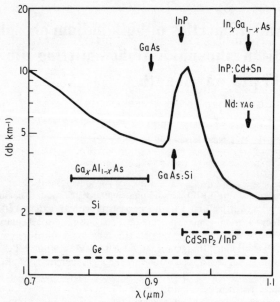

Figure 1. Total loss against wavelength of a germanium doped BTL fibre (courtesy of P D Lazay).

a laser diode with a single mode fibre, a consideration of lifetime makes further developmental work with respect to LED and multimode fibres equally important. Since material dispersion decreases considerably with increasing wavelength, LED operating near $1·05 \mu$m would be advantageous with respect to the dispersion limitation of pulse rate. Also, the total loss at $1·05 \mu$m is always smaller than at $0·9 \mu$m because scattering losses and possible contributions from the exponential tail of the UV absorption edge for SiO_2 decrease rapidly with increasing wavelength (Kapron *et al* 1972, Hubbard 1972, Pinnow *et al* 1973).

Two materials, $Ga_x In_{1-x} As$ (Nuese and Enstrom 1972, Nahory *et al* 1974, Nuese *et al* 1974a,b) and InP doped with Cd and Sn (Shay *et al* 1974) have so far produced efficient LED at room temperature near $1·05 \mu$m†. Present state-of-the-art $Ga_x In_{1-x} As$ diodes have the advantage of being narrower in width of emission than our InP diodes. However, the light emitted from $Ga_x In_{1-x} As$ diodes is self-absorbing while the emission of Cd + Sn/Zn doped InP diodes is shifted off the band gap of InP and can thus be coupled without absorption losses through the Zn-doped substrate to the optical fibre. In addition, the InP LED described in this paper have higher efficiency than present $Ga_x In_{1-x} As$ LED and both materials appear to be sufficiently promising to warrant further development.

Of the detector materials indicated in figure 1, Si is superior at $\lambda \leqslant 1 \mu$m and front illuminated Si-photodetectors of fast response with high quantum efficiency are available at $0·9 \mu$m. At $1·05 \mu$m the absorption coefficient of Si has become so small that

† Pulsed laser operation at room temperature near $1·06 \mu$m and CW lasing at 77 K has been recently reported for $Ga_x In_{1-x} As$ (Nuese and Enstrom 1972, Nahory *et al* 1974). Also, at $1·06 \mu$m the Nd-YAG laser is available, but requires optical pumping and is, therefore, outside the scope of this paper.

an excessive depletion layer thickness is required in such detectors and existing Ge-photodetectors operating from 0·6 to 1·6 μm would be superior. However, side illuminated devices could presumably extend the range of Si-detectors to 1·1 μm and recently discovered heterojunction devices may provide a further alternative to the existing technology (Shay *et al* 1974b, Wagner *et al* 1975).

Since the performance of InP LED and detectors is affected by the electrical properties and defect structure of the substrates, we evaluate in §2 of this paper the relative merits of a variety of crystal growth techniques with respect to defect density, chemical purity and electrical properties of the resulting crystals. Details on apparatus and on the actual crystal growth procedures are presented elsewhere (Bachmann and Buehler 1974, Bachmann *et al* 1975). In §3 we report on recent improvements in the preparation and properties of InP LED emitting near 1·05 μm wavelength.

2. The properties of bulk InP crystals grown by different crystal growth techniques

Bulk indium phosphide crystals suitable for use as substrate material in our LPE experiments have been grown by four techniques. (i) the gradient freeze technique, (ii) zone melting, (iii) the horizontal Bridgman method and (iv) liquid encapsulated Czochralski (LEC) pulling. InP crystals grown by techniques (i)–(iii) are all solidified in a boat, which is usually made from pyrolytic BN. Figure 2 shows the distribution of dislocations as revealed by etch pits on {111} slices cut from a gradient freeze-grown ingot scanning a cross section horizontally (full dots) and vertically (open circles), that is to say from the free surface to the bottom of the ingot where the InP was in contact with the BN boat. The general form of this distribution is characteristic for all crystals

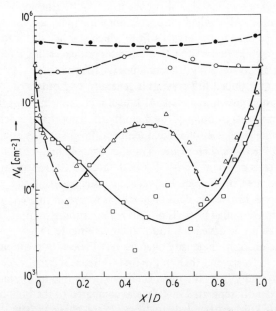

Figure 2. Dislocation density scans N_d against distance X measured in units of the crystal diameter D for InP crystals grown by the gradient freeze technique (circles) and by LEC pulling (triangles and squares), respectively.

grown in a boat and indicates that the strain, produced upon cooling by the differential thermal contraction of the InP and the boat material, governs the generation of defects. The coefficients of linear thermal expansion of pyrolytic BN are 10^{-6} °C^{-1} in the a-axis directions and $2 \cdot 5 \times 10^{-5}$ °C^{-1} in the c-axis direction†. For InP the thermal expansion was measured in the temperature range $25\,°C \leqslant t \leqslant 800\,°C$ using a dilatometer arrangement described by Sauer (1968). The expansion is linear in this range corresponding to $\alpha(\mathrm{InP}) = 5 \cdot 3 \times 10^{-6}$ °C^{-1}. Due to the fact that the c-axis of the BN tends to be oriented perpendicular to the walls of the boat, the side walls become bent inwards upon cooling, thus exerting a compressive force on the InP ingot which increases towards the upper edges of the boat. Vertically only small forces are experienced since the material can expand freely towards the free surface of the ingot. Because the InP melt does not wet the BN, only about one-half of the ingot is confined vertically in the boat and a maximum occurs in the vertical dislocation density distribution at $X/D \simeq 0 \cdot 5$. At a given vertical level the horizontal dislocation density distribution is quite uniform. The magnitude of the temperature gradient at the solid–liquid interface, which is typically $30\,°C\,\mathrm{cm}^{-1}$ in the case of zone melting as compared to $1\,°C\,\mathrm{cm}^{-1}$ for the horizontal Bridgman technique and $0 \cdot 2\,°C\,\mathrm{cm}^{-1}$ for gradient freeze crystal growth under almost isothermal conditions, is of little influence on the final dislocation density. Crystals containing a number of parallel 60° twist boundaries or twin lamellae in the first to freeze part have relatively low dislocation densities‡ ($10^4 \leqslant N_d \leqslant 10^5\,\mathrm{cm}^{-2}$) since these boundaries act as barriers and piling-up of dislocations on twist boundaries in the confined part of zone melted crystals has been observed (Bachmann *et al* 1975). Very non-uniform dislocation distributions and even cracks may develop in boat-grown ingots when the InP binds to the BN which is always the case when non-stoichiometric or impure melts are used and excess In metal or dirt collects at the BN/InP interface during the course of solidification. Also, when pure stoichiometric melts are used small phosphorus bubbles may be occluded in the crystals from which star-like dislocation clusters are punched out as shown in figure 3(b). Growth rates not exceeding a few $\mathrm{mm\,h}^{-1}$ are recommended for all crystal growth techniques (i), (ii) and (iii).

The dislocation density in LEC pulled InP crystals is generally 1–2 orders of magnitude smaller than in crystals grown in a boat. Although no confinement of the crystals by crucible walls exist in this technique, the majority of dislocations is generated by thermal stresses upon cooling since a sizable radial temperature gradient is established between the inside and the outer shell of the crystal in the cold high-pressure nitrogen atmosphere. For crystals grown at $3\,\mathrm{cm\,h}^{-1}$ growth rate and $\langle 001 \rangle$ growth axis a cylindrical region of compensated stress exists between the outer shell and the core of the crystal leading to a W-shaped dislocation density profile as shown in figure 2 (open triangles). However, in very slowly pulled crystals of $\langle 001 \rangle$ orientation also, profiles without inflection points in the inside have been found (compare figure 2).

In the growth direction the dislocation density changes very little with position and may be even lower in the last-to-freeze end than in the first-to-freeze section, except when the crystal is pulled-off rapidly from the melt. Increases in N_d due to thermal shock by a factor of 3 at the rapidly separated interface as compared to the bulk have been observed and this increased dislocation density decays over a characteristic length

† Union Carbide, Technical Information Bulletin No. 713-204EF, p2.
‡ The symbol N_d is used for dislocation density while N_D refers to the donor concentration.

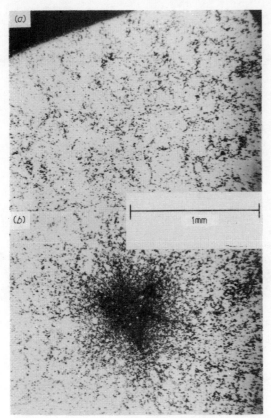

Figure 3. (*a*) Dislocation etch pits near to the edge of a (111) wafer cut from a gradient freeze-grown InP crystal. (*b*) Dislocation cluster around a micro-inclusion in the same wafer.

of about 1·5 mm. In the seeding region thermal shock has very little impact on dislocation density which is probably due to a sizable melt back of the seed after making contact to the melt. Figure 4 shows etch pits on a (111)-slice cut across the seed to crystal interface of an LEC pulled crystal. The dislocation density in the interface region is almost identical to N_d measured well off the interface in the bulk of the seed.

The levels of residual impurities in nominally undoped InP crystals as revealed by spark source mass spectrometric analyses are shown in table 1, which contains also a typical analysis of the starting material prepared from the elements at 27·5 atmosphere phosphorus pressure (Bachmann and Buehler 1974b). The major impurities are Si, O and C† at a level of several ppm. In synthesis experiments and during crystal growth via techniques (i), (ii) and (iii), where the decomposition of InP is prevented by establishing equilibrium with a phosphorus vapour atmosphere of appropriate pressure, the fused silica envelope confining the phosphorus vapour provides a source for both Si and O according to the reaction

$$10 \ SiO_2(s) + P_4(g) \rightleftharpoons 10 \ SiO(g) + P_4O_{10}(g). \tag{1}$$

† An investigation by Mullin *et al* (1972) showed that C is electrically inactive in InP.

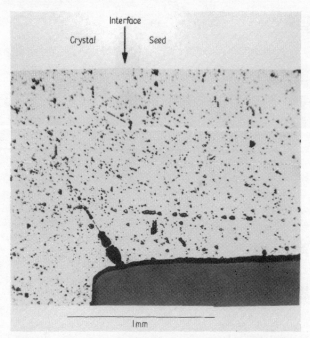

Figure 4. Dislocation distribution in the vicinity of the seed–crystal interface of an LEC pulled InP crystal.

Table 1. Spark source mass spectrographic analyses of nominally undoped indium phosphide crystals. ND indicates concentrations less than or equal to 0·1 ppma

Technique	Melt composition x_P	Part of the ingot	Impurity concentration (ppma)						
			C	O	N	S	Cu	Zn	Si
Gradient freeze (synthesis)	0·5	Middle	1·1	0·3	0·1	ND	ND	ND	1·4
Zone melting	0·5	Middle	2·3	0·9	0·4	1·2	0·6	0·4	3·0
Horizontal Bridgman	0·4	Middle	0·7	0·4	ND	0·5	ND	ND	1·1
LEC pulling	0·5	First to freeze	1·3	0·5	0·2	0·9	0·5	ND	1·4
LEC pulling	0·4	First to freeze	3·6	2·9	0·7	2·6	1·0	ND	5·0

The equilibrium constant of this reaction can be derived from data given in the JANAF thermochemical tables (1971) and the phosphorus pressure at the liquidus of the In/P system is known from our previous study (Bachmann and Buehler 1974b) leading to an estimated value of the partial pressure of SiO over a stoichiometric InP melt $p_{SiO} \simeq 1 \times 10^{-7}$ Torr. The vapour pressure of pure SiO at 1335 K is known from the work of Emons and Theisen (1972) to be $p_{SiO}^{O} = 2 \times 10^{-2}$ Torr. Assuming a Henry–Dalton relationship between the dissolved SiO and the vapour phase one obtains for the molar fraction of SiO in the melt, $10^{-6} \lesssim X_{SiO} \lesssim 10^{-5}$, which agrees approximately with the detected Si-concentration. In qualitative accordance with the SiO-transport hypothesis the levels of Si and O in different crystals correlate and the Si- and O-concentrations are reduced in crystals grown by the horizontal Bridgman technique at low phosphorus pressure. Also, in nominally undoped zone melted InP crystals the Si is uniformly distributed indicating continuous exchange with the vapour phase. The distribution coefficient of Si in InP is unclear at present since values differing by several orders of magnitude have been reported in the literature (Mullin *et al* 1972, Astles *et al* 1973). For a detailed discussion of impurity segregation in InP we refer the reader to our paper (Bachmann *et al* 1975). However, we would like to emphasize that reactions involving the other residual impurities (eg, residual H_2O) and the boat material (Weiner 1972) may contribute to the production of both SiO and P_4O_{10} and that a complete quantitative evaluation of SiO transport is not possible at present because the kinetics of incorporation of Si and O at the solid–liquid interface and the nature of the end products are unknown.

The electrical properties of nominally undoped InP crystals are tabulated in table 2. The highest mobility is observed in LEC pulled crystals grown at small growth rates from both stoichiometric and In-rich melt compositions. A comparison of tables 1 and 2

Table 2. Electrical properties of nominally undoped indium phosphide crystals at room temperature

Growth technique	Growth rate (mm h^{-1})	Melt composition (x_P)	Crystal part	$N_D - N_A$ (cm^{-3})	μ_e (cm^2 V^{-1} s^{-1})	ρ (Ω cm)
Gradient freeze	10	0·5	Middle	$1·1 \times 10^{16}$	3550	0·16
Zone melting	7	0·5	Middle	$2·5 \times 10^{15}$	3580	0·71
Horizontal Bridgman	7	0·4	Middle	$1·6 \times 10^{16}$	3290	0·12
LEC pulling	3	0·5	First to freeze	9×10^{15}	4130	0·17
LEC pulling	6	0·4	First to freeze	$6·6 \times 10^{15}$	4510	0·21

shows that the analytical concentration of uncompensated donor impurities exceeds by far the net carrier concentration $N_D - N_A$ and that the crystals of highest mobility and of lowest $N_D - N_A$ have not the lowest Si concentration. One possible explanation for this discrepancy is that a major part of the Si is precipitated, presumably as a silicon–oxygen complex.

Since epitaxial InP layers grown via liquid phase epitaxy (LPE) from Sn-rich solutions are always n-type, p-type substrate crystals are required which were produced by doping with Zn or Cd. The highest mobility of holes was $\mu_h = 120$ cm^2 V^{-1} s^{-1} at room temperature, measured in a Cd-doped LEC pulled crystal with $N_A - N_D = 1·7 \times 10^{-6}$ cm^{-3}.

3. The preparation and properties of InP LED with peak emission near 1·05μm wavelength

In a recent letter (Shay *et al* 1974a) we have reported preliminary results on InP diodes consisting of a heavily Sn-doped n-type epilayer which is partly compensated by Cd on a Zn-doped substrate of (100) orientation. These layers were grown by a liquid phase epitaxial tipping process from Sn-solutions of In, Cd and P and emitted with 1% external efficiency (10% internal efficiency) at room temperature in the regime $0·95 \leqslant \lambda \leqslant 1·1 \mu m$, depending on growth conditions. For a given solution composition and growth rate the electroluminescence spectra of the diodes were reproducible, for example, at a cooling rate of $3·7\,°C\,h^{-1}$ deep emission with peak intensity a $1·08\,\mu m$ was observed for all tipping experiments using a solution of 2·8% InP, 0·2% CdSnP$_2$, 95% Sn and 2% P. However, difficulties were encountered in tailoring the peak emission and improving the properties of the LED which we relate to insufficient control of growth rate in the tipping experiments.

Therefore, we now use a sliding boat apparatus as shown in figure 5, which represents a modified version of the arrangement used by Panish (1973a,b) for LPE growth of Sn-doped $Ga_xAl_{1-x}As$ structures. An essential feature of our sliding boat is a more uniform heat distribution, implemented by a closed well which prevents cooling of the surface of the solution by the hydrogen ambient protecting it from oxidation. In an open well sliding boat arrangement a sizable fraction of the precipitating constituents may crystallize on InP nuclei floating at the surface of the solution, since the density of

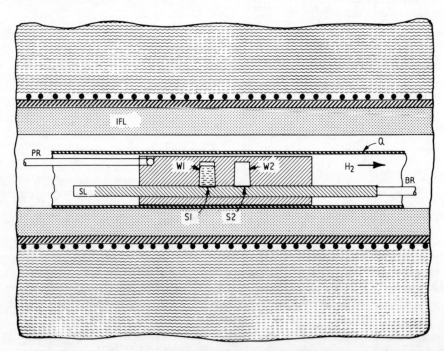

Figure 5. Sliding boat arrangement for LPE growth of InP LED. IFL = isothermal furnace liner; Q = fused silica tube; PR = push rod; BR = blocking rod; SL = carbon slider; S1, S2 = InP seeds; W1, W2 = wells.

InP is much smaller than that of the metal solvents (In, Sn) used in our LPE experiments. The resulting changes in solution composition cause irreproducible epitaxial layer growth on the bottom of the well. Also, convective currents may distort the epilayer growth when a vertical temperature gradient is established in the solution by cooling of its surface in the H_2-stream. In addition, a closed well system minimizes the loss of volatile constituents (Cd, P) from the solution and improves thus the reproducibility of the experimental conditions. Adapting Dawson's (1974) procedure for GaAs to our work, a nominally undoped InP seed S1 was placed under the solution for equilibration at constant temperature before the p-type seed S2 was slid into the growth position. After a short etching period generating a virgin surface for the epitaxial deposition, S2 was kept at constant temperature for 15 h and epitaxial layer growth was then initiated by cooling. A typical temperature–time programme is shown in figure 6.

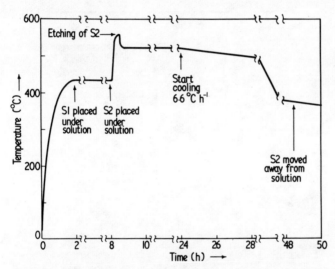

Figure 6. Temperature–time programme for the growth of InP epilayers yielding peak emission at 1·06μm with 3% external efficiency at room temperature.

An as-grown epilayer and a cross section of it, stained with a solution of 25 ml HNO_3 + 25 ml H_2O + 1 ml HF, are shown in figure 7. The epilayers are typically 30 μm thick which is smaller by a factor of about 3 than the thickness of epilayers usually produced in the previously described tipping process (Shay *et al* 1974a,b).

LED with peak emission at 1·06 and 1·08 μm have been prepared by the sliding boat process. All diodes cleaved from such LPE wafers had excellent electrical characteristics as indicated by figure 8 which shows a typical $I-V$ curve. The room temperature external efficiency of diodes grown in the sliding boat is 3% corresponding to 30% internal efficiency. It should be noted that these numbers still refer to layers produced by single epitaxy directly onto LEC pulled substrates and further improvements may be expected from double epitaxy experiments. The presence of Cd is essential for obtaining the deep emission. Since the Sn-doping level for layers grown from Sn-solutions is very

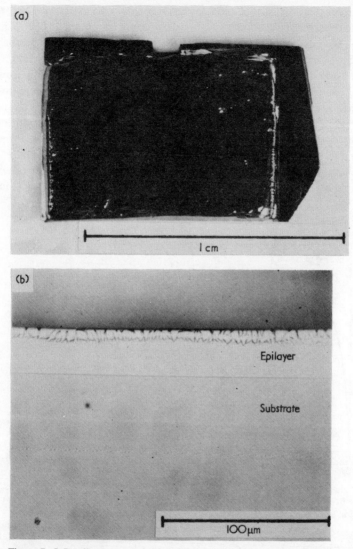

Figure 7. InP epilayer grown in the sliding boat arrangement on a (100)-substrate wafer: (*a*) top view, (*b*) cross section.

high (about $5 \times 10^{19} \, \text{cm}^{-3}$), relatively highly resistive substrates ($N_A - N_D = 10^{16} \, \text{cm}^{-3}$) had to be used to prevent excessive tunnelling currents. As it is well known from the work of Rosztoczy *et al* (1970) the Sn-concentration in epitaxial InP reduces with the mole fraction of Sn in mixtures of In + Sn used as solvents and a number of experiments with solutions of relatively small Sn-content were performed. Unfortunately, higher growth temperatures must be used in this case since the solubility of InP in In is much less than in Sn. It is most revealing that no deep emission is observed in layers grown at temperatures above the peritectic temperature at which $CdSnP_2$ is formed (565 °C). This indicates that the deep emission observed in diodes prepared below 565 °C is due

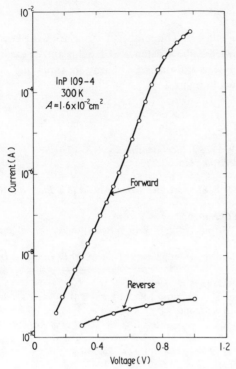

Figure 8. *I–V* characteristic of diodes prepared from the wafer shown in figure 7.

to the incorporation of Cd and Sn into the InP lattice as an ordered complex. The ability of obtaining highly efficient Cd + Sn doped LED near 1·05 μm wavelength is, in our opinion, related to the unique property of Cd and Sn to provide under the above growth conditions simultaneously charge and size compensation in the InP lattice. A more detailed discussion of the performance of our InP diodes as a component of a preliminary source/fibre/detector system will be presented elsewhere (Shay *et al* 1974).

4. Summary

Bulk InP crystals were prepared by tne gradient freeze technique, zone melting, the horizontal Bridgman technique and LEC pulling. The purest crystals as judged by chemical analysis were grown by the horizontal Bridgman technique from In-rich melt compositions. We propose that two major impurities, Si and O, are introduced into the InP melt via SiO vapour transport, but do neither limit the electrical properties nor determine the density of defects in present state-of-the-art InP crystals. The best crystals with respect to defect density ($N_d \simeq 10^3 \, \text{cm}^{-2}$) and mobility of electrical carriers ($\mu_e = 4510 \, \text{cm}^2 \, \text{V}^{-1} \, \text{s}^{-1}$ for electrons, $\mu_h = 120 \, \text{cm}^2 \, \text{V}^{-1} \, \text{s}^{-1}$ for holes at room temperature) were obtained by LEC pulling. Cd + Sn-doped n-type epilayers were grown on Zn-doped p-type substrates by LPE in a sliding boat arrangement. LED prepared from these wafers emit near 1·05 μm wavelength with 3% external efficiency at room temperature.

132   *K J Bachmann, E Buehler, J L Shay and D L Malm*

Acknowledgments

We would like to thank J W Fleming for his cooperation in measuring the coefficient of expansion of InP. Also, we acknowledge the assistance of L Schiavone in measuring the electrical properties of our InP crystals and thank P D Lazay for supplying us with the total loss—wavelength curve shown in figure 1.

References

Astles M G, Smith F G H and Williams E W 1973 *J. Electrochem. Soc.* **120** 1750
Bachmann K J and Buehler E 1974a *J. Electron. Mater.* **3** 279
—— 1974b *J. Electrochem. Soc.* **121** 835
Bachmann K J, Buehler E, Clark L, Malm D L, Shay J L and Strnad A R 1975 *J. Electron. Mater.*
Dawson L R 1974 *J. Crystal Growth* in press
Emons H-H and Theisen L 1972 *Monatsh Chem.* **103** 62
French W G, MacChesney J B, O'Connor P B and Tasker G W 1974 *Bell Syst. Tech. J.* **53** 951
Hubbard W M 1972 *Appl. Optics* **11** 2495
JANAF Thermochemical Tables ed D R Stull and H Prophet NSRDS-NBS **37** (Washington DC: NBS)
Kapron F P, Maurer R D and Teter M P 1972 *Appl. Optics* **11** 1352
MacChesney J B, O'Connor P B, DiMarcello F V, Simpson J R and Lazay P D 1974 *Proc. 10th Int. Congr. Glass Kyoto* (Ceram. Soc. Japan) vol 6 pp6—40
Mullin J B, Royle A, Straughan B W, Tufton P J and Williams E W 1972 *J. Crystal Growth* **13/14** 640
Nahory R E, Pollack M A and DeWinter J C 1974 *Appl. Phys. Lett.* **25** 146
Nuese C J and Enstrom R E 1972 *IEEE Trans. Electron Dev.* **ED-19** 1067
Nuese C J, Enstrom R E and Ettenberg M 1974a *Appl. Phys. Lett.* **24** 83
Nuese C J, Ettenberg M, Enstrom R E and Kressel H 1974b *Appl. Phys. Lett.* **24** 224
Panish M B 1973a *J. Appl. Phys.* **44** 2659
—— 1973b *J. Appl. Phys.* **44** 2667
Pinnow D A, Rich T C, Ostermayer F W and DiDomenico M 1973 *Appl. Phys. Lett.* **22** 527
Rosztoczy F E, Antypas G A and Casau C J 1970 *Proc. 3rd Int. Symp. Gallium Arsenide and Related Compounds* (London and Bristol: Institute of Physics) p86
Sauer H A 1968 *Rev. Sci. Instrum.* **39** 562
Shay J L, Bachmann K J and Buehler E 1974a *Appl. Phys. Lett.* **24** 192
—— 1974b *J. Appl. Phys.* **45** 1302
—— 1974c *Proc. IEEE Meeting Atlanta*
Wagner S, Shay J L, Migliorato P and Kasper H M 1975 *Appl. Phys. Lett.* to be published
Weiner M E 1972 *J. Electrochem. Soc.* **119** 496

Discussion

M Rodot (CNRS, Meudon)

When you spoke of InP doped with Cd and Sn, did you imply, as I understood it, that Cd and Sn are ordered in spite of their small concentration? What is this concentration? Can you comment about this?

Dr Bachmann

The Sn-concentration in diodes grown from Sn-solutions can be determined by both electrical measurements and analytical techniques and is 5×10^{19} cm^{-3}. The Cd concentration depends on growth conditions and determines the wavelength of peak emission. It is not easily evaluated by a direct analytical determination. However, from annealing experiments, in which the properties of epi-layers grown on a Cd-doped substrate from pure Sn and from Cd–Sn solutions before and after annealing are compared, we know at which concentrations diffusion of Cd occurs from the seed into the epi-layer and vice versa. The results of these experiments indicate Cd-levels in the 10^{18} cm^{-3} range. When I talked about Cd and Sn entering into the InP lattice as an ordered complex, I meant that at least part of the incorporated Cd and Sn atoms are paired as in the compound CdSnP$_2$. The distribution of these pairs within the InP host lattice may be at random and is presently unknown.

J Lebailly (RTC La Radiotechnique, Caen)

In this paper, as in the one of Ettenberg, we are dealing with crystals transparent to their own radiation from which light is escaping through all faces. The question refers to the application: what should be the light injection coefficient into a multimode optical fibre compared with the case of a classical planar electroluminescent diode in which band to band emission occurs and light is generated from one defined area?

Dr Bachmann

This question may be based on a misunderstanding. The emission from Ga$_x$In$_{1-x}$As is basically band to band emission while the emission in our diodes is shifted by more than 150 meV off the bandgap of InP. The 0·05 cm thick InP substrate, but not the epilayer, is therefore transparent to the emitted light while Ga$_x$In$_{1-x}$As is self-absorbing. It is true that by proper design of the gradation the problems of self-absorption in a diode of graded composition may be reduced but materials properties often limit the utilization of this window effect. (For a detailed discussion compare A A Bergh and P J Dean 1972 *Proc. IEEE* **60** pp208–10.) In order to prevent excessive losses through self-absorption, diodes with band to band emission have, therefore, to be sufficiently thinned down before mounting to the glass fibre (C A Burrus and B I Miller 1971 *Opt. Commun.* **4** 307). This is, in our opinion, a disadvantage in Ga$_x$In$_{1-x}$As diodes as compared to the InP diodes discussed in this paper which may be mounted without thinning directly to the fibre. For a discussion of the power coupled from a Burrus diode and from our InP diodes into a multimode fibre we refer to a recent review article of M DiDomenico Jr 1974 *Opt. Eng.* **13** and to J L Shay, K J Bachmann and E Buehler 1974 *Proc. IEEE Meeting, Atlanta.*

Synthesis of GaP by gallium solution growth in a horizontal sealed ampoule

G Poiblaud and G Jacob

RTC La Radiotechnique Compelec, 14000 Caen, France

Abstract. This paper describes a technique for preparing polycrystalline GaP suitable for use as starting material for liquid encapsulated Czochralski pulling and as source for liquid phase epitaxy. The synthesis of GaP is performed by a gallium solution growth method in a horizontal closed tube containing two boats of Ga placed in a symmetrical temperature gradient. GaP is formed by the reaction of Ga with phosphorus vapour resulting from sublimation of red phosphorus.

It is found that the GaP synthesis can be performed much faster by using a furnace system with two cold zones on either side of the growth zone and by loading the ampoule with a large excess of gallium. The reaction yield is discussed. The resulting crystals are evaluated from the point of view of electrical and optical properties. The purity is investigated; particular attention has been devoted to silicon and oxygen contamination as a function of the synthesis running time, the growth temperature and the boat materials. The oxygen content is determined by ^{3}He activation using the reaction ^{16}O(^{3}He, p)^{18}F and the silicon concentration by spectrochemical analysis. Experimental results show that the presence of nitrogen in the synthesized GaP reduces the residual doping and then improves the electrical properties of the material.

1. Introduction

Gallium phosphide is taking a predominant place in the light-emitting diode (LED) industry for the emission of a variety of colours, namely green, yellow and red. The fabrication of LED uses mainly the VPE and LPE technologies in which the GaP single-crystal substrates are grown by liquid encapsulation Czochralski (LEC) techniques, at present the only means of producing large area wafers with high yields (Nygren 1973).

Up till now GaP synthesis and single-crystal growth are not accomplished in the same process, thus necessitating for LEC pulling the presynthesis of the poly-crystalline starting material. Besides, the LPE procedure also needs a source of pure gallium phosphide.

To prepare the polycrystalline GaP, two techniques are commonly used: an open-tube flow method using the reaction of PH_3 with Ga (Grimmeiss and Kischio 1956, Dean *et al* 1968, Ringel 1971) and a high-pressure gallium solution growth method (Bass and Oliver 1968, Besselere and Le Duc 1968). In the first case the reaction yield in phosphorus and gallium is rather low; the second method has the disadvantage of needing complicated apparatus, high pressure and high temperature.

In this paper we describe a simple method of GaP synthesis by solution growth in a horizontal sealed ampoule, starting from pure gallium and phosphorus elements, operating at atmospheric pressure and relatively low temperature; then we report some physical properties of the resulting crystals.

2. Basic principle and experimental apparatus

In a previous publication (Poiblaud and Jacob 1973), we reported the growth of bulk gallium phosphide; the synthesis method described here is an extension of this technique. It occurs in a horizontal furnace system. Figures 1 and 2 are schematic diagrams of the two successive arrangements used, including the reactions involved in the procedure. A quartz ampoule, 110 cm long, 55 mm external diameter, 50 mm internal diameter, with walls 2·5 mm thick, is cleaned by 10% HF for 1 h followed by an *aqua regia* etching to dissolve fluorides and a water rinse of several hours. It is then dried under nitrogen and heated at 1000 °C under internal vacuum of 10^{-5} to 10^{-6} Torr. The ampoule is then loaded as sketched in figures 1 or 2 with commercial 99·9999% gallium and red phosphorus used as received without any treatment, and sealed under vacuum of 10^{-6} Torr obtained by an ion pump. The two boats of gallium are located in a symmetrical temperature gradient ($G = 20$ °C cm^{-1}) on either side of the centre of the furnace with $T_2 = 900$ and $T_3 = 1150$ °C.

In early work, the first experiments were conducted in the furnace shown in figure 1, characterized by only one cold zone at the level of the red phosphorus, using the sublimation equilibrium of red phosphorus to control the vapour pressure in the tube;

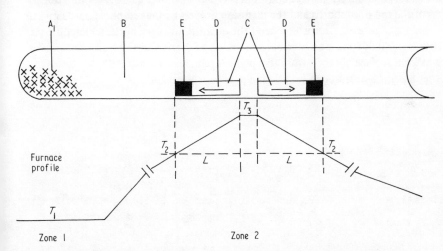

Figure 1. Principle of GaP synthesis by the method of gallium solution growth in a horizontal sealed ampoule, two-zone technique. A_1: red phosphorus sublimation P_4 (solid) $\rightleftharpoons P_4$ (vapour) at 500 °C. A_2: condensation of red phosphorus which has not reacted with Ga. B: equilibrium P_4 (vapour) $\rightleftharpoons 2P_2$ (vapour). C: reaction Ga (liq) + P_4 (vap) + P_2 (vap) $\rightleftharpoons$ (Ga + GaP) solution. D: diffusion of GaP in Ga following Fick's law, $J_{GaP} = -D(\partial C/\partial X)$. E: precipitation of GaP when the solubility limit in Ga is achieved. T_1: sublimation temperature of red phosphorus (500 < T_1 < 550 °C). T_4: temperature for controlling the phosphorus vapour pressure in the ampoule ($P = 1$ atmosphere for $T_4 = 420$ °C). $(T_3 - T_2)/L$: temperature gradient applied to the Ga boats ($T_3 = 1200$ °C; $T_2 = 900$ °C; $L = 15$ cm).

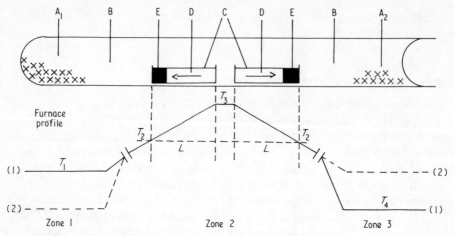

Figure 2. Three-zone technique.

the phosphorus is evaporated by heating at a temperature $T_1 = 420\,°C$ calculated from the relationship giving the vapour pressure versus temperature:

$$\lg P_{\text{atm}} = -0.656\,\frac{104}{T(\text{K})} + 9.46.$$

During the crystallization an equilibrium is established between the sublimation process of the red phosphorus and the diffusion process of the red phosphorus vapour in the gallium (figure 3). On the one hand the quantity of phosphorus which diffuses into the Ga per unit time is an increasing function of vapour pressure until saturation of the gallium surface C_s, then from this point it remains constant with pressure. On the other hand the quantity of evaporated phosphorus per unit time decreases with

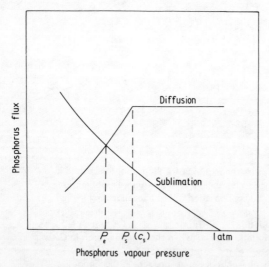

Figure 3. Evaporated and diffused phosphorus flux as a function of the pressure in the synthesis ampoule.

pressure in the ampoule becoming equal to zero when $P = 1$ atm. The intersection of these two curves gives the equilibrium pressure P_e within the ampoule. Growth experiments revealed that this equilibrium pressure was always less than 1 atm, resulting in a slow crystallization of GaP and a deformation of the synthesis ampoule which became slightly flat; this phenomenon shows the note of diffusion is greater than the note of sublimation.

In fact the sublimation kinetics of red phosphorus can vary between limits which are poorly defined owing to:

(a) variations of the physical form of the red phosphorus such as powder or pieces of variable dimensions

(b) variations of allotropic forms of the red phosphorus from one batch to another (mixture of several species) giving different equilibrium pressure for a given temperature. The non-reproducibility of the growth rate (variations of 20 to 30% observed between experiments supposed identical) can be attributed to this phenomenon as mentioned elsewhere (Poiblaud and Jacob 1973, Bachmann and Buehler 1974).

To avoid these disadvantages we use the experimental arrangement of figure 2, consisting of a furnace with two cold zones (zones 1 and 3). Here the phosphorus is heated to a temperature $500 < T_1 < 550\,^{\circ}C$, so that the quantity of evaporated phosphorus is greater than the quantity of diffused phosphorus. The excess phosphorus condenses in the second cold end A_2 of the ampoule, which is maintained at a temperature $T_4 = 420\,^{\circ}C$ to ensure an internal pressure of 1 atmosphere. When all the phosphorus is evaporated from point A_1, the temperatures T_1 and T_4 are inverted and so on.

This process allows an increase in the kinetics of the reaction; the reaction is here limited by the diffusion of P into Ga and not by the sublimation of the red phosphorus as was the case in the first arrangement.

3. Experimental results

3.1. Crystal morphology

Figure 4 represents a boat after synthesis. The GaP produced is of two kinds: first a porous polycrystalline ingot with many large gallium inclusions localized essentially in the grain boundaries (figure 5); the GaP grains small at low temperature are increasing

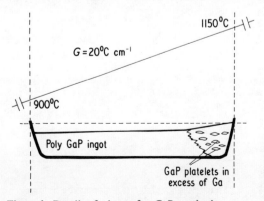

Figure 4. Details of a boat after GaP synthesis.

Figure 5. Polycrystalline GaP ingots grown from Ga solutions in horizontal closed tube (scale in mm).

with temperature. The two last centimetres of the boat are filled with gallium containing GaP platelets (figure 6) formed during the decrease of temperature. The excess of gallium is decanted; the polycrystalline ingot is broken into small pieces. The whole is etched by an $HF-H_2O_2$ mixture followed by a wash and a treatment with HCl to eliminate possible fluorides absorbed on the surface of the crystals. The small crystals so obtained are practically without gallium inclusions. In comparison the GaP platelets are small single crystals with growth faces oriented (111).

3.2. Reaction yield

The reaction yield of gallium and red phosphorus depends on various parameters: mainly temperature gradient in gallium, synthesis running time, and P:Ga ratio of elements introduced in the synthesis ampoule.

Figure 6. GaP platelets grown at the hot end of the boat (scale in mm).

First, we are going to show the importance of the phosphorus transport procedure using a furnace with two cold zones (figure 2) with respect to the synthesis procedure using only one cold zone (figure 1). Figure 7 shows the weight of GaP produced in the two procedures versus the synthesis running time, with constant loading conditions (100 g red P and 2 × 225 g Ga).

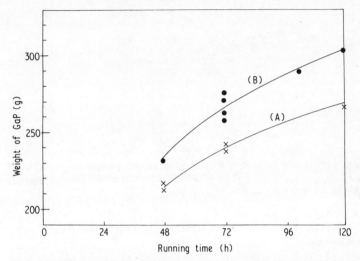

Figure 7. Influence of the phosphorus transport procedure. Curve A: furnace with one cold zone (T_1 = 420 °C). Curve B: furnace with two cold zones (T_1 = 550 °C) and T_4 = 420 °C following the set-up illustrated in figure 2 (ampoules loaded with 100 g of red P and 2 × 225 g of Ga).

The importance of the second procedure is clearly apparent; at first the reaction kinetics is high, then it is slowing progressively when the contact surface between the phosphorus vapour and the gallium is decreasing; the curve tends to an asymptotic value from which we deduce a phosphorus yield of 90–95% and a gallium yield of 45–48%, with a synthesis of 120 h.

However, to reduce the contamination by silica it is necessary to decrease the synthesis running time: this is achieved by loading boats with a high excess of gallium with respect to the phosphorus introduced. Figure 8 shows the weight of GaP produced as a function of the synthesis running time for different loading conditions (A: 100 g P and 2 × 300 g Ga; B: 100 g P and 2 × 225 g Ga). With a large excess of gallium, the reaction is practically complete after 36–48 h. The phosphorus yield is near 100%, but the gallium yield drops to 35–38%. If we take into account the fact that most of the unreacted gallium can be used again after purification, these last conditions remain very favourable.

3.3. Electrical properties

The electrical properties of the crystals were evaluated from Hall measurements made on four leaved clover shaped specimens etched by sand-blasting. The samples for measurements were chosen among single-crystal GaP platelets with two (111) parallel faces having a diameter of $\phi \geqslant 4$ mm. We must keep in mind that this GaP originates

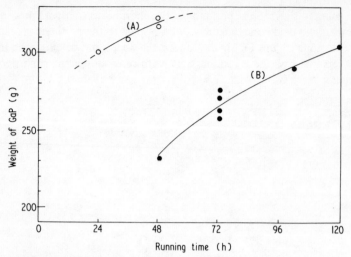

Figure 8. Influence of the loading ratio of gallium to red phosphorus on GaP yield. Curve A: ampoule loaded with 100 g of red P and 2 × 300 of Ga. Curve B: ampoule loaded with 100 g of red P and 2 × 225 g of Ga.

at the highest temperature of the boat (see figure 4); we can assume that it is the more contaminated material.

Typical results obtained on standard undoped crystals are shown in table 1.

We occasionally obtained much better results in apparently the same experimental conditions. At 300 and 77 K respectively resistivities of 10^2 and $10^4 \,\Omega$ cm, mobilities of 200 and 2000 cm² V⁻¹ s⁻¹...

We occasionally obtained much better results in apparently the same experimental conditions. At 300 and 77 K respectively resistivities of 10^2 and $10^4\,\Omega$ cm, mobilities of 200 and 2000 $cm^2\,V^{-1}\,s^{-1}$, carrier concentrations of 10^{15} and $5 \times 10^{11}\,cm^{-3}$ were measured.

Cathodoluminescence measurements made on the respective crystals suggest that these improvements in electrical properties can be associated with the presence of nitrogen in the crystals.

Table 1. Electrical properties of polycrystalline GaP

T (K)	Type	ρ (Ω cm)	μ (cm² V⁻¹ s⁻¹)	$N_D - N_A$ (cm⁻³)
300	n	0·2–2	100–150	6×10^{16}–3×10^{17}
77	n	30–250	400–750	4×10^{13}–6×10^{14}

3.4. Cathodoluminescence

We studied undoped GaP samples which had been previously used for Hall measurements. The cathodoluminescence spectra were excited by an electron beam of $25\,\mu A \times 20\,keV\,mm^{-2}$.

Figures 9 and 10 show the spectra obtained on typical materials having the electrical characteristics of table 1. At 300 K we note an intense green emission at 2·24 eV and a broad red line at 1·95 eV characteristic of silicon-doped material.

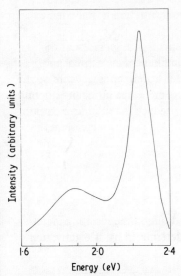

Figure 9. Cathodoluminescence spectrum of GaP crystals (as measured in table 1). $T = 300$ K.

Figure 10. Cathodoluminescence spectrum of GaP crystals (as measured in table 1). $T = 95$ K.

At 95 K the spectrum consists of the S_0 line (2·3 eV) of an exciton bound to sulphur, the D_0 line (2·23 eV) corresponding to the transition sulphur donor—valence band and the broad red peak at 1·90—1·98 eV, not well defined, generally associated with Si—Si pairs (Dean *et al* 1968). There is thus evidence of residual doping by silicon and sulphur.

On GaP having the improved electrical characteristics indicated earlier, no major differences are observed at 300 K; at 95 K, we obtain the spectra of figure 11. There is evidence of nitrogen doping; the spectrum consists of the N line (recombination of an exciton bound to an isolated nitrogen atom and its phonon replicas), the NN_1 line and its phonon replicas. From the absorption data and the intensity ratio of the NN_1/N—LO

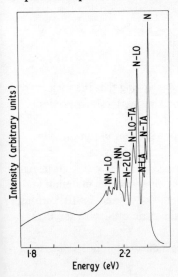

Figure 11. Cathodoluminescence spectrum of GaP crystals with improved electrical properties. $T = 95$ K.

lines (A T Vink, private communication), the nitrogen concentration is evaluated at $5 \times 10^{17} - 10^{18} \, \text{cm}^{-3}$.

To explain the presence of nitrogen in these experiments it is supposed that the red phosphorus has contained nitride compounds. By combining the electrical and optical measurements it is suggested that the presence of nitrogen tends to decrease the number of electrons in the conduction band by a trapping process; a nitrogen–sulphur competition on phosphorus sites could also contribute to reduce the sulphur incorporation thus N_D. This is compatible with the decrease of $N_D - N_A$ and the increase in the mobility.

3.5. Chemical purity

Results of chemical or spectrochemical analysis carried out on the resulting materials are listed in table 2. The total oxygen content has been determined by ^{3}He activation using the reaction $^{16}O(^3He, p)^{18}F$ (Kim 1971, 1974).

Table 2. Residual impurities in polycrystalline GaP determined by chemical or spectrochemical analysis

Elements	(ppm wt)
B	< 1
O	See figure 12
Na	≤ 0·1
Mg	0·3 – 3
Al	≤ 3
Si	See figure 13
S	≤ 1
Mn	< 0·2
Fe	0·2 – 5
Cu	≤ 0·5

It is found that Si and O are the predominant impurities showing that the major contaminant in the system is the silica material used for making synthesis ampoules and boats.

In figures 12 and 13, the oxygen and silicon contents of a number of samples are plotted, including polycrystalline GaP and resulting LEC GaP single crystals.

When we use silica boats, total oxygen concentrations of $3 \times 10^{18} - 2 \times 10^{19}$ atoms/cm^3 and $10^{18} - 4 \times 10^{18}$ atoms/cm^3 are respectively observed in polycrystalline materials and LEC crystals. The silicon level in the polycrystal is between 8×10^{17} and 5×10^{18} atoms/cm^3, slightly decreasing in the LEC crystals due to the interaction of Si with the B$_2$O$_3$ encapsulant.

It will be noted that mass spectroscopy analysis always indicates inferior values with a factor of up to ten.

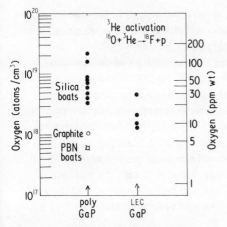

Figure 12. Oxygen concentration of GaP crystals determined by ³He activation.

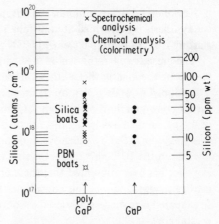

Figure 13. Silicon concentration of polycrystalline GaP determined by different techniques.

In our system we assume a process of contamination by an SiO_2–Ga interaction following the reactions:

$$3SiO_2 + 4Ga \longrightarrow 2Ga_2O_3 + 3Si$$

or

$$SiO_2 + 2Ga \longrightarrow Ga_2O + SiO.$$

Analysis made on samples along polycrystalline ingots show that synthesis temperatures from 925 to 1150 °C have no effect on the oxygen content whereas the silicon content determined by spectrochemical analysis remains low (less than 3 ppm) until 1050 °C before increasing notably in the last part of the boat.

Results of Si and O analyses are not sensitive to variations of the synthesis running time from 24 to 96 h. When pyrolytic boron nitride boats were used the total oxygen content was found to be less than 10^{18} atoms/cm³ and silicon less than 8×10^{17} atoms/cm³. The use of graphite boats also reduces the oxygen content but not the silicon content. The effects of temperature and running time were not studied in the case of synthesis in graphite and PBN boats.

Undoped LEC single crystals pulled from the polycrystalline materials so prepared are n-type with carrier concentrations in the range of 4×10^{15} to 10^{17} cm⁻³ from top to the bottom of ingots; on these undoped crystals mobilities higher than 150 cm² V⁻¹ s⁻¹ are determined. Undoped GaP LPE layers grown from GaP sources obtained in PBN boats are n-type with carrier concentration less than or equal to 2×10^{16} cm⁻³.

4. Conclusion

A method has been developed to synthesize GaP from a gallium solution growth technique in a horizontal sealed silica ampoule at atmospheric pressure using a phosphorus vapour transport procedure in a furnace with two cold zones.

This method offers a number of advantages: it does not require high expenditure, the synthesis is performed at atmospheric pressure, at relatively low temperature, starting from pure simple elements (GN grade Ga and P) in one day.

The present apparatus can be used to produce larger quantities of GaP per run by increasing the load in the synthesis ampoule.

Undoped crystals grown by this process in silica boats are n-type with average carrier concentrations of 10^{17} cm^{-3}. Chemical analysis of crystals revealed the presence of oxygen and silicon in significant concentrations: this contamination from quartz was greatly reduced by using pyrolytic boron nitride boats: residual doping was thus lowered in the 10^{16} cm^{-3} range. Crystals of improved electrical quality were obtained: evidence has been presented that these crystals were associated with the presence of nitrogen suggesting that this contributes to reducing the residual doping. The material so prepared was successfully used as source for LPE and as starting materials for LEC crystal growth.

Acknowledgments

The authors would like to thank M. J L Le Cann for technical assistance; MM. Gaffre and Varon for cathodoluminescence measurements. Thanks are also due to M. Haroutiounian (RTC Suresnes) for chemical analysis, M. Hendriks (Philips Labs, Eindhoven) for spectrochemical analysis and M. Engelman (CEA Saclay) for activation analysis.

References

Bachmann K J and Buehler E 1974 *J. Electrochem. Soc.* **121** 835
Bass S J and Oliver P E 1968 *J. Crystal Growth* **3** 286
Besselere J P and Le Duc J M 1968 *Mater. Res. Bull.* **3** 797–806
Dean P J, Frosch C J and Henry C H 1968 *J. Appl. Phys.* **39** 5631–46
Grimmeiss H G and Kischio N 1956 *Philips Techn. Rev.* **26** 136
Kim C K 1971 *Anal. Chim. Acta* **54** 407–14
—— 1974 *J. Appl. Phys.* **45** 243–5
Nygren S F 1973 *J. Crystal Growth* **19** 21–32
Poiblaud G and Jacob G 1973 *Mater. Res. Bull.* **8** 845–58
Ringel C M 1971 *J. Electrochem. Soc.* **118** 609

A simple LPE process for efficient red and green GaP diodes

C Weyrich, G H Winstel, K Mettler and M Plihal

Forschungslaboratorien der Siemens AG, München, West Germany

Abstract. A simple LPE process for both red and green GaP light emitting diodes is described. The main features of this process are:

(i) a low gallium consumption owing to the use of thin melts
(ii) the formation of the p−n junction by overcompensation of the gallium melt and
(iii) the introduction of all dopants via vapour phase.

Using thin melts the gallium can be saturated by dissolving GaP substrate material. This allows the gallium to be brought into contact with the substrate at room temperature: this facilitates boat construction and renders the process very suitable for large scale production. The influence of process parameters, for example substrate orientation, impurity concentrations, annealing processes etc on the efficiency of the luminescent diodes is discussed. The highest efficiencies obtained for encapsulated diodes are 4% for the red and 0·4% for the green at 0·5 A cm^{-2} and 100 A cm^{-2}, respectively.

As compared to conventional technologies the process described very easily allows for subsequent growing a red and a green light emitting junction on the opposite sides of the same GaP substrate. With such a structure we obtained a device which emits light of red, green or yellow hues depending on whether one or both p−n junctions are operated.

1. Introduction

The highest efficiencies reported for red and green GaP LED have been obtained by liquid phase epitaxy (LPE). High efficient red Zn, O-doped GaP diodes are made by double epitaxy (Saul *et al* 1969: 7·3%; Solomon and DeFevere 1972: 15·1%), that is to say, two epitaxial layers, a Te-doped n-layer and a Zn, O-doped p-layer are subsequently grown on a liquid encapsulated Czochralski (LEC) GaP crystal. Many attempts have been made to circumvent this two-step epitaxy process: the direct deposition of a Zn, O-doped p-layer on an n-type LEC substrate leads to efficiencies up to 2% (Hackett *et al* 1970), but the reproducibility seems to be questionable. Much higher efficiencies of up to 10·1% are obtained by growing the p-layer on a so-called SSD (Synthesis Solute Diffusion) crystal (Kaneko *et al* 1973). However, the morphology of the epitaxial layer is poor due to the polycrystallinity of the SSD material. Casey *et al* (1972) made red GaP diodes by Zn-diffusion into Te, O-doped layers grown by LPE. After appropriate annealing processes the efficiencies increased up to 1·5%.

Highly efficient green n-doped GaP diodes have been obtained by double LPE (Logan *et al* 1971: 0·6%), by double bin LPE (Ladany and Kressel 1972: 0·7%) or by an overcompensation method (Lorimor *et al* 1973: 0·3%; Weyrich 1973: 0·15%; Roccasecca *et al* 1974: 0·33%†). The advantage of this latter process is that the p−n

† This value is measured at 7 A cm^{-2} and should correspond to a maximum quantum efficiency of about 0·65%.

junction is formed in one epitaxy process by overcompensating the Ga melt adding Zn to the melt during cool-down procedure. This technique has also been applied to multislice LPE growth (Saul and Roccasecca 1973).

Despite the high efficiency of LPE devices a rather small percentage of the commercially available LED are LPE grown, the larger part being vapour phase epitaxy devices with diffused p–n junctions. Apart from cost, one reason for this is the technical difficulty involved in the LPE growth diminishing the typical performance of GaP LED for one order of magnitude as compared to research results.

In this paper a simple epitaxy process is presented for both red and green GaP LED which is also believed to be very suitable for large scale production. The main features of the process are (i) a low Ga consumption due to the use of thin melts, (ii) the formation of the p–n junction by overcompensation and (iii) the introduction of all dopants via the vapour phase (see also Saul and Roccasecca 1973).

2. Experimental

A schematic view of the LPE apparatus is shown in figure 1. Besides the boat construction the system is similar to that described by Saul and Roccasecca (1973).

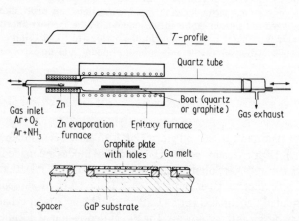

Figure 1. Epitaxy growth apparatus and cross section of the graphite boat.

However, contrary to the slider boat used by these authors our boat does not contain any moving parts since the 1 mm thick Ga melt is saturated by dissolving part of the substrate. This allows the Ga to be brought into contact with the substrates at room temperature. A cross section of the graphite boat as used for the growth of green LED is also shown in figure 1. In case of red LED quartz is used as boat material.

For the growth of layers with uniform thickness proportionality between Ga amount and substrate area is necessary (Peters 1973) which becomes important in multi-slice epitaxy. In our boat we can easily fit the volume of the Ga melt to the area of the substrate using spacers with different inner diameters. Furthermore this leads to a less severe dump growth at the edges of the substrate compared to slider boat experiments with different substrate diameters. The deposition efficiency is almost 100%, that means that nearly all of the dissolved material is regrown on the substrate which is typical for the growth from thin melts (Bergh *et al* 1973).

Figure 2 shows the temperature program of the epitaxy process for overcompensated red and green GaP diodes. Ar is used as carrier gas in both cases. Starting the process the boat is moved into the furnace kept at 600 °C and then heated up to 980—1080 °C.

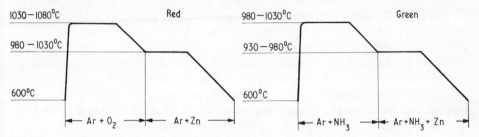

Figure 2. Temperature—time program of the LPE process for red and green GaP diodes.

About 25—50 μm of GaP are dissolved to accomplish melt saturation. Wetting problems do not appear if the substrates are carefully prepared prior to contact with Ga. Tapered dissolution can be avoided by close temperature control during the heat-up period. In case of LPE process for red diodes O_2 is added to the carrier gas to obtain O-doped layers, Te- or S-doping is accomplished in the experimental status of the work by adding Te or S in elemental form to the melt. However, this can also be done via the vapour phase. After 30 min for equilibration of the melt the temperature is lowered for 50 °C at a rate of 2—5 °C min^{-1} to form the Te, O-doped n-layer. At this time the temperature is held constant and Zn is added to the gas by evaporating elementary Zn in a separate furnace (figure 1). Oxygen must be purged out of the system prior to Zn-evaporation to prevent oxidation of Zn. After 30 min for equilibration the temperature is lowered to 600 °C to form the p-layer on top of the n-layer. At this temperature the boat is removed from the furnace. For the growth of epitaxial layers for green LED NH_3 is added instead of O_2 to the carrier gas in order to obtain N-doped layers. No intentional donor doping is necessary for green diodes as will be seen in §3.2.1.

For quantum efficiency measurements the epitaxial wafers were wire-sawn into elements of 1×1 mm^2 area with 0·3 mm thickness. The uncontacted elements were clamped between pressure contacts and operated at pulsed currents to prevent heating. Light output was measured with the diode crystal in a parabolic mirror at various diode currents. For electrical measurements AuZn and AuGe alloys were evaporated on the p and n side, respectively, and sintered at 500 °C to form ohmic contacts.

3. Diode characteristics

3.1. Red GaP diodes

3.1.1. p—n junction formation. Due to the Zn diffusion into the Te, O-doped layer at high temperatures the junction of overcompensated red diodes does not coincide with the interface corresponding to the overcompensation temperature. Similar behaviour was observed by Peaker *et al* (1972) in double epitaxial GaP: ZnO diodes over a range of Zn concentrations. Evidence for this was obtained by two simple experiments: first no O_2 was added during n-layer growth but during p-layer growth. Diodes made of this wafer exhibited quantum efficiencies which were two orders of magnitude lower than those of diodes with O-doped n-layers. This implies that the p–n junction is formed in the n-layer which did not contain oxygen in this experiment. In the second experiment oxygen was purged out of the melt after normal n-layer growth in pure Ar ambient for varying times. No decrease in efficiency was measured even after 3 h of purging. This indicates that the penetration depth of Zn into the n-layer is at least higher than one electron diffusion length which is about $0.4\,\mu m$ in our red diodes (see table 1). Another indication for the Zn-indiffusion was obtained by cathodoluminescence measurements using a scanning electron microscope (SEM). Figure 3 shows the dependence of room temperature cathodoluminescence intensity on distance perpendicular to the p–n junction of a red overcompensated diode and a cathodoluminescence image of the same diode. The position of the p–n junction was determined by measuring the induced

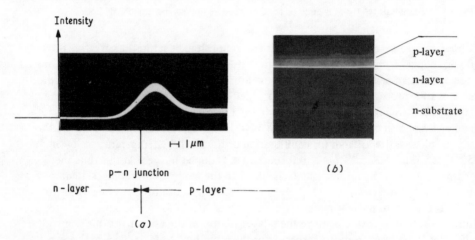

Figure 3. Cathodoluminescence intensity near the p–n junction of an overcompensated red GaP diode (*a*) and cathodoluminescence image of the same diode (*b*). The position of the p–n junction was determined by measuring induced junction current.

junction current. The maximum cathodoluminescence intensity occurs in the p-layer about $1\,\mu m$ distant from the p–n junction. The low intensity in the remaining p-layers is attributed to a decreased oxygen concentration as a consequence of oxygen losses of the Ga melt during overcompensation.

3.1.2. Quantum efficiency and Te concentration in the Ga melt for different crystal orientations. Figure 4 shows the dependence of quantum efficiency on Te concentration in Ga melt for (111)A-, (111)B-, and (100)-oriented substrate crystals. The partial pressures of Zn and O_2 were kept constant in all experiments. As the diffusion coefficient of Zn in GaP is high, implying bulk equilibrium incorporation and therefore no orientation-dependent incorporation (Jordan 1971) the curves in figure 4 must be explained in terms of Te and O incorporation. Considering only one crystal orientation quantum efficiency increases with increasing Te concentration at low doping levels and decreases

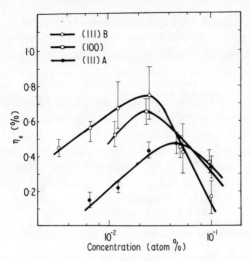

Figure 4. External quantum efficiency (at 1 A cm⁻²) of encapsulated red GaP diodes versus Te concentration in Ga melt for (111)A, (111)B, and (100) substrate orientations.

towards higher concentrations at high doping levels. A similar behaviour was observed by Kasami *et al* (1969) in n on p epitaxial diodes and by Casey *et al* (1972) in Zn-diffused diodes. These structures are comparable to our overcompensated diodes since the p−n junction is formed in both cases by Zn-diffusion into Te, O-doped layers. Following Kasami *et al* (1969) the increasing part of the curve results from the higher Zn concentration and thus from Zn−O pairs in the p-region adjacent to the p−n junction. We attribute the decrease in quantum efficiency at higher Te concentrations (in our case for $N_D - N_A > 3 \times 10^{18}$ cm⁻³) to an increasing formation of nonradiative centres which are correlated with the Te (Fabre *et al* 1974). Considering the curves of figure 4 for different substrate orientations the shift implies an orientation-dependent incorporation of Te in GaP. The order of the incorporation coefficients $k_{(111)B} > k_{(100)} > k_{(111)A}$ imposed by figure 4 is consistent with a surface controlled incorporation of the slow diffusing impurity Te in GaP (Zschauer and Vogel 1971).

3.1.3. Annealing experiments. Annealing at moderate temperatures (500–750 °C) normally increases the quantum efficiency of red GaP diodes. Saul *et al* (1969) observed a 3 to 4 fold increase in their double epitaxial diodes during annealing, Casey *et al* (1972) measured in diffused diodes a 30 fold increase. The beneficial effect of the annealing consists in an increase of Zn−O pair concentration and in the elimination of nonradiative recombination centres which lead to higher space charge currents (Luther *et al* 1974) or dead layers between the p−n junction and the light emitting p-region (Calverly and Wight 1970). However, in our diodes we could not observe any improvement of the efficiency by annealing at 500–600 °C after p-layer growth. Quenching of the epitaxial wafers from 900 °C to room temperature resulted in a 50% decrease of efficiency which according to Onton and Lorenz (1968) could be re-established by annealing at 500–600 °C. From these annealing experiments it can be concluded that in our diodes equilibrium Zn−O pair concentration is already established after p-layer growth.

Although annealing at low temperatures has not been successful we found that annealing at high temperatures (2 h at 1000 °C) which has to be accomplished after a few micrometres of p-layer growth resulted in a twofold increase of efficiency. Encapsulated cube-shaped diodes made in this way have average maximum quantum efficiencies between 1·5 and 2% at 1 A cm^{-2}, the highest value being 4·0% at 0·5 A cm^{-2}. To understand the mechanism for this efficiency improvement we measured the following characteristics of the diodes: current–voltage dependence $I(V)$, capacitance voltage dependence $C(V)$ and net donor and acceptor concentrations by surface-barrier capacitance measurements on angle lapped surfaces, the electron and hole diffusion lengths L_e and L_h by measuring the induced junction current at cleaved surfaces in SEM (Hackett *et al* 1972), the decay time of red luminescence τ_D, and light output versus current density $L(I)$. The results are summarized in table 1.

Table 1. Characteristics of overcompensated red GaP diodes with and without annealing at 1000 °C (average values)

	Annealed	Unannealed
Current–voltage dependence $1 \sim \exp(qV/nkT)$	$n = 1\cdot70$	$n = 1\cdot67$
Impurity gradient at the p–n junction	$1\cdot60 \times 10^{22}\,\mathrm{cm}^{-4}$	$2\cdot12 \times 10^{22}\,\mathrm{cm}^{-4}$
Electron diffusion length L_e	$0\cdot43\,\mu\mathrm{m}$	$0\cdot38\,\mu\mathrm{m}$
Hole diffusion length L_h	$0\cdot23\,\mu\mathrm{m}$	$0\cdot24\,\mu\mathrm{m}$
Decay time of red luminescence τ_D	260 ns	170 ns
$I_m n_{em}$	$1\cdot57 \times 10^{-2}\,\mathrm{A\,cm}^{-2}$	$1\cdot68 \times 10^{-2}\,\mathrm{A\,cm}^{-2}$

All junctions exhibit linearly graded impurity distributions with average gradients of $1\cdot60 \times 10^{22}\,\mathrm{cm}^{-4}$ for annealed diodes and $2\cdot12 \times 10^{22}\,\mathrm{cm}^{-4}$ for unannealed diodes. The net donor concentration near the p–n junction as revealed by surface barrier capacitance measurements is about $1 \times 10^{18}\,\mathrm{cm}^{-3}$ and the net acceptor concentration is about $3 \times 10^{18}\,\mathrm{cm}^{-3}$. From the measured electron diffusion length it can be estimated that within one diffusion length of the junction the impurity distribution is always graded. The net Zn concentration increases in this region from 1·0 to $1\cdot7 \times 10^{18}\,\mathrm{cm}^{-3}$ in the case of annealed junctions and from 1·0 to $1\cdot9 \times 10^{18}\,\mathrm{cm}^{-3}$ in the case of unannealed junctions. Assuming the same oxygen concentration for both types of diodes the number of Zn–O pairs in this region should not differ for more than 10% as calculated from pairing theory (Wiley 1971).

The internal quantum efficiency of red GaP diodes as developed by Rosenzweig *et al* (1969) is given by

$$\eta_i = \frac{qL_e N_c}{I\tau_R} \ln\left(1 + \frac{n(0)}{n_t}\right) \tag{1}$$

where L_e is the electron diffusion length, N_c the Zn–O pair concentration, I is the diode current density, τ_R is the radiative lifetime of an electron trapped on a Zn–O pair, $n(0)$ is the excess electron density at the edge of the depletion layer and n_t is the excess electron density at 50% saturation of the Zn–O pairs. An important consequence of this model is an inverse proportionality between maximum quantum efficiency η_{em} and current density at maximum efficiency I_m for diodes with identical characteristics (Hackett *et al* 1969). The validity of this inverse proportionality could be shown for our diodes as can be seen in table 1. As maximum efficiency occurs at $n(0) = n_t$ the Zn–O pair concentration N_c can be calculated from equation (1)

$$N_c = \frac{\eta_{em} I_m \tau_R}{q L_e \eta_0 \ln 2}$$

using $\eta_{em} = \eta_{im}\eta_0$, η_0 being the optical coupling efficiency. Assuming $\tau_R = 1\,\mu s$ corresponding to an average hole concentration of $4 \times 10^{17}\,cm^{-3}$ (Jayson *et al* 1970) in the region within one diffusion length of the junction and assuming 60% external coupling efficiency the resulting Zn–O pair concentration is about $5 \times 10^{15}\,cm^{-3}$ for both annealed and unannealed diodes. This value is about 6 times lower than those calculated for our efficient red double epitaxial diodes. Using the same oxygen concentration in our Te, O-doped layers as measured by Luther *et al* (1973) of $0\cdot5-1 \times 10^{16}\,cm^{-3}$ which is at least one order of magnitude lower than in Zn, O-doped layers (Saul and Hackett 1972) and taking into account that 45% of oxygen donors are bound at nearest neighbour pairs with Zn as calculated from pairing theory (Wiley 1971) we obtain quite good agreement with the above mentioned value for the Zn–O pair concentration. The reduced oxygen incorporation in Te, O-doped layers compared to Zn, O-doped layers seems at present to be the most restricting factor in obtaining higher efficiencies than 4% with overcompensated red GaP diodes.

Finally the higher efficiency of high temperature annealed diodes can be explained by their longer decay time indicating a lower free hole concentration in the light generating p-region. Lower free hole concentrations lead to reduced nonradiative Auger recombination (Jayson *et al* 1970) and therefore to higher efficiencies. One indication for a lower free hole concentration is the smaller impurity gradient at the p–n junction of annealed diodes as compared to unannealed diodes. As n_t in equation (1) is inversely proportional to τ_D (Rosenzweig *et al* 1969) maximum quantum efficiency assumed to occur at $n(0) \equiv n_0 \exp(qV/kT) = n_t$ is shifted in unannealed diodes to higher current densities due to $I \sim \exp(qV/nkT)$. Thus according to $I_m \eta_{em} = $ constant, unannealed diodes have lower efficiencies as compared to annealed diodes.

3.2. Green GaP diodes

3.2.1. Quantum efficiency and donor concentration of the n-layer. Figure 5 shows the dependence of quantum efficiency on net donor concentration for (111)B-oriented substrates. Net donor concentrations were determined by surface barrier capacitance measurements and by p–n junction capacitance measurements. The ammonia concentration in Ar is about $0\cdot1$ vol %. The resulting N concentration in the GaP layer as measured by optical absorption according to Thomas and Hopfield (1966) and evaluated according to Lightowlers *et al* (1974) is $3 \times 10^{18}\,cm^{-3}$. A similar dependence as shown in figure 5 was obtained by Logan *et al* (1971). Net donor concentrations of $0\cdot5-1 \times 10^{17}\,cm^{-3}$ were obtained without intentional doping of the Ga melt. The main impurities in such layers are S and C as revealed by photoluminescence measurements

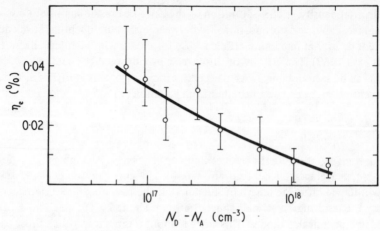

Figure 5. External quantum efficiency (at 10 A cm^{-2}) of encapsulated green GaP diodes versus net donor concentration in the n-layer.

at 2 K. Using (111)A- or (100)-oriented substrates the epitaxial layers are often initially highly compensated or even p-type and flip after a few micrometres to n-type growth. This is a consequence of the different orientation dependence (Zschauer and Vogel 1971) and temperature dependence of slow diffusing impurities in GaP.† Furthermore the quantum efficiencies of p—n junctions on (111)A- and (100)-oriented substrates grown in the same epitaxial run are 50% lower as compared to (111)B-oriented substrates. This is attributed to an orientation dependent incorporation of N in GaP as measured by Roccasecca *et al* (1974).

3.2.2. Quantum efficiency and temperature of overcompensation. The quantum efficiency of our overcompensated green diodes is dependent on the temperature at which overcompensation occurs as shown in figure 6. The starting temperature for epitaxial growth is always 50 °C higher than the overcompensation temperature. Partial pressures of NH$_3$ and Zn are kept constant in all experiments. Generally the efficiency increases with decreasing temperature. This is in agreement with Lorimor *et al* (1973) who measured a 50% higher efficiency in material grown essentially at 900 °C as compared to material grown at 1000 °C. They attribute the higher efficiency to an increase of electron and hole diffusion lengths. This seems also to be valid for our diodes. For diodes with overcompensation temperatures of 940 °C we measured electron and hole diffusion lengths of 1·5 and 3·0 μm, respectively, and quantum efficiencies up to 0·4% at 100 A cm^{-2}. The lower efficiency for diodes with overcompensation temperatures of 910 °C is caused by a reduced n-layer thickness since we found in epitaxial wafers with tapered n-layers a rapid decrease in electroluminescence efficiency for n-layer thicknesses less than 8 μm.

3.3. Two p—n junction device

By our simple epitaxy process we subsequently grew a red light emitting junction on the (111)A side and a green light emitting junction on the (111)B side of one GaP

† (For the temperature dependence of S and Te incorporation in GaP see Sudlow *et al* 1972 and Jordan *et al* 1973.)

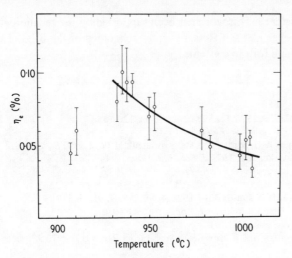

Figure 6. External quantum efficiency (at 10 A cm⁻²) of encapsulated green GaP diodes plotted against temperature of p–n junction formation.

substrate crystal. Contrary to experiments with double epitaxial red p–n junctions the epitaxy process for the green diode reduces the efficiency of the red overcompensated diode only slightly. Growing of the junctions in the inverse order results in a deterioration of the green electroluminescence during growth of the red junction. Mesa etching was accomplished to reveal the substrate for the common n contact. After evaporating contacts the wafers were sawn into individual elements which were mounted red p–n junction down on a header with three terminals. The resulting device emits light of red, green, and yellow hues depending on whether one or both junctions are operated.

4. Conclusions

Efficient red and green GaP LED were made by a simple epitaxy process. The characteristics of this process are the use of thin Ga melts, saturation of Ga by dissolving substrate material and p–n junction formation by overcompensation. All dopants can be introduced via the vapour phase. Red GaP diodes made by this method have efficiencies up to 4% which is 2–3 times lower than for the best double epitaxial diodes; this is attributed to a 6 times lower Zn–O pair concentration. This reduction is due to the lower oxygen incorporation in the Te, O-doped layers as compared to Zn, O-doped layers of double epitaxial diodes. For green diodes maximum efficiencies of 0·4% were obtained which is close to best values reported in literature.

Acknowledgments

The authors are indebted to K H Zschauer for many helpful discussions, to K Schwarzmichel and to K Zeuch for supplying the substrate material, to H D Wolf for photoluminescence measurements and to E Wolfgang for the SEM measurements. Thanks are due to M Komainda and W Kunkel who grew the epitaxial layers and to

154   *C Weyrich, G H Winstel, K Mettler and M Plihal*

G Meier for technical assistance. This work has been supported by the technological program of the Federal Department of Research and Technology of the FRG. The authors alone are responsible for the contents.

References

Bergh A A, Saul R H and Paola C R 1972 *J. Electrochem. Soc.* **120** 1558
Calverley A and Wight D R 1970 *Solid. St. Electron.* **13** 382
Casey H C, Luther L C, Lorimor A G, Jordan A S and Kowalchik M 1972 *Solid St. Electron.* **15** 617
Fabre E, Barghava R N and Zwicker W K 1974 *J. Electron. Mater.* **3** 409
Hackett W H, Rosenzweig W and Jayson J S 1969 *Proc. IEEE* **57** 2072
Hackett W H, Saul R H, Dixon R W and Kammlott G W 1972 *J. Appl. Phys.* **43** 2857
Hackett W H, Saul R H, Verleur H W and Bass S J 1970 *Appl. Phys. Lett.* **16** 477
Jayson J S, Barghava R N and Dixon R W 1970 *J. Appl. Phys.* **41** 4972
Jordan A S 1971 *J. Electrochem. Soc.* **118** 781
Jordan A S, Trumbore F A, Wolfstirn K B, Kowalchik M and Roccasecca D D 1973 *J. Electrochem. Soc.* **120** 791
Kaneko K, Ayabe M, Dosen M, Morizane K, Usui S and Watanabe N 1973 *Proc. IEEE* **61** 884
Kasami A, Naito M, Toyama M and Maedea K 1969 *Japan. J. Appl. Phys.* **8** 1469
Ladany I and Kressel H 1972 *RCA Rev.* **33** 517
Lightowlers E C, North J C and Lorimor O G 1974 *J. Appl. Phys.* **45** 2191
Logan R A, White H G and Wiegmann W 1971 *Solid St. Electron.* **14** 55
Lorimor O G, Hackett W H and Bachrach R Z 1973 *J. Electrochem. Soc.* **120** 1424
Luther L C, Harrison D A and Derick L 1973 *J. Appl. Phys.* **44** 4072
Onton A and Lorenz M R 1968 *Appl. Phys. Lett.* **12** 115
Peaker A R, Sudlow P D and Mottram A 1972 *J. Crystal Growth* **13/14** 651
Peters R C 1973 *Proc. 4th Int. Symp. Gallium Arsenide and Related Compounds* (London and Bristol: Institute of Physics) p55
Roccasecca D D, Saul R H and Lorimor O G 1974 *J. Electrochem. Soc.* **121** 962
Rosenzweig W, Hackett W H and Jayson J S 1969 *J. Appl. Phys.* **40** 4477
Saul R H, Armstrong J and Hackett W H 1969 *Appl. Phys. Lett.* **15** 229
Saul R H and Hackett W H 1972 *J. Electrochem. Soc.* **119** 542
Saul R H and Roccasecca D D 1973 *J. Electrochem. Soc.* **120** 1128
Solomon R H and DeFevere 1972 *Appl. Phys. Lett.* **21** 257
Sudlow P D, Mottram A and Peaker A R 1972 *J. Mater. Sci.* **7** 168
Thomas D G and Hopfield J J 1966 *Phys. Rev.* **150** 680
Weyrich C 1973 *Diskussionstagung 'Optoelektronik, Materialgüte und -entwicklung' Stuttgart March 1973*
Wiley J D 1971 *J. Phys. Chem. Solids* **32** 2053
Zschauer K H and Vogel A 1971 *Proc. 3rd Int. Symp. Gallium Arsenide and Related Compounds* (London and Bristol: Institute of Physics) p100

Peltier-induced growth of GaAlAs

J J Daniele and C Michel

Philips Laboratories, Briarcliff Manor, NY 10510, USA

Abstract. Peltier-induced liquid phase epitaxy (LPE) was performed on $Ga_{1-x}Al_xAs$ $(0·0 < x < 0·5)$ by passing an electrical current through the solid–liquid (S–L) interface. Cooling and heating at the S–L interface and growth rate modulation was achieved. Layers as thin as 800 Å and as thick as $18 \mu m$ were obtained and periodic structures of up to 20 layers with an average layer thickness of $0·2 \mu m$ have been constructed. The Peltier-induced LPE was superimposed on normal LPE growth layers.

Because of the small Peltier coefficient of the Ga–GaAlAs system, direct current was found to be unsuitable since it resulted in predominant Joule heating. A method based upon applying a constant Joule heating with AC on which a DC signal is superimposed was developed. This results in a net cooling or heating near the S–L interface of about $0·6 \,°C$ for a current of 12 A cm^{-2}.

By applying DC levels of $3-12$ A cm^{-2} for durations ranging from 0·25 to 9 min, Peltier-induced layers of $0·2-18 \mu m$ were grown. The initial Peltier-induced growth rate (v_{pelt}) of up to ten times the normal growth rate is followed by a decrease proportional to $t^{-0·6 \pm 0·2}$ during the first three minutes of growth. After three minutes, v_{pelt} levels off at a constant level of two to four times v_{normal}. The step-cool model of LPE growth adequately explains this behaviour.

It was also found that v_{pelt} is linearly dependent on the DC level. The Peltier coefficient P is about 0·35 V for Ga–GaAlAs at about 830 °C. Theory predicts that a modulation of the growth rate should result in a modulation of the aluminium concentration in the solid. Preliminary data suggest that a decrease of about 1% has been achieved during the initial accelerated growth.

1. Introduction

Variations in composition and growth are difficult to control and to predict during liquid phase epitaxial (LPE) growth. Generally, LPE growth proceeds near equilibrium conditions where the phase diagram dictates the composition, and the growth rate can be adjusted by changing the cooling rate. To change the composition of the solid abruptly, as is required for the growth of heterojunctions, the substrate must change melts. It is only recently that some of the problems associated with melt transfer have been recognized and alleviated (Dawson 1973).

To change the solid composition to values other than those obtained under equilibrium conditions, but without changing melts, requires the use of a step-cool process (Deitch 1970, Taynai *et al* 1971). In this process, the composition and growth rate depend in a complex way on the system parameters. Thus, the requirement remains for an LPE growth technique which allows microscopic control of the growth rate and composition. The Peltier-induced LPE technique described in this paper shows promise of filling this need.

When an electrical current passes through the solid–liquid (s–l) interface during normal LPE, heat is extracted at the interface due to the Peltier effect. The resulting temperature change modulates the growth rate. Because the heat is removed directly from the s–l interface and because the rate of removal is directly proportional to the current, a high level of external electrical control over the growth should be achievable. Since the composition of pseudobinary III–V compounds and the segregation of impurities are related to the growth rate (Taynai *et al* 1971, Burton *et al* 1953, Bridges 1956), dopant segregation and solid composition can in principle be modulated.

Peltier-induced crystal growth was first suggested by Ioffe (1956). Pfann *et al* (1957) applied the Peltier effect to zone melting of Ge and suggested its use for making p–n–p and p–n–p–n structures by modulation of dopant segregation. Singh *et al* (1968) used Peltier-induced markers to delineate the s–l interface during Czochralski growth of InSb. Lichtensteiger *et al* (1971) found that the impurity distribution in Czochralski growth of InSb was modulated by the Peltier effect.

The first application of the Peltier effect during LPE was by Kumagawa *et al* (1973) on InSb. They used a combination of DC and marker pulses to measure the microscopic Peltier-induced growth rate. The application of Peltier-induced marker pulses to study GaAs LPE was first made by Blom *et al* (1975). In this material, the Peltier coefficient is very small and Joule effects were found to predominate. Interface demarcation lines were, however, produced by Joule and Peltier heating causing a meltback and subsequent rapid regrowth.

In the present work, the Peltier-induced growth of GaAlAs is quantitatively characterized. The modulation of the growth rate of the epitaxial layer by externally applied electrical current pulses, using the pulse amplitude and duration as variables, is described. GaAlAs is especially interesting since Peltier modulation of the growth rate may result in modulation of the aluminium concentration in the solid.

2. Experimental

When a current passes through an s–l interface, several competing thermoelectric effects take place, that is to say, Joule heating, Peltier cooling (or heating) and the Thompson effect. The Thompson effect can be neglected except when very large thermal gradients exist (Bardeen and Chandrasekar 1958). Joule heating is the dominant process at any practical current in the Ga–GaAlAs system (Blom *et al* 1975) and without special precautions net heating occurs at all but the lowest currents.

To eliminate the effects of Joule heating during Peltier-induced growth, Pfann *et al* (1957) suggested a method that will be referred to here as the AC/DC method. In this method, a constant Joule heating is maintained with alternating current, and Peltier heating or cooling is produced with a superimposed DC signal. Simultaneous with application of the DC signal, the AC component is adjusted such that the total Joule heating is maintained constant. Thereby a new thermal equilibrium is obtained at a slightly higher temperature and the Peltier effect works about this new equilibrium temperature as if no Joule effect was present.

Liquid phase epitaxy (LPE) was carried out in a boat manufactured from graphite and boron nitride (figure 1). The graphite top half of the boat contained the melt chamber. A boron nitride slider electrically insulates the top and the bottom half of the

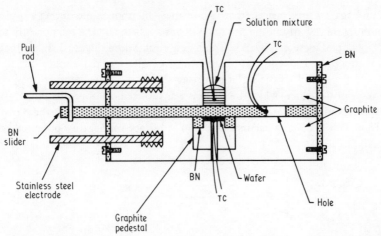

Figure 1. Schematic diagram of boat made of graphite and boron nitride used during Peltier-induced LPE. TC are thermocouples.

boat. The bottom half contains the substrate (0·015 in thick, Si-doped, n-type ($n = 3 \times 10^{18}$) GaAs having the (100) orientation) which is located directly under the melt chamber. A hole in the slider allows the melt to come into contact with the wafer at the appropriate time. Electrical contact to the back of the wafer is made with a Ga–Al liquid metal alloy in pressure contact. Electrical contact to the boat is made with stainless steel rods and current passes through the melt (because of the boat geometry the current density was lower near the centre of the wafer), through the s–l interface and the wafer and out through the bottom half. The temperature is monitored directly under the wafer, in the centre of the melt, and in the boron nitride slider next to the melt.

In a typical run, the boat was heated to 900 °C and subsequently cooled and equilibrated at 850 °C. After 1–2 h the slider was moved so that the melt contacted the wafer and normal LPE was initiated with a cooling rate of 0·25 °C min^{-1}. After 5–10 μm of GaAlAs was grown, a 1 kHz AC signal was applied. The accompanying Joule heating established a new thermal equilibrium. After an additional 5–10 μm of GaAlAs was grown by normal cooling, Peltier cooling was applied by means of the AC/DC method. For each Peltier pulse the DC and AC levels were adjusted using a voltage divider circuit to ensure an unchanged Joule heating. After all the pulses were applied, an additional 5–10 μm of GaAlAs was allowed to grow and the slider was moved back into its original position, isolating the melt from the wafer.

Subsequently, the wafer was cleaved and etched in either a diluted AB etch (5 g CrO$_3$: 5 ml HF : 40 mg AgNO$_3$: 20 ml H$_2$O, see Abrahams and Buiocchi 1965) or the aluminium sensitive etch (H$_2$O$_2$ at pH = 7·00, Logan and Reinhart 1973). Both etchants revealed the Peltier bands but the H$_2$O$_2$ etch resulted in a sharper definition. Photomicrographs were then taken of the cleaved edges using interference contrast microscopy.

3. Results

During each DC (Peltier) pulse, Peltier cooling or heating (depending on the polarity) of up to 0·6 °C was observed at the various thermocouples in the boat. A plot of temperature against time measured at two of these thermocouples during a typical run is shown in figure 2. Early in the run the AC was applied, and the temperature at all the thermocouples increased by 1·5 °C to a new equilibrium. Figure 2 shows that during the first Peltier pulse the temperature near the S–L interface decreased while the temperature at the back of the wafer increased. This is expected because one junction is the reverse of the other. When the current was reversed, an opposite temperature change was recorded indicating a true Peltier effect and the absence of the AC skin effect. Thus, Peltier cooling occurs at the S–L interface for negative potential on the Ga melt.

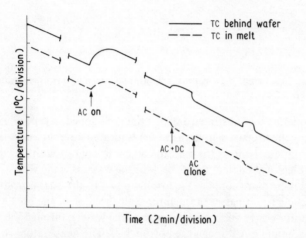

Figure 2. Temperature–time characteristic measured at the side of the melt and at the back of the substrate during Peltier-induced LPE.

The Peltier-induced growth rate was then studied as a function of pulse amplitude and pulse duration. In the first set of experiments, a constant level of DC + AC was applied using a DC pulse with pulse widths for the various experiments between 0·25 and 9·0 min and pulse separation of 6–10 min (throughout the paper the current density is defined as the total current divided by the area of the S–L interface). These pulses resulted in alternative regions of Peltier and normal growth. A cleaved and etched cross section of such a layer is shown in figure 3. The Peltier regions are clearly visible and range in thickness from 0·5 to 3·3 μm. The average layer thickness grown during each DC pulse was divided by the pulse duration to give the average Peltier-induced growth rate. The normal rate was similarly found. A plot of the average Peltier-induced growth rate, measured from the layer shown in figure 3, is shown in figure 4. The average growth rate starts at about six times the normal rate and typically decreases as $t^{-0·4}$ to $t^{-0·8}$ for the first 2–3 min of the pulse. After this initial decrease the growth rate reduces more slowly. Finally after 3–4 min the growth rate approaches a constant level. This constant level is about 2·5 times the normal growth rate in this specific case and ranges typically from 2 to 4 times the normal growth rate. Pulses of 3 to 9 min duration were also used and the results confirmed that the growth rate was constant after the first 3 min.

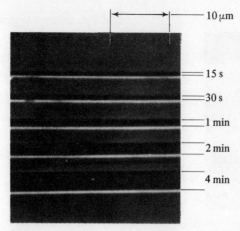

Figure 3. Photomicrograph of a cleaved and etched GaAlAs layer containing Peltier-induced bands. Current density was 10 A cm^{-2} for 0·25 to 4 min durations. Pulse separation was 8 min.

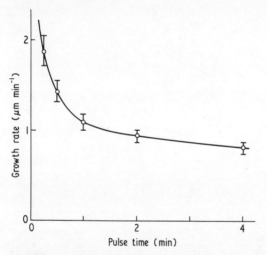

Figure 4. Plot of average growth rate versus pulse duration taken from figure 3. Normal growth rate was about 0·34 μm min^{-1}.

In the second set of experiments the effect of varying the DC pulse amplitude was studied. Pulses of constant duration ranging for the various experiments between 2–3 min and at intervals of 6–10 min were applied with various DC levels. Currents of 4 to 12 amps were used and the Peltier-induced and normal growth rates were measured. Figure 5 shows a photomicrograph of a layer containing Peltier-induced layers produced by 4, 8, and 12 A cm^{-2}, respectively. A typical plot of the observed average growth rate as a function of current is shown in figure 6. It is demonstrated that, for times less than 3 min, the Peltier-induced growth rate is a linear function of current and is added to the normal growth rate v_0.

Further experiments to determine the range of Peltier layer thicknesses demonstrated

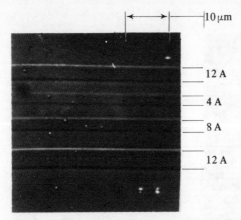

Figure 5. Photomicrograph of a cleaved and etched layer of GaAlAs containing Peltier-induced bands. Current density ranged from 3·3 to 10 A cm^{-2}, the pulse duration was 3 min, and the pulse separation was 6 min.

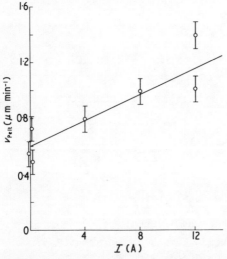

Figure 6. Plot (taken from figure 5) of average growth rate against current for pulses of constant duration. The normal growth rate v_0 (given by the distance between the Peltier bands divided by the time between pulses) varies systematically throughout the entire LPE layer as shown by the three points at $I = 0$. Since the 12 A pulses were made at opposite ends of the layer, their difference in growth rate reflects the total change of v_0 during growth. A direct correspondence can be made between the points above the line at $I = 0$ and $I = 12$ A and similarly for those points below the line at the same two currents.

that layers as thick as 18 μm and as thin as 0·2 μm were grown using pulse durations ranging from 0·17 to 9 min. Layered structures with as many as 20 layers of average thickness of 0·2 μm (estimated by averaging over a 4 μm region) were also grown by applying 10 s pulses at 20 s intervals. Still thinner layers could be formed by meltback and regrowth and layers as thin as 800 Å were made in this way.

4. Discussion

The time dependence of the Peltier-induced growth rate is separated into two parts: (i) for times less than 3 min, and (ii) for times longer than 3 min. For $t < 3$ min, an approximation to the Peltier-induced growth can be obtained from the step-cool growth method (Hsieh 1975, Ghez 1973). In this method, growth proceeds by diffusion when a supercooled GaAlAs solution is placed in contact with a GaAs seed. The layer thickness varies with time as $d \sim t^{1/2}$ giving a growth rate dependence of $v \sim t^{-1/2}$. The initial boundary conditions for the step-cool and the Peltier case are very similar, being a step function decrease in concentration the S–L interface ($x = 0$) (Hsieh 1974, Ghez 1973). Assuming growth by diffusion, it is then reasonable to expect a similar time dependence in both cases. The data in figure 4 show that for the first 3 min of growth the average Peltier-induced growth rate varied as $v \sim t^{-0.4}$ to $t^{-0.8}$ which is a reasonable result in light of the similarity between the Peltier and step-cool approximations.

For the step-cooled segment of the growth ($t < 3$ min), v_{Pelt} was calculated to be $0.33\,\mu m\,min^{-1}$ using an expression given by Hsieh (1974) and using an estimate of the Peltier-induced temperature change ($0.8\,^{\circ}C$) at the S–L interface (appendix). In this case, the temperature change and thus the growth rate are proportional to the DC current; in agreement with experiment.

For times longer than 3 min, the step-cool approximation no longer holds and a constant growth rate is maintained possibly by the Peltier-induced temperature and As concentration gradients at the interface.

In a separate calculation, a Peltier coefficient P of approximately 0.35 V was estimated from the measured temperature change at a specific distance from the interface (Carslaw and Jaeger 1959) (appendix). This value for P is in agreement with the published value of 0.5 V for GaAs (Carlson *et al* 1963).

A consequence of the Peltier-induced growth rate modulation of GaAlAs is the possibility of accompanying modulation of the aluminium concentration (X_{al}) in the epitaxial layer. Changes in the growth rate lead to changes in dopant segregation in Czochralski growth according to the theory of Burton *et al* (1953) and changes in aluminium segregation in previous experiments on GaAlAs by Taynai *et al* (1971). The H_2O_2 at pH = 7·00 etch, which is particularly sensitive to aluminium changes in GaAlAs, was found to very clearly reveal the Peltier bands. Moreover, when pure GaAs was grown under identical Peltier conditions as GaAlAs, no such bands were observed after etching with AB or H_2O_2 etch. This suggests that the Peltier bands are due to Al concentration changes in the epitaxial layer. More direct evidence comes from preliminary cathodoluminescence and photoluminescence measurements done on angle-lapped and cleaved samples.

5. Conclusions

Peltier-induced (cooling) LPE growth of $Ga_{1-x}Al_xAs$ ($0 < x < 0.5$) was demonstrated by using the AC/DC method to eliminate the predominant Joule effect.

The average Peltier-induced growth rate was found to depend linearly on the DC level and was in addition to the normal growth rate. The Peltier coefficient for GaAlAs was estimated to be about 0.35 V which is in the same range as that of GaAs (Carlson *et al*

1963). The average Peltier-induced growth rate was found to be time dependent. The initial Peltier-induced growth rate was 6 to 10 times the normal growth rate and was found to decrease at rates ranging from $t^{-0.4}$ to $t^{-0.8}$. After 3 minutes, the growth rate reached a constant level usually 2 to 4 times the normal growth rate and remained constant for times up to 9 min. By varying the duration of the Peltier (cooling) pulses, Peltier-induced layers as thick as $18\,\mu\mathrm{m}$ and as thin as $2000\,\mathrm{Å}$ were produced.

The Peltier-induced growth rate in the step-cooled segment for $t < 3$ min was calculated to be $0.33\,\mu\mathrm{m\ min}^{-1}$ (appendix) which is smaller than the measured value of about $1\,\mu\mathrm{m\ min}^{-1}$. This difference may be accounted for by the use of low temperature data for computing K_1, K_2 and K_P (appendix), an error in the estimate of the Peltier-induced temperature change at the $\mathrm{s-L}$ interface or possibly additional growth mechanisms such as ion migration.

Preliminary evidence suggests a Peltier-induced modulation of the aluminium concentration in the solid of typically 1%. Further measurements are needed to establish a quantitative relation between the Al concentration and the Peltier pulse parameters.

Acknowledgments

The authors gratefully acknowledge the work of Dr G A Acket in making the photoluminescence measurements referred to in this paper. The authors also acknowledge many helpful technical discussions with Dr G M Blom, Dr S K Kurtz, and Dr J Fitzpatrick, and the excellent technical assistance of E Brennan.

Appendix

A simple thermal conduction model was developed describing the observed Peltier cooling and heating and the Peltier coefficient for the Ga—As system was derived. In this planar model of the LPE growth process (figure A1) it is assumed that heat flows solely by conduction, and that there is no thermal contact resistance between materials. It is further assumed that the Thompson effect is negligible and that due to the use of

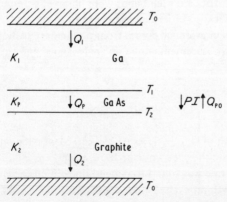

Figure A1. Planar model for heat flow in the Peltier LPE growth system.

the AC/DC method, electrical resistance heating can also be neglected. In addition, the heat absorbed by the growing crystal (0·02 W at $3 \, \mu m \, min^{-1}$) is negligible in comparison to the thermal heat flow (4 W) and can be neglected. The calculation is made for the steady state case and a thermal conductance is defined for each region as follows:

$$K_1 = K_{Ga}\left(\frac{A}{l}\right) = \frac{1 \cdot 36}{0 \cdot 54}(0 \cdot 8) = 2 \cdot 2 \, W \, {}^{\circ}C^{-1} \qquad (A1)$$

(Touloukian and Ho 1970),

$$K_2 = K_{graphite}\left(\frac{A}{l}\right) = \frac{1 \cdot 5}{0 \cdot 97}(1 \cdot 2) = 1 \cdot 86 \, W \, {}^{\circ}C^{-1} \qquad (A2)$$

(Poco Graphite Inc, Texas),

$$K_P = K_{GaAs}\left(\frac{A}{l}\right) = \frac{1 \cdot 5}{0 \cdot 0382}(0 \cdot 7) = 2 \cdot 75 \, W \, {}^{\circ}C^{-1} \qquad (A3)$$

(Tye 1969), where A is the area in square centimetres and l the thickness in centimetres.

When a current is passed through the system, heat is removed by the Peltier effect from the Ga–GaAs junction and deposited at the GaAs–graphite junction at a rate equal to PI. Since there is no mechanism for net gain or loss of heat, the temperatures T_1 and T_2 at junctions 1 and 2, respectively, will adjust themselves so that the heat flow is continuous throughout. The heat flow in each of the Ga, graphite, and GaAs regions, Q_1, Q_2 and Q_P, respectively, are defined as follows:

$$Q_1 = (T_0 - T_1)K_1 \qquad (A4)$$

$$Q_2 = (T_2 - T_0)K_2 \qquad (A5)$$

$$Q_P = PI - (T_2 - T_1)K_P. \qquad (A6)$$

Since the heat flow is continuous ($Q_1 = Q_2 = Q_P$), the expressions for $T_0 - T_1$ and $T_2 - T_0$, the Peltier-induced temperature changes at the interfaces, follow directly:

$$T_0 - T_1 = \frac{PI}{(1 + K_1/K_2)K_P + K_1} \qquad (A7)$$

$$T_2 - T_0 = \frac{PI}{(1 + K_2/K_1)K_P + K_2}. \qquad (A8)$$

Using equation (A8), a value for P is derived from the measured temperature change at the back of the wafer (where the measurement can be done more accurately due to a shorter thermal path), $T_2 - T_0 \simeq 0 \cdot 6 \, {}^{\circ}C$ (measured at about 2 mm from the back of the wafer) for a 12 A current. Substituting the above values into (A8) yields a Peltier coefficient of 0·35 V for Ga–GaAs at 850 °C. This value is in agreement with published results of 0·5 V for GaAs, $n = 1 \cdot 5 \times 10^{17}$ (measured at 600 °C) (Coulson *et al* 1963).

Because the thermal distribution is assumed linear, the temperature change at the S–L interface can be found by extrapolation, yielding $\Delta T \approx 0 \cdot 8 \, {}^{\circ}C$.

Using this temperature change, the step-cool growth rate is now calculated as follows (Hsieh 1974):

$$v_{Pelt} = \frac{\Delta T}{C_s m}\left(\frac{D}{\pi}\right)^{1/2}\frac{1}{t^{1/2}} = 0.33\,\mu\mathrm{m\,min^{-1}}\tag{A9}$$

where D is the diffusion coefficient of As in Ga at 830 °C (about $10^{-4}\,\mathrm{cm^2\,s^{-1}}$ according to Rode 1973), C_s is the concentration of As in the solid (50 atom %), m is the slope of liquidus line in the phase diagram (30 °C/atom % at 830 °C according to Panish and Ilegems 1972) and t is 30 s.

A substitution of equation (A8) into equation (A9) shows that v_{Pelt} for the step-cooled region is indeed proportional to the Peltier current.

References

Abrahams M S and Buiocchi C J 1965 *J. Appl. Phys.* **36** 2855
Afromowitz M A 1973 *J. Appl. Phys.* **44** 1292
Bardeen J M and Chandrasekar B S 1958 *J. Appl. Phys.* **29** 1372
Blom G M, Witt A F, Daniele J J and Kyros T 1975 to be published
Bridges H E 1956 *J. Appl. Phys.* **27** 746
Burton J A, Prim R C and Slichter W P 1953 *J. Chem. Phys.* **21** 1987
Carlson R O *et al* 1963 *J. Phys. Chem. Solids* **23** 422
Carslaw H S and Jaeger J C 1956 *Conduction of Heat in Solids* (London: Oxford University Press) chapter 2
Dawson L R 1973 *Electrochem. Soc. Meeting RNP325, Chicago, May 1973* (Princeton, NJ: Electrochem. Soc.)
Deitch R H 1970 *J. Cryst. Growth* **7** 69
Ghez R 1973 *J. Cryst. Growth* **19** 153
Hsieh J J 1974 *J. Cryst. Growth* **27** 49
Ioffe A F 1956 *Zhur. Tekh. Fiz.* **26** 478
Kumagawa M, Witt A F, Lichtensteiger M and Gatos H C 1973 *J. Electrochem. Soc.* **120** 583
Lichtensteiger M, Witt A F and Gatos H C 1971 *J. Electrochem. Soc.* **118** 1014
Logan R A and Reinhart F K 1973 *J. Appl. Phys.* **44** 4172
Panish M B and Ilegems M 1972 *Progress in Solid State Chemistry* ed H Reiss and J O McCaldin vol 7
Pfann W G, Benson K E and Wernick J H 1957 *J. Electron.* **2** 597
Rode D L 1973 *J. Cryst. Growth* **20** 13
Singh R, Witt A F and Gatos H C 1968 *J. Electrochem. Soc.* **115** 112
Taynai J D, Deitch R H and Summers C J 1971 *J. Electron. Mater.* **1** 213
Touloukian Y S 1970 *Thermophysical Properties of Matter* (New York: IFI/Plenum) vol 10
Touloukian Y S and Ho C S 1970 *Thermophysical Properties of Matter* (New York: IFI/Plenum) vol 1
Tye R P 1969 *Thermal Conductivity* vol 2 (London: Academic Press) p235

Surface imperfections in GaAs–Al$_x$Ga$_{1-x}$As DH LPE layers

K K Shih, G R Woolhouse, A E Blakeslee and J M Blum

IBM Thomas J Watson Research Center, Yorktown Heights, New York 10598, USA

Abstract. Structural defects in GaAs–Al$_x$Ga$_{1-x}$As double heterostructure epitaxial layers have been studied. Several distinct surface features, which may play a role in laser degradation, have been characterized: surface pits, meniscus lines, stacking faults and unwetted regions. The structure of these features has been elucidated using TEM, SEM and optical microscopy. The origin of these defects and their relation to growth conditions are discussed.

1. Introduction

The double heterostructure (DH) GaAs–Al$_x$Ga$_{1-x}$As room temperature CW laser is on the verge of becoming a widely-used optoelectronic semiconductor device. However, one thing that has been lacking in this device is reliability. Although the lifetime has been extended from minutes to thousands of hours, the technology required to reproducibly produce long-lived lasers must still be brought under control. The improvement thus far realized in lifetime has been attributed to such factors as the reduction of diode mounting strain, as reported by Hartman and Hartman (1973). Dark line defects have been observed by DeLoach *et al* (1973) and Yonezu *et al* (1974); large dark spots have been observed by Johnston (1974); and impurity effects have been reported by McMullin *et al* (1974). Implicit in all this work is the fact that if dislocations can be avoided in the laser multi-layer structures, laser lifetime should be significantly increased. For this reason, it is important to determine the relationship between morphological and structural defects and the presence of dislocations. The purpose of this paper is to describe the characteristics of certain defects observed and to discuss their possible origin and relation to laser degradation.

2. Experimental

2.1. Material preparation

A widely-used growth technique for preparing DH laser material by the liquid phase epitaxy (LPE) method is the multiple bin boat with slider as described by many authors (Miller *et al* 1972, Nelson 1971, Panish *et al* 1971). The small-volume thin-solution approach developed previously (Blum and Shih 1971, Donahue and Minden 1970, Lockwood and Ettenberg 1972) has been modified for the growth of multiple layers.

A schematic drawing of the modified liquid phase epitaxial apparatus used for this work is shown in figure 1. The graphite boat consists of six wells; the first five are used to grow five epi-layers and the last well is used for wiping any Ga residue left on the

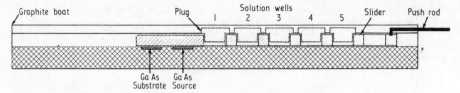

Figure 1. Apparatus for the growth of multiple LPE layers.

surface of the wafer. Each well has a graphite plug to confine a small volume of thin solution (about 1·5 mm high). An equilibration substrate (Blum and Shih 1971, Dawson 1973) has been used to maintain near-equilibrium growth conditions.

The starting growth temperature is about 840 °C, and a constant cooling rate of 0·2 °C min^{-1} is used. The substrate is an n-type (100) GaAs single crystal Si-doped with electron concentration about $1-2 \times 10^{18}$ cm^{-3}. For a typical GaAs–Al$_x$Ga$_{1-x}$As DH laser, the first layer is n-type GaAs doped with Te or Sn with electron concentration about 1×10^{18} cm^{-3}. This layer, about 8 μm thick, is grown as a buffer layer to smooth out the irregularities of the substrate surface. The next layer is n-type Al$_x$Ga$_{1-x}$As ($x \sim 0·2-0·3$), about 2 μm thick, doped with Sn or Te with electron concentration about 1×10^{17} cm^{-3}. An 0·3 μm-thick GaAs active layer doped with Si or Ge with carrier concentration about 5×10^{17} cm^{-3} is then grown. This is followed by a 1 μm-thick Ge-doped p-type Al$_x$Ga$_{1-x}$As ($x \sim 0·2-0·3$) layer with carrier concentration approximately 5×10^{17} cm^{-3}. The last layer is a 1 μm-thick Ge-doped p-type GaAs layer with carrier concentration about 2×10^{18} cm^{-3}. In addition to these typical laser structure wafers, other special samples were grown having different layer thicknesses and/or different doping concentrations.

2.2. Defect observation technique

Because of the microscopically flat and smooth surface layers that can routinely be grown with this modified multiple bin LPE technique, the surface morphology and defect structure of laser layers can be critically studied by various techniques. Bright and dark field optical microscopy, scanning electron microscopy (SEM) and transmission electron microscopy (TEM) were used to characterize the various as-grown defect structures seen in these layers.

An etching technique was also used to reveal the nature of the defects seen in these layers. The AB etchant (Abrahams and Buiocchi 1965), modified by slightly reducing the CrO$_3$ in the etching solution was used to etch the samples at room temperature for a short time interval (30 seconds to a few minutes). This modification provided the slow etching rate required in order not to destroy the thin epi-layers while at the same time revealing the defect structure.

TEM specimens were prepared as follows: first the bulk of the substrate was polished away until the wafer was about 100 μm thick, then squares 1–2 mm on a side were prepared by cleaving. These were mounted with epi-layers down on glass slides. The edge was masked with grease leaving a central window about 0·5 mm in diameter exposed. The samples were then etched to perforation in a solution whose composition was 8:1:1 H$_2$O$_2$:H$_2$SO$_4$:H$_2$O. Perforation was detected using an optical microscope with transmitted light. This yielded substantial (several hundred micrometres in diameter) electron transparent areas (at 100 kV) of material close to the original wafer surface. This technique enabled surface observation of the wafer to be precisely correlated with crystalline defects visible in the electron microscope.

3. Results and discussions

The wafers usually have a smooth and mirrorlike appearance to the naked eye. When examined by optical microscopy, it was found that there are at least four distinct features of interest on the as-grown surface: pits, lines, stacking faults and unwetted regions.

3.1. Surface pits

3.1.1. Pits on the as-grown surface. Small surface pits with density in a few cases as high as 10^7 cm^{-2} have been observed on the surface of the wafers using dark field optical microscopy as shown in figure 2. It was found by the replica TEM technique that the

20μm

Figure 2. Photograph of surface pits on a (100) as-grown surface of DH layers, dark field optical microscopy.

pits are typically 2000 Å in diameter and about 500 to 2000 Å deep. TEM samples showed that some of these pits do not contain dislocations as shown in figure 3(*a*).

3.1.2. Pits developed after etching. The modified AB etchant was used to reveal the structure of the surface pits. All pits disappeared when the top two or three layers had been etched away.

Many special samples with the same doping as the laser structure but with a thicker top layer were then grown. After the surface of these wafers was etched with AB etchant, defect structures occurring as line segments were observed, as shown in figure 4(*a*) and (*b*) using dark and bright field optical microscopy, respectively.

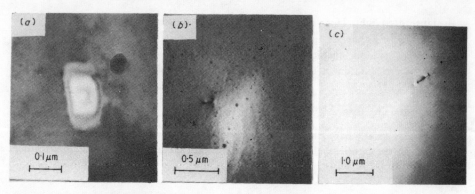

Figure 3. Transmission electron micrographs of pits (*a*) dislocation-free surface pit on as-grown surface, $g = 022$, (*b*) 'heavy dot' pit with dislocation in etched sample, $g = 222$, (*c*) linear pit with dislocation in etched sample, $g = 022$.

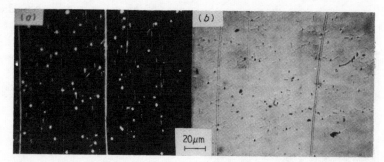

Figure 4. AB etch pits on (100) surface of thick top layer of DH multi-layer sample. (*a*) Dark field, (*b*) bright field. Etched in AB etchant for 4 min at room temperature.

Lines oriented in three directions occur in approximately equal numbers in these patterns. One set is short lines which are parallel to one of the two possible $\langle 110 \rangle$ directions in the (100) surface; the other two are parallel to $\langle 210 \rangle$ directions. The lines occur singly or in combinations with or without a heavy dot in the centre. One frequently occurring combination, that of two lines with a heavy dot in the centre, can be clearly seen in figure 4. These lines and dots are similar to the linear pits observed in $GaAs_{1-x}P_x$ and GaP epi-layers by Stringfellow and Greene (1969) and Stringfellow *et al* (1974) which are said to represent threading dislocations oriented along $\langle 211 \rangle$ and [100], respectively.

TEM investigations showed that the 'heavy dot' pits were associated with dislocations which were almost normal to the wafer surface. In such cases the sample had to be tilted through a large angle (about 42°) to obtain a clear dislocation image as shown in figure 3(*b*). The linear pits, on the other hand, were associated with dislocations making a large angle with [100] as shown in figure 3(*c*). It seems likely that the 'heavy dot' pits are due to [100]-oriented dislocations and the linear pits to $\langle 211 \rangle$-oriented dislocations.

The factors affecting the dislocation density, such as doping and other growth conditions, are still under investigation, but it seems probable that inclusions and the damage they inflict during wiping (scratching) are the important ones. Optical micrographs supporting this view are shown in figure 5.

The importance of dislocations in the laser degradation process has been recently very vividly demonstrated by Petroff and Hartman (1973). The dislocations observed on AB etched pits may be harmful for laser lifetime.

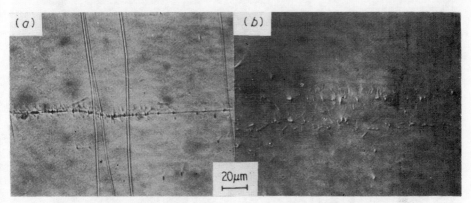

Figure 5. AB etch pits on (100) surface of a DH multi-layer sample. (*a*) Pits produced by a scratch, (*b*) pits which may have been produced by inclusions along wiping direction.

3.2. Meniscus lines

Another very intriguing effect in the growth of LPE GaAs and Al$_x$Ga$_{1-x}$As layers is the existence of a set of approximately parallel and equally spaced lines on the top surface of the wafers as shown in figure 6. These lines are most easily seen as in this figure under dark field illumination. They have been seen on virtually every sample examined. Figure 6 displays several interesting features of these lines. They have a prominent direction, which is generally normal to the direction of movement of the graphite slider and is independent of the crystal orientation. They may 'hang up' around obstacles on the surface as well as at the edge of the wafer. There are some regions which are completely free of these lines, and these regions are bounded by heavy lines formed from the convergence of many other lines. It is suggested that these lines may be due to the impression left by the meniscus of the solution when the solution was withdrawn from the wafer, hence the name 'meniscus lines'. They are similar to those observed by Crossley and Small (1972) and H G B Hicks (1974 private communication). These observations closely parallel our own. It appears that Hicks may have been the originator of the term 'meniscus lines'. TEM studies have shown that dislocations are sometimes associated with meniscus lines (figure 7). Evidence that inclusions can be trapped by them has also been seen.

It was deduced by interferometry and by SEM that the profile of these lines has an S- or Z-shaped character consisting of a peak adjacent to a trough, with no net change in surface height across the whole configuration. Typical peak-to-valley distance is about 500 Å but can be much larger or much smaller. Gross features are sometimes observed when the solution is introduced to the substrate without having been in contact with a source substrate.

Figure 6. Photograph of meniscus lines on a (100) as-grown surface (dark field). The size of the wafer is about 1.2×1.5 cm.

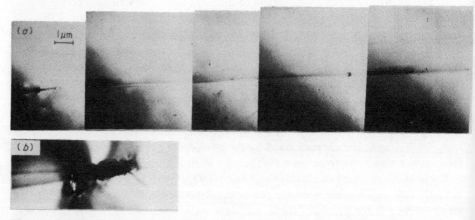

Figure 7. Transmission electron micrographs of a meniscus line (*a*) dislocations associated with the line, (*b*) dislocation network further along the line, $g = 022$.

The existence of adjacent peaks and troughs strongly suggests localized simultaneous etching and accelerated growth, but why such an effect should occur along these contours is not at all obvious. There is some evidence that the effect may be related to variations in surface tension, but there is no evidence to date that the lines have any influence on the rate of laser degradation although associated dislocations presumably will have such an effect. However, it was found that these lines do exist at least at some subsurface interfaces and perhaps at all of them. One must consider the possibility that, when the more basic problems of active layer thickness and planarity control are well in hand, a series of jogs in the heterojunction planes, each of which is a sizeable fraction of the thickness of the active layer itself, might present some kind of resistance to the wave guiding of the laser modes.

3.3. Stacking faults

Another surface feature which was sometimes observed on the as-grown surface is the short line defect, of which some can be seen in figure 8(*a*). These short lines sometimes have different lengths but always extend along ⟨110⟩ directions. It was confirmed by TEM analysis (figure 8(*b*)) that they are stacking faults similar to those observed by Abrahams and Buiocchi (1965) on GaAs. These faults are presumably generated by

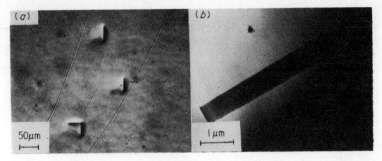

Figure 8. (*a*) Photograph of stacking faults on a (100) as-grown surface, (*b*) transmission electron micrograph of a stacking fault, $g = 022$.

foreign particles on the substrate surface and are particularly frequent when the wall of the quartz tube is contaminated with residual impurities from previous runs, such as phosphorus, zinc, etc. This type of defect can have a detrimental effect on laser lifetime. Fortunately, its presence can be easily avoided by keeping the system clean.

3.4. Unwetted regions

Sometimes there are areas, especially near the edge of the wafer, where either one or all the layers were not continuous and large pits containing droplets of Ga are left on the surface. It is generally believed that the unwetted regions are caused by contamination during LPE growth. The number of this kind of defect can be reduced by cleanness and careful handling of substrate, source material and boat during the LPE process.

4. Summary and conclusions

A modified horizontal sliding boat using small-volume thin solutions has been used to grow multiple epitaxial layers with smooth surface and uniform layer thicknesses.

The defect structure of LPE multi-layers has been studied using dark field optical microscopy, SEM, TEM and AB etching techniques. It was found that there are at least four significant defects on the surfaces of the layers: surface pits, meniscus lines, stacking faults and unwetted regions.

Stacking faults due to foreign particles in the system and unwetted regions due to discontinuous growth are detrimental to laser lifetime but they can be easily controlled by keeping the system clean. Meniscus lines left by the meniscus of the solution during sliding motion are probably unavoidable in a sliding boat LPE system but they may not be directly harmful to laser lifetime. The surface pits do not correspond one to one with dislocations but dislocations have been observed in the AB etched pits after etching. These pits which contain dislocations may be harmful to laser lifetime.

Acknowledgments

The authors wish to acknowledge gratefully the technical assistance of J F DeGeloromo and R M Potemski and helpful discussions with J C McGroddy, M B Small, A W Smith and P G McMullin. They also appreciate the SEM and replica analysis made by C F Aliotta and SEM work by O C Wells and C G Bremer.

References

Abrahams M S and Buiocchi C J 1965 *J. Appl. Phys.* **36** 2855
Blum J M and Shih K K 1971 *Proc. IEEE* **59** 1498
Crossley I and Small M B 1972 *J. Cryst. Growth* **15** 275
Dawson L R 1973 *J. Electrochem. Soc.* **120** 181C
DeLoach B C, Hakki B W, Hartman R L and D'Asaro L A 1973 *Proc. IEEE* **61** 1042
Donahue J A and Minden H T 1970 *J. Cryst. Growth* **7** 221
Hartman R L and Hartman A R 1973 *Appl. Phys. Lett.* **23** 147
Johnston W D 1974 *Appl. Phys. Lett.* **24** 494
Lockwood H F and Ettenberg M 1972 *J. Cryst. Growth* **15** 81
McMullin P G, Blum J, Shih K K, Smith A W and Woolhouse G R 1974 *Appl. Phys. Lett.* **24** 595
Miller B I, Pinkas E, Hayashi I and Capik R J 1972 *J. Appl. Phys.* **43** 2817
Nelson H 1972 *US Pat.* No. 3,565.702
Panish M B, Sumski S and Hayashi I 1971 *Metall. Trans.* **2** 795
Petroff P and Hartman R L 1973 *Appl. Phys. Lett.* **23** 469
Stringfellow G B and Greene P E 1969 *J. Appl. Phys.* **40** 502
Stringfellow G B, Lindquist P F, Cass T R and Burmeister R A 1974 *J. Electron. Mater.* **3** 497
Yonezu H, Sakuma I, Kamejima T, Ueno M, Nishida K, Nannichi Y and Hayashi I 1974 *Appl. Phys. Lett.* **24** 18

Discussion

D J Stirland (Plessey)

The absence of a dislocation at some of the etch pits is not surprising perhaps; if the ⟨001⟩ direction dislocations are those giving the larger pits after the modified AB etc, is it not probable that the thinning treatment used to prepare the TEM specimens will preferentially attack and hence remove these dislocations? Figure 3(a) seems to represent a completely holed specimen, as if the thinning etch has in fact removed everything from the centre of the pit region.

Dr Shih

It is very possible that dislocations in the TEM specimens may have been removed by the thinning treatment. However, we have found that even in the same TEM sample, there are some pits which contain dislocations and other pits which do not.

Degradation of GaAs–(Al,Ga)As double heterostructure light emitting diode

Kenji Ikeda, Toshio Tanaka, Makoto Ishii and Akiko Ito

Central Research Laboratories, Mitsubishi Electric Corporation, Itami, Hyogo, Japan

Abstract. The degradation mechanism of GaAs–(Al, Ga)As double heterostructure light emitting diode has been investigated. Two kinds of degradation are found. One is associated with $\langle 010 \rangle$ dark lines on (001) plane which are observed in degraded LED aged under current density of higher than 1000 A cm^{-2}. Another is associated with $\langle 110 \rangle$ dark lines on (001) plane, which are observed in degraded LED operated under current density of less than 1000 A cm^{-2}. The directions of dark lines can be explained by considering that the dark lines come from formation of a (111) non-radiative plane. Degradation of this type can be reduced by using diode chips thicker than 190 μm or using In solder in bonding the LED chip on a metal stem. Lives of the diodes can be expected to become longer than 10 000 h.

1. Introduction

Recently, practical use of optical communication systems and other electro-optical equipment have been considered in real earnest. As a candidate for the light source of the system, some kinds of light emitting diode LED have been investigated. In particular double heterostructure (DH) LED can be expected to be highly efficient and fast in response as proposed by Lee *et al* (1973), Hsieh and Rossi (1974) and Ettenberg *et al* (1973). For these LED, not only high radiance and high-speed response but also long life are required. We have successfully developed a GaAs–(Al,Ga)As DH-LED, which has advanced characteristics in efficiency and frequency response as was expected. On the other hand, semiconductor lasers with a double heterostructure generally present a difficult problem of short life in CW operation at room temperature. Under these circumstances, it is very important to clarify the degradation mechanism of the DH-LED. In this paper, we have concentrated our attention on the degradation of double heterostructure light emitting diodes.

Two kinds of degradation have been found. One of these is associated with $\langle 010 \rangle$ dark lines on (001) plane which are observed in LED degraded by current density higher than 1000 A cm^{-2}. Another is associated with $\langle 110 \rangle$ dark lines on (001) plane, which are observed in degraded LED operated under current density of less than 1000 A cm^{-2}. Degradation of this type can be reduced by using a diode chip thicker than 190 μm or using In solder in bonding the LED chip on a metal stem. Lives of the diodes can be expected to become longer than 10 000 h.

2. Experimental technique

GaAs–(Al, Ga)As double heterostructure crystals were grown on a n-GaAs (001) substrate by liquid phase epitaxy. Sliding boat technique was employed. The Al contents of the n and p layers were 0·25 mol%. The thickness of the p-GaAs layer, where injected carriers radiatively recombined, was controlled to be 0·8 μm. The dopants of n and p layers were Te and Zn, respectively. Diodes were fabricated into a rectangular form ($L = 1·0$ mm, $W = 0·5$ mm and 0·1 mm thickness) as shown in figure 1. The side

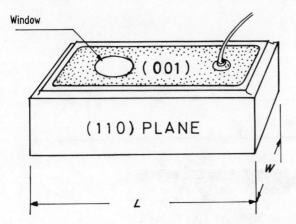

Figure 1. Schematic profile of DH-LED chip. $L = 1·0$ mm and $W = 0·5$ mm.

surfaces of the diodes are (110) or its equivalent plane. In order to avoid a bonding stress, wire bonding of an electrode for p-(Al, Ga)As was made at some distance from the window through which the light output was obtained. The diameter of the window is 0·2 mm. The diode chip was mounted on a TO-39 metal header by Au–Si alloyed solder sealed in a dry N_2 atmosphere by a metal cap with a glass window. The emission spectrum of this LED at room temperature has a peak located at 8600 Å which corresponds to the bandgap energy of p-GaAs crystal.

Life tests for 50 diodes were carried out as follows. The stress current was 100 mA DC which corresponded to about 300 A cm^{-2} in the current density for a window area. The junction temperature should be about 65 °C when the stress current is applied. Output power was periodically monitored by a Si solar cell.

3. Results and discussion

Figure 2 shows degradations of the DH-LED. External efficiency of the LED reduces with ageing time. At the same time, the reverse current, measured as −3·0 V, increases. The other characteristics, for example, emission spectra and forward voltage, are not affected. After ageing for about 2000 h, the degraded diodes were observed by using a microscope with an infrared television camera so as to be able to detect the infrared image. Figure 3(a) is a photograph showing a window of the diode under external visible light illumination. No change corresponding to degradation can be found in the figure. Figure 3(b) is an IR photograph showing the same part of figure 3(a) without external

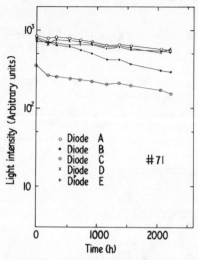

Figure 2. Time dependences of light intensities of DH-LED.

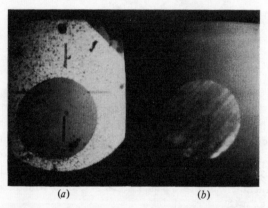

Figure 3. Microphotographs of the window of DH-LED after about 2000 h ageing obtained by using an infrared TV camera. (*a*) is obtained under external visible light illumination. (*b*) shows the same part of the diode as that in (*a*).

illumination. One can see in the figure that there are some dark lines in the window. Moreover, the direction of the dark lines is ⟨110⟩ on the basis that the lines are perpendicular to the side plane of the diode (110). Biard *et al* (1966) found dark lines corresponding to (110) slip planes in (111) samples of homojunction LED. On the other hand, in semiconductor lasers having a GaAs–(Al, Ga)As double heterostructure similar essentially to the DH-LED mentioned here, degradation has been investigated by Yonezu *et al* (1973) and Petroff and Hartman (1973). Dark lines found in laser diodes have ⟨100⟩ direction. As far as the authors are aware no-one has reported ⟨110⟩ dark lines.

In order to know where the dark lines lie in a multilayer crystal, we cleaved the diode chip along a long side of the diode chip (110). After cleavage, the chips were mounted on stems again. The cross sections of the diodes have been examined by scanning electron microscopy (SEM). In figure 4, a picture obtained by a cathodo-luminescence method in a cross section of the degraded LED is shown. Dark lines can be found in the cleaved face of the diode in the figure. The dark lines continue into the GaAs substrate from the surface of the p-(Al, Ga)As layer. The dark lines in the cleaved cross section intersect (001) junction plane by $54°$. These experimental results suggest that this kind of dark line is on the (111) crystal plane.

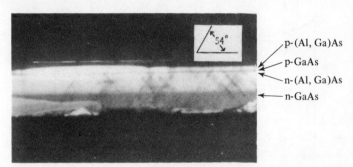

Figure 4. Cathodoluminescence pattern of the cross section of degraded DH-LED by using an SEM system. Dark lines intersect the junction plane by $54°$.

In order to investigate the cause of the degradation, we have examined internal strain in the diode chip by the photoelastic technique described by Hartman and Hartman (1973). In this effect, a part with strong stress is observed to be light, and a stress-free portion is observed to be dark. Photographs obtained by the photoelastic technique for LED soldered by In and by Au–Si are shown in figure 5(*a*) and (*b*), respectively. It is

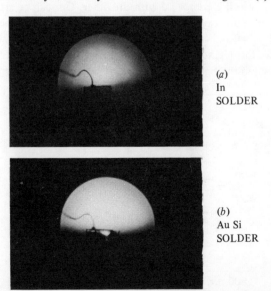

Figure 5. Observation of internal strain field in the diode chips by the photoelastic effect. (*a*) and (*b*) show those obtained for the diode mounted by Au–Si solder and by In solder, respectively.

clearly shown in the figure that the strain field of the latter is stronger than that of the former. Comparison between optical power changes measured in LED mounted by In solder and in those soldered by Au–Si has been made. Ten samples were used for each group. The result is given in figure 6. As can be seen in the figure, the group of diodes

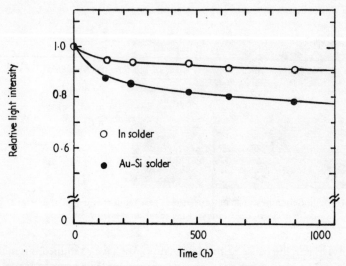

Figure 6. Degradation curves for two groups of DH-LED. Open circles show optical power measured in a group of LED mounted by In solder. Full circles represent the optical powers measured in a group of LED mounted by Au–Si solder.

using In solder has a long half power life compared with the other. This can be attributed to the difference in hardness between In and Au–Si or to the difference in the melting points. In both cases, it would be natural to expect that making the chip thicker would increase the life. Figure 7 shows the degradation of LED as a function of diode thickness (90, 150 and 190 μm). As shown in the figure, the life time is longer when thicker diode

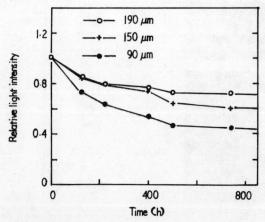

Figure 7. Degradation of LED as a function of diode thickness. Each plot shows the average of relative power obtained by ten samples.

chips are used. Few dark lines of ⟨110⟩ can be observed when the thickness of diode is 190 μm. By taking into consideration these experimental data, we can expect the half power life of LED to be several 10 000 h, when diode chips thicker than 190 μm are used. Moreover, by using In solder, we might expect longer life.

There is the other degradation mode when operation current density is in excess of 1000 A cm^{-2}. Figure 8 shows a picture of the window of the LED observed by infrared TV. The diode is operated by a pulse current density of 5000 A cm^{-2} whose duty ratio

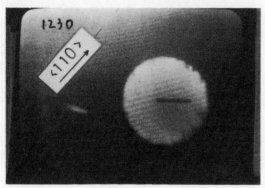

5000 A/cm

Figure 8. Microphotograph of the window of an LED degraded by pulse current density of 5000 A cm^{-2}.

is 10^{-2}. One can find a few dark lines near the electrode in the window. The direction of the dark lines are ⟨010⟩ and its equivalent directions, which is similar to those observed in laser diodes. These kinds of dark lines were also formed, even when ageing was done by DC current and/or even when the chip thickness was about 200 μm. So these may come not only from the difference of thermal expansion coefficients between GaAs crystals and metals. Some of the other forces coming from high current density may be applied.

4. Summary

The degradation mechanism of GaAs–(Al, Ga)As double heterostructure light emitting diode has been investigated. We have found two kinds of degradations. One of them is associated with ⟨010⟩ dark lines on (001) plane which are observed in degraded LED operated under current density of higher than 1000 A cm^{-2}. These ⟨010⟩ dark lines are previously reported in semiconductor lasers. Another is associated with ⟨110⟩ dark lines on (001) plane, which are observed in degraded LED operated under current density of less than 1000 A cm^{-2}. The directions of the dark lines can be explained by considering that the dark lines observed come from the formation of (111) non-radiative planes. Degradation of this type can be reduced by using diode chips thicker than 190 μm or using In solder in bonding the LED chip on the metal stem. The lives of the diodes can be expected to become longer than 10 000 h. The cause of the formation of ⟨110⟩ dark lines can be attributed to the difference of thermal expansion coefficients between diode chip and solder or stem material.

When chips thicker than 190 μm are used, we can expect the half power life of the LED to be several 10 000 h. Moreover, by employing In solder, we might expect longer life, for example, 100 000 h even when diodes are operated by about 1000 A cm^{-2} in current density.

When chips thicker than 190 μm are used, we can expect the half power life of the LED to be several ten-thousand hours. Moreover, by employing In solder, we might expect longer life, for example, 100 000 h even when diodes are operated by about 1000 A cm^{-2} in current density.

Acknowledgments

We wish to express our thanks to Mr H Namizaki for his kindness in the course of experiments on the photoelastic effect, and also to Mr F Iwasaki for his help in the SEM measurements.

References

Biard J R, Pittman G E and Leezer J F 1966 *Proc. Int. Symp. GaAs* (London: Institute of Physics) p113
Ettenberg M and Kressel H 1973 *IEDM Techn. Digest* p317
Hartman R L and Hartman A R 1973 *Appl. Phys. Lett.* **23** 147
Hsieh J J and Rossi J A 1974 *J. Appl. Phys.* **45** 1834
Lee Tien-Pei, Burrus C A and Miller B I 1973 *Proc. IEEE Quantum Electron.* **QE-5** 820
Petroff P and Hartman R L 1973 *Appl. Phys. Lett.* **23** 469
Yonezu A *et al* 1973 *J. Japan Soc. Appl. Phys.* **43** Suppl 59

Characterization of GaAs−AlGaAs double-heterojunction laser structures using optical excitation

G A Acket, W Nijman, R P Tijburg and P J de Waard

Philips Research Laboratories, Eindhoven, The Netherlands

Abstract. Experiments carried out on three-layer GaAs−AlGaAs double heterojunction structures with a top p-AlGaAs layer which is transparent to HeNe radiation, are described. The active GaAs layer is excited using DC or pulsed HeNe lasers. Scanning photoluminescence, photocurrent and optical phase shift (lifetime) measurements are performed. Furthermore, laser diodes (both pulsed and long-lived CW lasers) are made from the very same structures.

It is found that the photoluminescence intensity occasionally varies strongly with position on the top surface, but that this variation is usually *not* related to variations of internal radiative efficiency of the active layer, but to the presence of a 'loss' of electrons at the depletion layer. Furthermore, simultaneous measurements of photoluminescence intensity and photocurrent frequently evince a completely correlated and 'complementary' behaviour. These phenomena can be explained by the band-gap discontinuity at the p-GaAs−n-AlGaAs interface occurring mainly in the conduction band and being more or less rounded off due to grading. From these data the external quantum efficiency of the active p-GaAs layer was estimated. Results of structures grown with and without a 'wash melt' between the n-AlGaAs and p-GaAs melts are shown. The results are compared with the results obtained on the laser diodes made from these structures. It is found that the threshold current densities show a correlation with the quantum efficiency of the p-GaAs active layer.

1. Introduction

Since the first reports on laser operation at low current densities of diodes made from GaAs−AlGaAs double heterojunction structures (Alferov *et al* 1970a, Hayashi *et al* 1970, Kressel and Hawrylo 1970) much effort has been devoted to the study of such diodes with special emphasis on long-lived CW lasers. It is sometimes observed that diodes made from structures grown with identical melt compositions show different threshold currents and external differential quantum efficiencies. Even diodes prepared from the same wafer sometimes exhibit considerable differences. In such cases it is vital to know which is the critical parameter in causing such variations and it would be helpful if methods would be found for determining the laser capability of a wafer before laser diode fabrication. For such purposes techniques seem suitable in which electron− hole pairs are created selectively within the active region of GaAs−AlGaAs double hetero- junction structures. In fact, one would expect the photoluminescence intensity to be a good criterion for determination of the radiative efficiency of the active p-GaAs region. However, it will be shown below that the photoluminescence intensity alone usually does not provide enough information for estimating the radiative efficiency of the active region and that additional information obtained from photocurrent measurements must be included. Therefore, a systematic study was made of the variations of photoluminescence intensity and photocurrent on a number of three-layer double heterostructures. From the combined variations of photoluminescence and photocurrent, the variation of radiative efficiency of the active region with position on the wafer surface was obtained. This efficiency is related to the minority carrier lifetime. Therefore, we also measured the minority carrier lifetime in the active region of the wafers studied. The results of the experiments performed also yield information concerning the grading of the n-AlGaAs–p-GaAs heterojunction. The data were compared with the behaviour of laser diodes fabricated from the same wafers.

2. Experimental methods

The structures investigated were three-layer n-AlGaAs–p-GaAs–p-AlGaAs structures on a n-type GaAs substrate in which the thickness of the active p-type GaAs layer was $0 \cdot 1 - 2 \, \mu m$ (the usual p-GaAs contacting layer is absent in these structures). The aluminium arsenide content of the p-AlGaAs top layer was at least 42% so that this layer is transparent to HeNe laser radiation ($\lambda = 6328$ Å). Contacts were applied to both the p-AlGaAs top layer and the n-GaAs substrate.

CW as well as pulsed (mode-locked) HeNe lasers were used to excite the p-GaAs active region through the transparent p-AlGaAs top layer. The DC photoluminescence intensity was measured either as a function of wavelength, exciting a fixed position on the surface of the structure or as a function of position close to a cleaved edge at a fixed wavelength. In order to obtain a good spatial resolution the laser was focused to a spot of about $5 \mu m$ diameter on the sample surface using a ×10 microscope objective. The sample was mounted on a micrometer translation stage equipped with a displacement transducer. The photoluminescence measurements were performed both with the sample open-circuited as well as with a reverse bias voltage applied to the n-AlGaAs–p-GaAs junction. With the latter junction reverse-biased also the photocurrent flowing across it was measured. The experimental situation is schematically depicted in figure 1. The variation in the integral photoluminescence intensity over the whole area of the wafer was inspected visually by exciting it with a diverging HeNe beam and inspecting the luminescence under a microscope with a silicon vidicon TV camera equipped with suitable filters. This procedure was used to select regions of sufficient interest for more detailed examinations. In order to obtain information concerning the variation of aluminium content near the n-AlGaAs–p-GaAs interface, we also carried out photoluminescence measurements at various positions on structures angle-lapped at $3°$.

Apart from these DC measurements we also measured the electron lifetime in the p-GaAs active regions by exciting it through the transparent p-AlGaAs top layer with a mode-locked HeNe laser oscillating at $75 \cdot 3$ MHz and measuring the phase shift between the AC photoluminescence radiation and the HeNe radiation. The measuring procedure has been described earlier (Acket *et al* 1974).

Laser diodes were fabricated from the wafers studied. Oxide-stripe lasers were made. In order to reduce the contact resistance of the p-AlGaAs top layer a skin diffusion of Zn (Hwang and Dyment 1973) was applied. The lasers had a length of about $300 \mu m$.

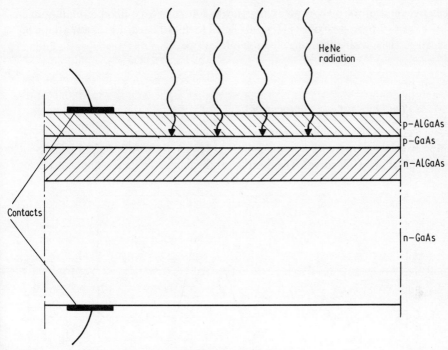

Figure 1. Schematic representation of experimental situation.

The lasers had a stripe-width of about 15 μm. Laser diodes to be used in pulsed experiments were mounted on modified TO-18 headers. Those diodes to be used CW were soldered with lead–tin or indium upside down on copper heat sinks.

3. Materials preparation

The three-layer double heterojunction structures were grown by the well-known multiple-bin liquid phase epitaxy technique as described by Panish *et al* (1971). The graphite boats used were also provided with a substrate support, whose vertical position can be controlled by means of a wedge operated from outside the furnace as described by van Oirschot and Nijman (1973). This construction is chosen among other things in order to improve the control of the wipe-off process. Often, the bin containing the melt from which the first (n-AlGaAs) layer was grown and the bin containing the melt for the second (p-GaAs) layer were separated by a bin containing a melt having the same composition as that for the p-GaAs region which served as a 'wash' solution to reduce drag-over of aluminium from the n-AlGaAs melt (see also Kressel *et al* 1973). On other occasions the source-seed method (Rode 1973) was used. The growth temperature was about 825 °C.

The n-$Al_x Ga_{1-x}$As region ($x \simeq 0.35$) had a thickness of 6–8 μm. It was Sn-doped to an electron concentration of $1–3 \times 10^{17}$ cm^{-3}. The p-GaAs region was doped with Ge and had a hole concentration of $2.5–3 \times 10^{18}$ cm^{-3}. The p-AlGaAs, which had an aluminium arsenide content of at least 42%, was also Ge-doped and had a hole concentration of about 5×10^{17} cm^{-3}. The thickness of the p-AlGaAs layer was $1–1.5$ μm.

The thickness of the p-GaAs active region was determined by cleaving, etching in the selective hydrogen peroxide (pH ≃ 7·05) etch (Schwartz *et al* 1972) and studying the cleavage plane with a scanning electron microscope.

4. Results

4.1. *Photoluminescence and photocurrent*

A typical front surface photoluminescence spectrum obtained on a double hetero-junction structure grown with a 'wash' solution between the n-AlGaAs and p-GaAs melts is shown in figure 2 (full curve). The position of the luminescence peak indicates that

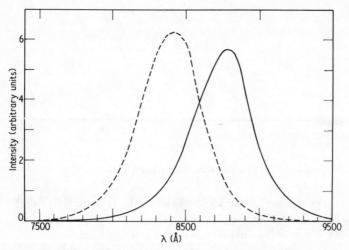

Figure 2. Typical photoluminescence spectra from p-GaAs active layer of a p-AlGaAs–p-GaAs–n-AlGaAs structure grown with (full curve) and without (dashed curve) 'wash' solution.

practically no important drag-over of aluminium towards the active region has occurred. The spectra obtained on layers grown without 'wash' solution have practically the same shape but the luminescence peak is often shifted towards shorter wavelengths by 200–300 Å (figure 2, broken curve). It should be mentioned that a similar shift was also found in some samples grown with a 'wash' solution and having a very thin (300–500 Å) active region. This may indicate that some growth already takes place in the 'wash' solution region.

The photoluminescence intensity sometimes varies markedly over the area of the structure (see figure 3, curve A). At the positions of highest intensity the response is very similar to the photoluminescence output of thick p-GaAs layers of the same composition (also provided with a p-AlGaAs top layer) grown on a GaAs substrate (where the internal photoluminescence efficiency was estimated to be at least 0·5 and where the lifetime against hole concentration characteristic indicated that radiative recombination was the dominant recombination mechanism (Acket *et al* 1974)). Hence, at the positions of maximum luminescence intensity, the internal efficiency of the active

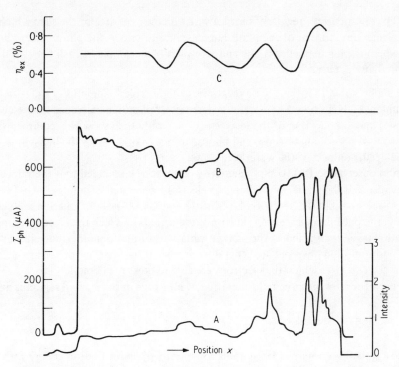

Figure 3. Variation of photoluminescence intensity at luminescence peak (curve A) and photocurrent (B) as a function of position parallel to a cleaved edge of a double hetero-structure (DH 48). The efficiency η_{ex} obtained after correction for diffusion losses is plotted in curve C.

region in our n-AlGaAs–p-GaAs–p-AlGaAs structures is also high and should be at least 50%. This is in agreement with the fact that the minority carrier lifetimes measured on regions with a high DC photoluminescence are as high as 3–4 ns. This is quite high for the hole concentration present in the active region ($p = 2 \cdot 5 - 3 \times 10^{18}$ cm^{-3}) (see Hwang and Dyment 1973 and Acket *et al* 1974).

The observation that the DC photoluminescence is often much lower in other regions of the wafers would suggest at first glance that there the internal radiative efficiency of the active region should be low. The electron lifetime measured on these regions turns out to be lower too, namely $0 \cdot 7 - 1 \cdot 5$ ns. However, it will be shown below that this behaviour is mainly produced by the presence of the depletion layer at the n-AlGaAs–p-GaAs junction. When an external reverse bias voltage is applied to contacts made onto the p-AlGaAs top layer and the n-GaAs substrate the following facts are usually observed:

(i) The photoluminescence intensity is often hardly influenced by closing the external circuit and applying a reverse bias voltage. This is observed at regions with high photo-luminescence as well as regions with low photoluminescence intensity.
(ii) The electron lifetimes in these regions are often hardly changed by application of reverse bias.

(iii) The photocurrent regularly shows a variation over the area of the slice which is 'complementary' to that of the photoluminescence intensity, the photocurrent being high where the luminescence is low and vice versa. An example of such behaviour is seen in figure 3, curve B.

It should be noted that there are exceptions to these 'rules'. Sometimes wafers are encountered which show a considerable decrease of photoluminescence (and electron lifetime) upon application of the reverse bias, and only then show the 'complementary' behaviour of photoluminescence and photocurrent. In other cases the behaviour mentioned occurs on part of the wafer area only.

From observation (iii) (complementarity) it must be concluded that the photoluminescence is high in regions where the electrons generated within the p-GaAs regions are less able to cross the depletion layer of the n-AlGaAs–p-GaAs junction and, conversely, the photoluminescence is low on those regions in which the electrons cross the depletion layer easily. Hence, the marked variations of photoluminescence intensity with position often observed are generally *not* related to variations of radiative efficiency but to the extent to which the electrons are able to cross the depletion layer. In the presence of reverse bias, the net generation of electrons in the excited region is given by

$$G = \frac{P_{abs}}{h\nu} - \frac{I_{ph}}{q} \tag{1}$$

where P_{abs} is the amount of HeNe radiation absorbed, $h\nu$ the photon energy for the HeNe radiation (1·96 eV), I_{ph} the photocurrent in the external circuit and q the electron charge. If P_{lum} stands for the luminescence intensity the external efficiency η_{ex} is given by

$$\eta_{ex} = \frac{P_{lum}(x)}{G(x)} \tag{2}$$

(where x is the position on the slice). This is indicated in figure 3, curve C, for the results of that layer. The quantity η_{ex} can be expressed as an external quantum efficiency since the luminescence output has been compared with that of some samples in which the total luminescent power emitted was determined using an elliptical mirror and a calibrated photodetector (van der Does de Bye 1969). The values for η_{ex} obtained on various structures are listed in table 1. Here it should be remarked that due to total internal reflection an internal efficiency of 100% corresponds to an external efficiency of about 1·4% (Gooch 1973), which is indeed close to the highest experimental values. The highest values of η_{ex} are obtained for the structures with a relatively thick active region, namely 0·8–2·0 μm. The structures with thinner active layers have values of η_{ex} which are somewhat lower and show more pronounced spatial variations.

Our data indicate that the ability of the electrons to cross the depletion layer may vary over the area of the wafers. This can be explained as follows. The electron concentration of the n-type GaAlAs is about one order of magnitude lower than the hole concentration, so that the depletion layer is practically completely within the n-AlGaAs. The band-gap of the AlGaAs is greater than that of GaAs and there is evidence that most of the discontinuity of the bandgap takes place in the conduction band (Alferov 1970b). If the band-gap discontinuity is sharp (see figure 4(a)) the electrons created

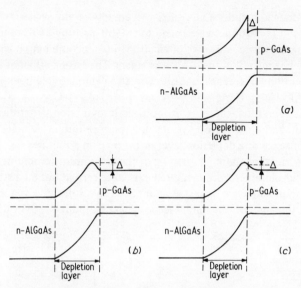

Figure 4. Possible variation of band edges as a function of position near the n-AlGaAs–p-GaAs heterojunction: (*a*) Abrupt junction (Alferov 1970b). (*b*) Graded n-Al$_x$Ga$_{1-x}$As–p-GaAs junction (Womac *et al* 1972). (*c*) Graded n-Al$_x$Ga$_{1-x}$As–p-Al$_y$Ga$_{1-y}$As junction.

within the active p-GaAs layer will not be able to cross the depletion layer, but if considerable grading occurs the discontinuity in the conduction band is rounded off (Womac and Rediker 1972) and the electrons are more likely to reach the depletion layer and, in the presence of a reverse bias voltage, a high photocurrent is to be expected (see figure 4(*b*)). Hence, the variations in photocurrent and photoluminescence observed may be related to variations in grading of the n-Al$_x$Ga$_{1-x}$As–p-GaAs heterojunction. However, an additional possibility, and indeed a very likely one, is that some drag of aluminium takes place towards the p-GaAs melt, so that the actual heterojunction is more an n-Al$_x$Ga$_{1-x}$As–p-Al$_y$Ga$_{1-y}$As junction. If y rapidly decreases when moving into the active layer, a gradient in the position of the conduction band occurs, preventing electrons created within the active region from reaching the depletion layer (figure 4(*c*)). Indeed we will see below that the experiments suggest that the situation sketched in figure 4(*c*) is the one obtained most commonly.

It has been mentioned above that the presence of the depletion layer can lead to a loss of electrons generated in the active region and a concurrent reduction of photo-luminescent intensity, even in the absence of contacts and current leads, provided the electrons are able to reach the depletion layer. This can be explained as follows. With the sample open circuited and optical power incident on the sample the p–n junction becomes forward biased such that the forward current, produced by this photovoltage, V_{ph}, just compensates the photocurrent produced by the electrons created in the active region. If the forward current corresponding to the voltage V_{ph} is an injection of electrons over the p–n junction into the p-type material, no net loss of electrons will occur. However, if at this voltage the forward current is mainly due to electron–hole recombination in the depletion layer (Sah *et al* 1957) a net loss of electrons from the

p-GaAs does take place. Indeed junctions, in which a severe loss of electrons was
observed under open-circuit conditions, were found to exhibit a strong $2kT$ branch
in the current voltage characteristic indicative of electron–hole recombination within
the depletion layer (which is mainly in the n-AlGaAs layer). The strong effect of the
depletion layer on the photoluminescence intensity was also demonstrated in experiments in which both p-AlGaAs–p-GaAs–p-AlGaAs and p-AlGaAs–p-GaAs–n-AlGaAs
were studied. Here it was found that the p–p–p double heterojunction structures
exhibited a much stronger and more uniform photoluminescence output, which
obviously is due to the absence of a depletion layer in the p–p–p structures.

In order to obtain information about the grading of the aluminium content in the
active region, detailed photoluminescence measurements were carried out on a structure
grown without washing melt and angle-lapped at about $3°$. The results are given in
figure 5. Far away from the junction the luminescence peaks at about 8250 Å but when

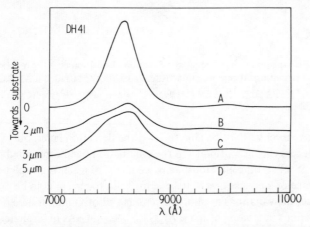

Figure 5. Photoluminescence spectra at various positions on the surface of a $3°$ angle-
lapped double heterojunction wafer. Positions on the bevel surface are indicated.
Position 0 is on the surface of the active layer about 5 μm away from the n-AlGaAs layer.

the junction is approached the intensity of the luminescence at shorter wavelengths
(7900–8000 Å) increases considerably although luminescence still remains present
around 8250 Å. This indicates the presence of a gradient of aluminium concentration
of several per cent over a depth of 0·1 μm in the active region, the continuing presence
of the longer wavelength contribution around 8250 Å being probably due to electrons
moving down the gradient of the conduction band and recombining further away from
the junction.

4.2. Results of the laser diodes

The behaviour of the diodes made from the three-layer structures was quite similar
to those made from four-layer wafers. Furthermore, the relatively high aluminium
content of the p-AlGaAs did not significantly affect the laser behaviour. Suitably
mounted diodes having an active layer thickness of less than 0·5 μm were easily operated

CW. The data (table 1) relate to diodes in which a stripe contact was defined by photo-etching an SiO_2 layer grown on top of the structure. The stripe width usually was 15 μm, whereas the thickness of the top p-AlGaAs layer ($p \simeq 5 \times 10^{17} cm^{-3}$) is about 1·5 μm. Hence, current spreading yields values for the threshold current density which are too high by a factor of about three (Dumke 1973). This factor was determined experimentally for our diodes by measuring the distribution of the spontaneous radiation along the junction plane just below threshold. A typical distribution thus obtained is shown in figure 6. The threshold current density below the stripe contact was then easily derived from this distribution together with the total (pulsed) threshold current.

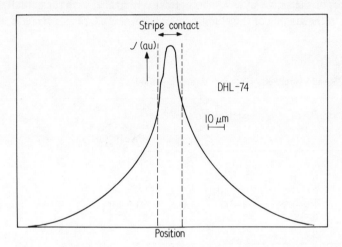

Figure 6. Distribution of current density J in the junction plane in a direction perpendicular to the stripe contact as obtained from the distribution of spontaneous radiation below threshold.

5. Discussion

The spatial variations of photocurrent observed are interpreted in terms of a variation of the height of the conduction band barrier the electrons have to overcome in order to reach the depletion layer. The incident HeNe power used was 2·5 mW, which corresponds to a maximum obtainable photocurrent $I_{max} = 870 μA$. The minimum photocurrent observed was about 30 μA so $I_{max}/I_{ph} \leqslant 30$. If we assume that

$$\frac{I_{ph}}{I_{max}} \simeq \exp -(\Delta/kT) \tag{3}$$

we find $\Delta \leqslant 0·08$ eV. This indicates that the situation of figure 4(a), where $\Delta \simeq 0·40$ eV, is never obtained. The above estimate (0·08 eV) is really an upper limit, much smaller values ($\leqslant 0·05$ eV) being more frequently encountered. The latter value reasonably agrees with the photoluminescence peak shift shown in figure 5. So we see that figure 4(c) with $\Delta \leqslant 0·05$ eV on the average describes the situation rather well. This also applies to samples grown with a 'wash' melt, since here the layers of 300–500 Å in thickness also showed a strong shift of the photoluminescence peak. It should be noted that the band diagram of figure 4(b) is less successful in explaining the variations of photocurrent observed since there the barrier is situated within the depletion layer so that the reverse voltage used in the photocurrent measurements should cause this barrier to vanish. The fact that the photoluminescence intensity was often rather insensitive to the applied reverse voltage also indicates that the controlling barrier should be situated in the neutral part of the active layer.

Now we compare the results obtained on the wafers with those of the laser diodes. Table 1 shows that the normalized threshold current densities of DH 12 and especially DH 48, which have been grown without a 'wash' melt, are lower than those of the DHL layers. This is probably due to the presence of some drag-over of aluminium which may lead to a thickness of the active layer which is effectively smaller than the actual thickness owing to the grading of the aluminium concentration (see also Nakashima *et al* 1974).

Table 1. Results obtained on laser structures and diodes

Wafer	η_{ext} (%)	τ_{exp} (ns)	d (μm)	J_{th}/d Oxide stripes (kA cm^{-2}μm^{-1})
DH 12	1·3	3–10	1·6	3·9 ± 1·0
DH 48	0·6–0·9	2·9–3·5	1·1	3·4 ± 0·2
DHL 53	0·6–1·2	2·0–3·6	0·8	6·9 ± 0·7
DHL 56	1·0–1·9	1·7–3·0	1·0	6·1 ± 1·0
DHL 74	0·3–0·7	2·6–3·5	0·4	7·3 ± 1·5
DHL 77	0·1–0·8	1·6–2·9	0·20	9 ± 2

DH, wafer grown without 'wash' melt.
DHL, wafer grown with 'wash' melt.

Inspection of the results of the DHL layers (table 1) shows that the structures with thinner active regions have higher normalized threshold current densities. Moreover, the efficiencies of these layers are on the average smaller than those with thicker active layers. This suggests that the variations of normalized threshold current density should be largely due to variations of efficiency of the active layer. It is not yet clear what is responsible for the lower efficiency in the thinner structures. Our previous experience on the p-GaAs–p-AlGaAs junctions (Acket *et al* 1974) indicates that no negative effect on the minority electrons in the p-GaAs is produced by this junction, so that it seems probable that the n-AlGaAs–p-GaAs heterojunction is more critical.

In conclusion, it has been found that photoluminescence intensity alone is not sufficient to characterize the efficiency of the active layer because of possible loss of electrons at the depletion layer of the n-AlGaAs–p-GaAs heterojunction. The determination of the external photoluminescence efficiency with appropriate correction for this loss of electrons (equations (1) and (2)) appears to be a better way to characterize the laser quality of the wafers studied. It has also been shown that the measurements described provide information concerning the grading at the n-AlGaAs–p-GaAs heterojunction and that the band diagram of figure 4(*c*) best described the data.

Acknowledgments

The authors are grateful to W Leswin for growing many of the structures investi-gated, T van Dongen for making the diodes, W Schoenmakers for measurements on the laser diodes and H Wielink for performing many of the photoluminescence and photo-current measurements. P Boers kindly made available the photoluminescence inspection equipment. J L C Daams performed the determination of the thickness of the active regions by scanning electron microscope.

References

Acket G A, Nijman W and 't Lam H 1974 *J. Appl. Phys.* **45** 3033
Alferov Zh I, Andreev V M, Garburov D Z, Zhilgaev Ju V, Morozev E P, Portnoi E L and Trofim V G
    1970a *Fiz. Tekn. Poluprovodn* **4** 1826
Alferov Zh I, Andreev V M, Zimogorova N S and Tret'yakov D N 1970b *Sov. Phys.–Semicond.* **3**
    1373
van der Does de Bye J A W 1969 *Rev. Sci. Instrum.* **40** 320
Dumke W P 1973 *Solid St. Electron.* **16** 1279
Gooch C H 1973 *Injection Electroluminescent Devices* (London: Wiley)
Hayashi I, Panish M B, Foy P W and Sumski S 1970 *Appl. Phys. Lett.* **17** 109
Hwang C J and Dyment J C 1973 *J. Appl. Phys.* **44** 3240
Kressel H and Hawrylo F 1970 *Appl. Phys. Lett.* **17** 169
Kressel H, Lockwood H F, Nicoll F H and Ettenberg M 1973 *IEEE J. Quantum Electron.* **Q**E-9
    383
Nakashima H, Chinone N, Taguchi Y and Nakada O 1973 *J. Appl. Phys.* **44** 2688
van Oirschot T G J and Nijman W 1973 *J. Crystal Growth* **20** 301
Panish M B, Sumski S and Hayashi I 1971 *Metall. Trans.* **2** 795
Rode D L 1973 *J. Crystal Growth* **20** 13
Sah C, Noyce R N and Shockley W 1957 *Proc. IRE* **45** 1228
Schwartz B, Dyment J C and Haszko S E 1972 *Proc. 4th Int. Symp. Gallium Arsenide and Related
    Compounds* (London and Bristol: Institute of Physics) p187
Womac J and Rediker R H 1972 *J. Appl. Phys.* **43** 4148

(AlGa)As double heterojunction CW lasers: the effect of device fabrication parameters on reliability

I Ladany and H Kressel

RCA Laboratories, Princeton, New Jersey 08540, USA

Abstract. Experiments are reported which were designed to examine the effect of two common device process steps on the life of CW laser diodes. These include the effect of Zn diffusion used to either improve the ohmic contact to the p-side of the diode or to provide lateral current confinement, and the method used to define the laser diode sidewalls. It is shown that excessively deep Zn diffusion (resulting in disturbances close to the recombination region) leads to accelerated degradation. Mechanical damage of the diode edges such as introduced by wire sawing is shown to have similar detrimental effects. Using identical starting material, orders of magnitude differences in CW laser life are observed and directly attributable to the above process steps. Lasers emitting reliably at $\lambda_L \sim 8100$ A for periods of at least 2000 h have been made by proper control of the device fabrication.

1. Introduction

Two major laser diode failure modes have been well characterized in the past. The first, usually termed 'catastrophic' degradation depends on the optical flux density and the failure cause is mechanical damage of the facet in the emitting region (Kressel and Mierop 1967). The second, denoted 'gradual' degradation, due to internal diode changes, is directly related to the electron—hole recombination in the active region of the diode (Kressel and Byer 1969). The effect does not appear related to the optical flux density and existing evidence suggests that the internal changes in the diode occur similarly whether the diode is operating in the lasing or the spontaneous mode.

Gradual degradation phenomena are known to be associated with the formation or growth of metallurgical defects in the recombination region which increase the internal absorption of the radiation and reduce the internal quantum efficiency. Some of the initially present defects which contribute to accelerated degradation in GaAs diodes have been identified (for a review see Kressel and Lockwood 1974). These include dislocations (Kressel *et al* 1970), precipitates and contaminants (Cu for example). In addition, there are differences in degradation rate when different dopants are used in the recombination region. An important recent finding is that the addition of Al to the GaAs diode recombination region greatly increases the resistance to gradual degradation, other growth, operating and structural parameters being constant (Ettenberg *et al* 1973, Yonezu *et al* 1973a).

In this paper we report on experiments designed to study the effect on gradual degradation of CW laser diodes due to two key device process steps. The first deals with the role of Zn diffusion commonly used either to improve the diode ohmic contacts or to provide isolation in stripe geometry diodes. The second deals with the effect of the diode sidewalls (ie, the two sides perpendicular to the cleaved facets forming the mirror ends of the Fabry–Perot cavity). We show that identical diode materials can exhibit operating lifetimes ranging from a few hours to thousands of hours depending on the fabrication process used.

2. Experimental

The diodes used in these experiments, designed for emission at 8000–8300 Å were $Al_xGa_{1-x}As-Al_yGa_{1-y}As$ double heterojunction structures with $x \simeq 0.1$ in the recombination region and $y \simeq 0.3$ in the adjoining p- and n-type regions. The choice of the present material composition and diode structure is based on the availability of glass fibres with very low absorption losses in the 8000 Å spectral range (Keck *et al* 1973), and on the fact that previous work has shown that diodes containing about 10% Al in the recombination region are much more resistant to gradual degradation than similar diodes with GaAs in the recombination region (Ettenberg *et al* 1974).

The diode material was grown by liquid phase epitaxy using the thin melt technique (Lockwood and Ettenberg 1972) in a horizontal growth apparatus on (100) GaAs substrates. The width of the n-type recombination region was $0.2-0.3\,\mu m$. Germanium was used to dope the p-type regions, while Sn or Te were used for the n-type regions. The recombination region was not deliberately doped, and contained an electron concentration of $5 \times 10^{16}\,cm^{-3}$. To minimize the thermal and electrical resistance of the diodes, the distance between the recombination region and the surface of the diode was typically $2-3\,\mu m$ and included a p-type heavily doped GaAs layer to improve the ohmic contact. Metallization was accomplished by evaporation of tin on the n-side and deposition of Ni or evaporation of multiple metal layers on the p-surface. Figure 1 shows a scanning electron micrograph of a typical diode cross section. In subsequent

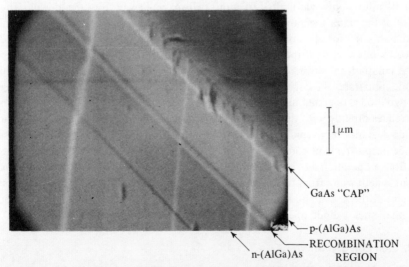

Figure 1. Scanning electron micrograph showing the cross section of a typical $Al_xGa_{1-x}As-Al_yGa_{1-y}As$ double heterojunction diode.

discussion we will refer to the GaAs surface layer as the 'cap' layer and to the p-type (AlGa)As layer as the 'p-wall'.

The basic design concepts of double heterojunction diodes of this type having very narrow recombination regions have been discussed previously (Kressel *et al* 1971). These diodes can combine very low threshold current densities with moderate beam divergence because of the controlled spread of the radiation from the recombination region into the low loss adjoining p- and n-type regions. When processed in conventional broad area form, the threshold current density values of the present diodes were typically 1100–1500 A cm^{-2} (with a diode length of 500 μm). The differential quantum efficiency was 30–40% and beam divergence in the direction perpendicular to the junction plane (at the half-power point) was 35–45°.

All diodes were assembled p-side down on Cu heat sinks using In solder. The broad area diode electrical and thermal resistances were typically 0·4 Ω and 10 °C W^{-1}, respectively.

3. Results

3.1. Zn diffusion

The effect of Zn diffusion on the diode life was determined from experiments in which deep-diffused, shallow-diffused and undiffused portions of the same wafer were compared. The total distance between the surface and the recombination region was 1·5–2 μm in the wafer selected. The p-type surface of the wafer was protected with SiO$_2$ (Dyment and D'Asaro 1967) except for 13 μm wide stripe windows. One section of the wafer was then diffused at 750 °C to a depth of approximately 1 μm using an excess of Zn and As in a sealed ampoule. The diffusion front as revealed by etching penetrated down to the thin 0·5–0·8 μm 'p-wall', and possibly even deeper, although this is not easily determined. In the case of the shallow-diffused section of the wafer, the diffusion depth was limited to about 0·1 μm; well away from the 'p-wall'. The final third of the wafer was not diffused. All sections of the wafer were similarly metallized and assembled for operation on copper heat sinks. The typical threshold current for 500 μm long narrow stripe diodes was 550–600 mA for all sections of the wafer.

A comparison between the operating lifetime of the deep-diffused and undiffused diodes illustrates the large diffusion-induced difference seen. Figure 2 shows the output of two diodes operated just below their lasing threshold where the spontaneous output correlates directly with the change in the internal quantum efficiency. The deep-diffused diode degrades very rapidly in the first 40 hours of operation, while the undiffused diode output remains constant.

Stable operation was also obtained in shallow-diffused diodes where the diffusion front is kept away from the recombination region. Figure 3 is illustrative of the small change in the power emission against current curves seen after 2000 h of continuous lasing of such a diode operated in a dry air atmosphere.

3.2. Junction definition

The effect of the diode edges on the operating life was determined by preparing (from the same wafer) diodes having a width of 100 μm, formed by conventional wire

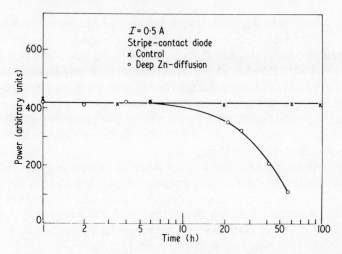

Figure 2. Power output as a function of time for two diodes fabricated from the same wafer but processed differently. One diode had a deep Zn diffusion (about $1\,\mu m$) from an excess Zn source prior to ohmic contact application, while the other diode was not diffused.

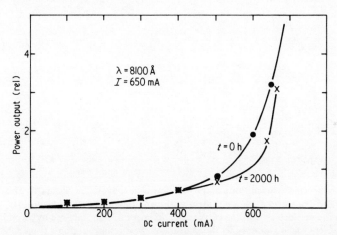

Figure 3. Power output against current of a CW laser diode with a shallow Zn diffusion (about $0\cdot1\,\mu m$; one not affecting the recombination region) before and after 2000 h of continuous operation.

sawing, and comparable area diodes constructed using a $100\,\mu m$ wide stripe contact obtained by SiO_2 isolation. The metallization and assembly procedures were similar for both, and a shallow Zn-diffused contact region was used in both structures. Both diode geometries yielded devices capable of CW operation at room temperature at current densities below $1600\ A\ cm^{-2}$. In the stripe contact diode, the walls of the structure were a distance of about $140\,\mu m$ from the edge of the stripe. An examination of the near-field emission pattern clearly showed that the current was restricted to the area of the metallic contact and hence well removed from the diode edges.

We consider first the results of the operating life test of the sawn-side diode which was operated DC at 1875 A cm^{-2}, somewhat above the CW threshold current density of 1550 A cm^{-2}. Figure 4(*a*) shows the curves of power output (one side of the laser) as a function of current initially and after 42 minutes of operation, during which time significant degradation in the characteristics occurred as reflected in the increased threshold current. The operating current was then increased to 0·85 A (2125 A cm^{-2}) in order to continue operating in the lasing mode. As shown in figure 4(*a*), a further rightward displacement of the curve occurred indicating further degradation at the end of only 17 additional minutes of operation.

To further demonstrate that degradation of the diodes occurred due to internal effects and not as a result of changes in the thermal or electrical resistance of the diodes, which would increase the CW threshold current, measurements were also made at each stage of degradation using pulse excitation (10^{-4} duty cycle). The pulsed power–current curves shown in figure 4(*b*) at the various stages, corresponding to the CW data of figure 4(*a*), indicate similarly increasing threshold current.

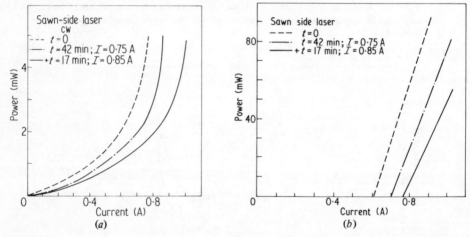

Figure 4. Power output against current for a broad area, sawn-side, diode operating CW. The diode area is 4×10^{-4} cm^2. The diode degradation is evident as shown, both in the DC mode (*a*), and the pulsed mode (*b*), by the displacement of the curves after an initial 42 min followed by an additional 17 min at a higher current made necessary by the increase in the threshold current. The operating current density was about 1875 A cm^{-2}.

The fact that internal laser damage occurred was confirmed by two other observations. First, the diode near-field emission became nonuniform as shown in figure 5. This effect is characteristic of gradual degradation (Kressel and Byer 1969) and is the result of the nonuniform nature of the degradation process. It is noteworthy that the diode emission drops off particularly strongly at the diode edges indicating the progression of defects into the diode for the sawn sidewalls.

Second, the power output, as measured under pulsed conditions, becomes unstable, showing spiking behaviour as shown in figure 6, whereas before degradation the output was uniform. This observation suggests internal changes in the lasing behaviour due to the introduction of defective regions, although a conclusive model is still lacking.

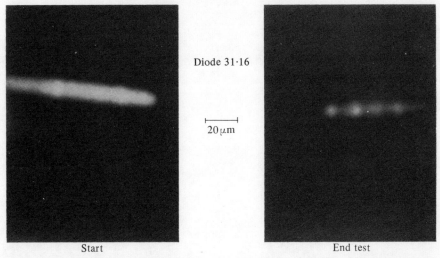

Diode 31·16

|—————|
20 μm

Start   End test

Figure 5. Near-field emission patterns of a sawn-side laser diode initially and after gradual degradation showing the decrease in the emission uniformity, particularly at the diode edges (near the sawn sidewalls).

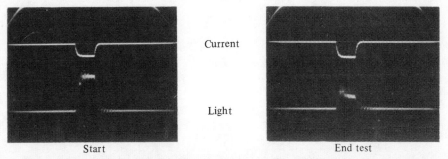

Current

Light

Start   End test

Figure 6. Comparison of the pulsed current and light output waveform before and after degradation of a sawn-side laser diode. Spiking in the light pulse is evident after degradation. Vertical scale 2 A/division, horizontal scale 500 ns/division.

Life test results obtained with a diode identical to the above except for the use of the 100 μm wide stripe contact configuration produced altogether different results. Figure 7 shows the cw power—current curve of such a diode at the start of the test and after 1200 h of operation at a current of 0·8 A (1300 A cm^{-2}). Very little change in the diode characteristics occurred during this period of operation, in remarkable contrast to the degradation of the diodes described above where the junction was defined by sawing. It should be noted that the difference in life between the two types of diodes cannot be attributed to the somewhat higher operating current density of the sawn-side diode (1875 as against 1300 A cm^{-2}) since other laser tests conducted at lower current densities gave essentially similar results (rapid degradation for the sawn-side units) while comparable material prepared using planar stripe configuration always exhibited lifetimes well in excess of 1000 h.

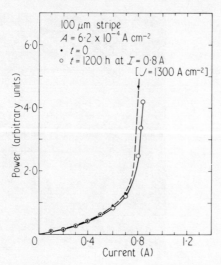

Figure 7. Power output plotted against current (DC) initially and after 1200 h of continuous laser operation at room temperature of a 100 μm-wide stripe contact diode made from the same material used for the sawn-side diode (figure 4). The diode was operated at 1300 A cm⁻². Very little degradation is observed in comparison to the data shown in figure 4.

Finally, we note that the large differences in the diode life cannot be explained by significantly different junction temperatures, since the thermal and electrical resistances of devices made by the two processes (when adjusted for area differences) are comparable.

4. Discussion and conclusions

The experimental results described here have focused on two fabrication processes which can lead to orders of magnitude differences in cw laser diode lifetime. We have shown that Zn diffusion *per se* need not be detrimental to the laser life as long as it does not penetrate too close to the diode recombination region. The effect of excessive Zn diffusion can at this time be only surmised as leading to the initial formation of defects, including Zn interstitial atoms, vacancies or possibly Zn precipitates. In the course of diode operation, these defects contribute to the degradation by promoting the rapid formation of nonradiative recombination centres and impurity clusters. From the present results, it is evident that great care is needed in the use of Zn diffusion steps, particularly in structures where the diffusion is used as the isolation method to form stripe-geometry lasers (Yonezu *et al* 1973b) because these require the diffusion front close to the recombination region for maximum current confinement. A lack of precise control of the diffusion is likely to lead to erratic operating lifetimes.

We have also shown that the exposed sidewalls of laser diodes can be a cause of rapid degradation. In the example chosen, the sidewalls were formed by sawing with a wire saw, a method known to generate a highly damaged surface. The experimental results, including the fact that the near-field pattern emission intensity reduction during operation is most severe in the diode edge region, strongly suggest that defect generation or multiplication is initiated in these damaged surface regions. It is known that dislocation multiplication by a climb mechanism can occur which gives rise to localized regions containing a high density of nonradiative and absorbing centres (Petroff and Hartman 1973). The presence of dislocations originally introduced by sawing would provide sources which can grow in the course of operation.

Our results have important technological implications. They suggest the need for careful annealing in proton-bombarded lasers, where a damaged region is deliberately placed at the edge of the emitting area. Furthermore, great care is needed to minimize surface damage whenever exposed sidewalls are present, particularly if the width of the diode is narrow as is the case in the fabrication of stripe lasers by etching narrow mesa areas (Tsukada *et al* 1973).

Acknowledgments

We are indebted to D Marinelli, V Cannuli, D B Gilbert, J Alexander and M Harvey for technical assistance, to Dr H F Lockwood and Dr M Ettenberg for valuable discussions, and to Dr E Levin for the SEM of figure 1. This research was partly supported by the Office of Naval Research, Arlington, Virginia, and NASA, Langley Air Force Base, Hampton, Virginia, USA.

References

Dyment J C and D'Asaro L A 1967 *Appl. Phys. Lett.* **11** 292
Ettenberg M, Kressel H and Lockwood H F 1974 *Appl. Phys. Lett.* **25** 82
Ettenberg M, Lockwood H F, Wittke J and Kressel H 1973 *Technical Digest, IEEE Int. Electron Dev. Meeting, Washington DC,* p317
Keck D B, Mauer R D and Schultz P C 1973 *Appl. Phys. Lett.* **22** 307
Kressel H and Byer N E 1969 *Proc. IEEE* **57** 25
Kressel H and Mierop H P 1967 *J. Appl. Phys.* **38** 5419
Kressel H, Byer N E, Lockwood H F, Hawrylo F Z, Nelson H, Abrahams M S and McFarlane S H 1970 *Metall. Trans.* **1** 635
Kressel H, Butler J K, Hawrylo F Z, Lockwood H F and Ettenberg M 1971 *RCA Rev.* **32** 393
Kressel H and Lockwood H F 1974 *J. Phys. (Paris)* Suppl No. 4 **35** C3–223
Lockwood H F and Ettenberg M 1972 *J. Crystal Growth* **15** 81
Petroff P and Hartman R L 1973 *Appl. Phys. Lett.* **23** 469
Tsukada T, Ito R, Nakashima H and Nakada O 1973 *J. IEEE Quantum Electron.* **QE-9** 356
Yonezu H, Kobayashi K, Minemura K and Sakuma I 1973a *Technical Digest, IEEE Int. Electron Dev. Meeting, Washington DC* p324
Yonezu H, Sakuma I, Kobayashi K, Kamejima T, Urno M and Nannichi Y 1973b *Japan. J. Appl. Phys.* **12** 1585

Devices based on electroabsorption effects in reverse-biased GaAs–GaAlAs double heterostructures

J C Dyment, F P Kapron and A J SpringThorpe
Bell-Northern Research, Ottawa, Ontario, Canada

Abstract. In this paper, three novel optical devices are discussed: (i) an integrated emitter—modulator in which an LED and an adjacent junction modulator share a common optically guiding layer but are electrically isolated from each other, (ii) a multiple-section photodetector in which each section responds only to selectively absorbed optical wavelengths, and (iii) a variable bandpass optical filter which incorporates the EA effect with a multilayer dielectric coating.

The performance of the three devices is evaluated with particular consideration of the dependence of wavelength discrimination on absorption edge steepness.

1. Introduction

Under the influence of an electric field, the optical absorption edge in a semi-conductor can shift to lower photon energies. This phenomenon, known as the Franz—Keldysh effect (Franz 1958, Keldysh 1958), causes a large electroabsorption (EA) of those optical frequencies just below the zero-field bandgap, particularly in reverse-biased double heterostructure (DH) p–n junctions where electric field strengths exceeding 10^5 V cm^{-1} are attainable. Recent papers (Reinhart 1973, Wilson and Reinhart 1974) have demonstrated the use of the EA mechanism to make light intensity modulators with extinction ratios up to 20 dB.

In this paper, we first present data for individual DH modulator structures which demonstrate how the EA effect can be optimized with respect to material parameters. We then discuss three novel optical devices:

(i) an integrated emitter—modulator in which the LED emitter and adjacent modulator share a common optically guiding layer but are electrically isolated from each other,
(ii) a multiple-section photodetector in which each section responds only to selectively absorbed optical wavelengths, and
(iii) a variable bandpass optical filter which incorporates the EA effect with a special multilayer dielectric coating.

The performance of each device is determined by the strong EA effects which occur in DH waveguides.

2. Characterization of individual modulator structures

Figure 1 (inset) shows the standard DH modulator structure which was characterized. Four LPE layers were grown using a conventional carbon slider system with a source—

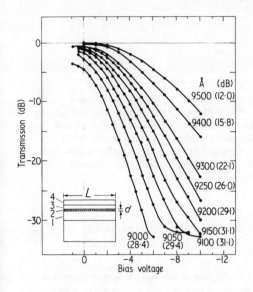

Figure 1. Typical data from a DH modulator (see inset) of length $L = 0.58$ mm showing light transmission against bias voltage for several wavelengths. Layers 1 and 3 are, respectively, n-type and p-type $Ga_{1-x}Al_xAs$ ($x = 0.35$) and layer 4 is p-type GaAs. The higher index guiding layer 2 has thickness d and is either p-type or n-type $Ga_{1-y}Al_yAs$ (with y between 0 and 0.10). Bracketed numbers after each wavelength are extinction ratios. $N_A = 1.5 \times 10^{17} \, cm^{-3}$, $d = 0.7 \, \mu m$.

seed technique (Miller and Casey 1973). The important parameters for the EA optimization are the carrier concentration (N_A or N_D) and thickness d of the guiding layer 2, and, to a lesser extent, the carrier concentration and Al contents in the lower index layers 1 and 3 above and below layer 2. Light from a monochromator, 30 Å wide, was coupled into the waveguide and the light output was measured after passage through a length L of the waveguide. Coupling was accomplished by using a specially designed modulator package which permitted high power objective lenses to be positioned close to both the input and output cleaved facets. Phase sensitive detection was employed in conjunction with a photomultiplier and slit combination. Typical data, for a device with a low doping level in layer 2, are plotted in figure 1.

The extinction ratio (usually expressed in dB) is defined as the ratio of light output, at any reverse bias V, relative to the output at $V = 0$. As the centre wavelength of the light packet is decreased, the extinction ratio first increases to some maximum value (31.1 dB at $\lambda = 9100$ Å and 9150 Å for the figure 1 device) and then decreases again for shorter wavelengths. These decreases are due to the higher insertion losses very near the bandgap which reduce the initial amplitude at $V = 0$ (see figure 1). The maximum value of the extinction ratio, normalized to $L = 0.5$ mm, can be used as a figure of merit to compare different modulator structures. Figure 2 is a plot of this figure of merit for waveguide carrier concentrations from $10^{16} \, cm^{-3}$ to greater than $10^{18} \, cm^{-3}$. Under the length normalization, the 31.1 dB from figure 1 has been reduced to 26.7 dB.

A rapid fall-off in modulator performance is observed for guiding layer carrier concentrations not less than $3 \times 10^{17} \, cm^{-3}$. This occurs primarily because the depletion region width, across which the high electric field exists, does not always extend across the entire waveguiding thickness. For example, we calculate (Sze 1969) that in going from no bias to a reverse bias of -10 V for a doping level of $10^{17} \, cm^{-3}$, the depletion width grows from 0.13 to $0.37 \, \mu m$; the corresponding figures for $10^{18} \, cm^{-3}$ are 0.04 to $0.11 \mu m$. Clearly, for the larger carrier concentration, a smaller fraction of light guided in a submicron layer is coupled to the high field region (Wilson and Reinhart 1974,

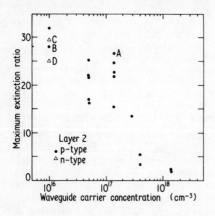

Figure 2. Maximum extinction ratio as a function of carrier concentration in guiding layer 2, with data normalized to $L = 0.5$ mm. The point labelled A corresponds to the modulator of figure 1. Three other modulators B, C, D are indicated for future reference in figure 4.

Dyment *et al* 1974) and hence smaller extinction ratios are expected. Below 3×10^{17} cm^{-3}, ratios often exceeding 25 dB/0·5 mm were measured and the best modulation typically occurred at N_A or $N_D \simeq 10^{16}$ cm^{-3} and $d \leqslant 1\mu$m. This improvement may not continue indefinitely since at very low doping levels the maximum junction electric field falls (Sze 1969) though the depletion width stays fixed at a value approximately equal to d. We further point out that although p-type guiding layers comprised the majority of data, some n-type layers were studied with equally good results.

The performance of each of the types of devices to be discussed depends strongly on the details of the absorption edge shift under applied voltage. To be better able to discuss wavelength selection properties, we have replotted the data of modulator A in figure 1 to show transmission versus wavelength in figure 3. As the reverse voltage is

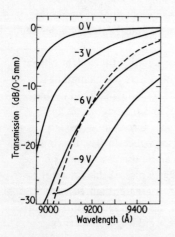

Figure 3. The data of figure 1 have been replotted for selected reverse biases to obtain light transmission against wavelength (normalized to $L = 0.5$ mm). The dashed line corresponds to -3 V but with the length increased to $L = 1.3$ mm in order that the transmission equals the -6 V ($L = 0.5$ mm) value of -15 dB at $\lambda = 9170$ Å.

increased the transmission edge becomes less steep and also shifts to longer wavelengths. The rates of shift are shown in figure 4 for the modulators A, B, C, D identified earlier in figure 2. In general, modulators with the lowest guiding layer carrier concentrations (about 10^{16} cm^{-3}) showed more even shifts at low transmission levels than did more heavily doped guiding layers. This is a desirable feature for the EA devices we will now discuss.

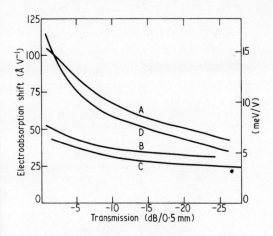

Figure 4. Rate of EA shifts as a function of the transmission level, obtained from data similar to figure 3. Modulators A, B, C, D from figure 2 are analysed.

3. Integrated emitter–modulator

An example of this device is shown schematically in figure 5(*a*) for the case of an LED emitter with adjacent modulator sections juxtaposed to two opposite faces. In side view it can be seen that these three sections share a common optical cavity but are electrically isolated from one another. The electrical isolation is achieved by using selective etches to remove successively layer 4 and layer 3. Etching stops at layer 2 which must be n-type (about 10^{16} cm^{-3}) in this design in order that the emitter and modulator junctions are electrically isolated and can be oppositely biased. The etching procedure is periodically monitored with probes and is terminated when the breakdown voltage between any two sections exceeds approximately 25 V.

The light spectrum, after passing through one of the modulator sections, is shown in figure 5(*b*). Note that as the reverse voltage is applied, the intensity decreases rapidly (by 20 dB in some devices) and the peak position shifts to longer wavelengths. This

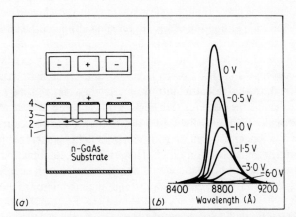

Figure 5. (*a*) Top and side views of an integrated emitter–modulator design. Positive bias (+) indicates the single LED emitter and negative bias (−) indicates the two modulator sections. Layers are as described in figure 1 with layer 2 being n-type. (*b*) Typical spectra taken after light passes through one of the modulator arms biased at the indicated values. Guiding layer 2 has $N_D \simeq 10^{16}$ cm^{-3} and overall length 1·1 mm.

latter observation is explained by the fact that the absorption edge moves to longer wavelengths when the reverse bias voltage is increased (as in figure 3). These points are more fully brought out in figure 6 where we show experimental spectra obtained when two different LED sources are coupled through modulator A (of figure 1). In the case of LED A (which is selected from the same wafer as modulator A), the transmission edge is already well into the tail of the spectrum at $V = 0$. The light emerging

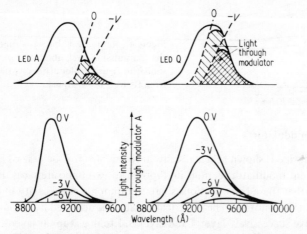

Figure 6. Upper portion: schematic illustrating the relative positions of transmission edge of modulator A at biases of 0 and $-V$ within the light spectrum of LED A (from the same wafer as modulator A) and of LED Q (which is from a different wafer) chosen so that a large portion of the light passes through the modulator at $V = 0$. Single and double cross-hatching represent light transmitted through A biased at voltages of 0 and $-V$, respectively. Lower portion: actual experimental data for modulator A of figure 1 at $V = 0, -3, -6, -9$ V.

from the modulator (shaded region) is of small amplitude which further decreases rapidly with reverse bias. Thus high extinction ratios are possible although the insertion loss is very high since only a small portion of the total light is transmitted. A much lower insertion loss was measured for the second source (LED Q from a different wafer) which emitted at longer wavelengths so that a much larger portion of the light passed through modulator A at $V = 0$. In this case, the output spectra are wider and the extinction ratios are lower than in the previous case because the absorption edge shift is not large enough to absorb all of the LED light. In order to reduce the insertion loss in the device of figure 5, a longer wavelength LED emitter might be achieved by selectively diffusing Zn into the emitter region only†. The resultant recombination from band-to-acceptor rather than band-to-band would decrease the photon energy of the LED output so that a greater fraction would pass through the modulator section.

4. Multiple-section photodetector with selective wavelength absorption

A four-section photodetector device is shown schematically in figure 7(a) and in the photograph of figure 8. The crystal structure and fabrication techniques are similar to

† Such devices are currently being evaluated.

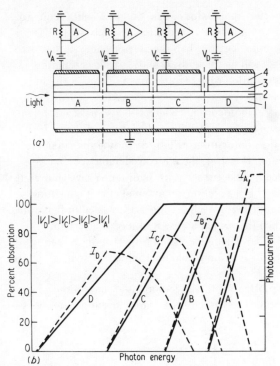

(a)

(b)

Figure 7. (a) Side view of a four-section photodetector with layers as described in figure 5. The various sections are biased successively more negative so that $|V_A| < |V_B| < |V_C| < |V_D|$. In each section, a resistor R and amplifier A could be arranged to ensure circuit response only to a unique range of wavelengths (see discussion of figure 9). (b) Schematic representation of the absorption edge shifts in sections A, B, C, D (solid lines which assume decreasing slopes) and corresponding calculated photocurrents (broken curves). Shifts in peak photocurrent response to lower photon energies can be seen.

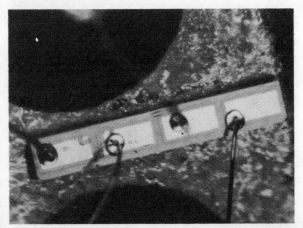

Figure 8. Photograph of a four-section photodetector device (as in figure 7(a)) mounted on a header. Overall device length is 1·5 mm. All sections share a common n-type optical waveguide with $N_D \simeq 10^{16}$ cm^{-3}.

those previously described for the emitter–modulator design. Suppose light is coupled into the waveguide of the detector from the left and encounters sections A, B, C, D with each section biased successively more negative than the previous one. As discussed in figures 3 and 4, the absorption edges will be successively shifted to lower photon energies, resulting in the composite absorption edge picture drawn schematically in figure 7(b). Each section will then selectively absorb light energies which fall within a narrow band. This, in turn, will generate photocurrents which will also have a different spectral response in each section as seen in the dashed line calculations of figure 7(b). The assumption has been made that the photocurrent will be directly proportional to the light absorbed in a particular section. However, there is considerable spectral overlap of the photocurrents generated in adjacent sections. In order to have an overall response to a unique range of wavelengths in a section, an external circuit consisting of resistor R and amplifier A could be used with each section as illustrated in figure 7(a). The amplifier would respond only when the voltage across R exceeded a certain threshold level. The amplifier threshold level could be varied from section to section and its magnitude plus the photocurrent spectrum would determine the circuit response in a given section.

In figure 9, experimental data are plotted for three sections and four sections of a photodetector. The amplitude and spectral response of the photocurrent depends very

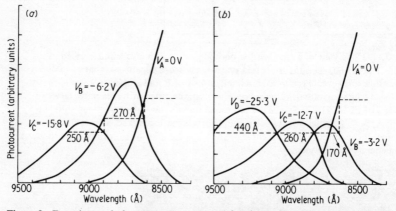

Figure 9. Experimental photocurrents measured for the device of figure 8 in which three sections are biased (a) or four sections are biased (b) at the indicated voltages. If the external circuits of figure 7(a) were set to respond to photocurrents *above* the dashed horizontal lines, then each section would respond to a unique set of wavelengths with the indicated bandwidths.

critically on the bias voltages of the previous sections. Suppose that the external amplifiers A of figure 7(a) were set to respond to voltages produced when photocurrents greater than the horizontal dashed lines flowed through R. Then a unique set of wavelengths would be detected in each external circuit with the indicated bandwidths varying from 170 to 440 Å for this device.

5. Variable bandpass optical filter

This device has not been experimentally evaluated, but figure 10 illustrates its construction. The guiding layer 2 can be either p-type or n-type in contrast to the two

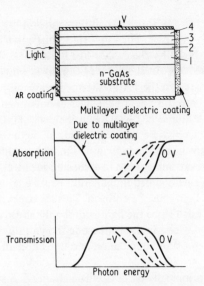

Figure 10. Schematic of a variable bandpass optical filter with layers as in figure 1 and layer 2 either p-type or n-type. The multilayer dielectric coating determines the filter cut-off at lower photon energies while the EA effect provides a 'variable' cut-off at higher energies. An optional anti-reflection coating can be used at the input cleaved facet.

previous devices where n-type was necessary. The filter has an optional anti-reflection coating at the input end to reduce insertion losses and a multilayer dielectric coating at the output end to define the absorption edge. A bandpass filter is formed with the high energy side being variable and determined by the EA effect while the low energy side is fixed and is determined by the properties of the dielectric coating. Such a filter can be used to select a narrow band of optical frequencies from a broadband light source or it can be used to reject unwanted optical signals which fall outside of the bandpass. The variable feature can be extremely useful where improved resolution is desired without having to physically change filters. In addition, its small size and potential use in integrated optics are important considerations.

6. Discussion

The amount of absorption edge shift with voltage is an important consideration for all three devices discussed here. In many devices, reverse voltages in the range of -10 to -20 V produce shifts of 700 Å. This will then be the amount of possible bandpass variation in the filter and, in the case of the emitter–modulator, this shift will be sufficient to eliminate most of the LED light and produce high extinction ratios. For the photodetector, absorption edge spacings of 200 to 300 Å between adjacent sections are possible with incremental voltages of -3 to -6 V.

The 10 to 90% absorption edge bandwidth for a 0·5 mm length ranges from 150 to 350 Å at zero bias; it increases along with the applied reverse bias. For longer lengths, however, the bandwidth will be smaller for the same bias. This has important implications for the resolving power of the detector in figure 7. For example, at -3 V in figure 3, a 0·5 mm length of modulator A has a −10 to −20 dB transmission bandwidth of 95 Å compared to about 170 Å for the −6 V case. The −6 V (0·5 mm) curve could be replaced by a −3 V (1·3 mm) curve as shown by the dashed line in figure 3. The two curves have identical transmission at λ = 9170 Å but the −3 V (1·3 mm) curve yields a smaller bandwidth of about 130 Å compared to the 170 Å obtained previously.

Detectors in which the light is coupled into the sides of the devices have been reported by others (Stillman *et al* 1974, Mathur *et al* 1970, Krumpholz and Maslowski 1971). Since the photons are absorbed in the high-field region, nearly all the created electron—hole pairs will be collected as photocurrent which results in high quantum efficiencies and fast response speeds. These advantages can be expected in the detector of figure 7(*a*). A potential disadvantage of side-coupled illumination may be the small effective cross section area into which the light must be coupled. This should not be a serious drawback in an integrated optics design since the light is probably already propagating in a narrow waveguide. Using devices similar to the one shown in figure 8, we have performed the functions of light emission, modulation and detection all in the same integrated chip.

Modulator response speeds are limited principally by device capacitance and series resistance. The large-area devices discussed in figures 5 and 7 have junction capacitances of 80—100 pF which can be reduced by a factor of 10 using a technique such as proton bombardment to decrease the junction area (Dyment *et al* 1972). This reduction should enable multiplexing rates in excess of 1 Gbit s^{-1} for these devices.

7. Conclusion

With reverse biases of typically −10 to −20 V, our best EA modulators exhibited 30 dB extinction ratios and 700 Å absorption edge shifts for 0·5 mm lengths. The corresponding guiding layer parameters (N_A or $N_D \simeq 10^{16} \mathrm{cm}^{-3}$, thickness $\simeq 1 \mu\mathrm{m}$) have been employed in the three devices discussed in this paper. The first device, an integrated emitter—modulator design, had extinction ratios approaching 20 dB, but the rather high modulator insertion loss would be reduced by using a longer wavelength emitter. The second device was a multiple-section photodetector which demonstrated that the wavelength dependence of photocurrent in any section was voltage controllable. Finally, a variable bandpass optical filter, with the long wavelength edge fixed by a multilayer coating and the other edge variable by the EA effect, has been proposed.

Acknowledgments

The authors thank G M Caskennette, H Nentwich and P M Garel-Jones for technical assistance and R R Fergusson for helpful discussions concerning the manuscript.

References

Dyment J C, D'Asaro L A, North J C, Miller B I and Ripper J E 1972 *Proc. IEEE* **60** 726
Dyment J C, Kapron F P and SpringThorpe A J 1974 *Paper presented at the Device Research Conference, Santa Barbara, California*
Franz W 1958 *Z. Naturf.* **13** 484

Keldysh L V 1958 *Zh. Eksp. Teor. Fiz.* **34** 1138 (Translation 1958 *Sov. Phys.—JETP* **7** 788)
Krumpholz O and Maslowski S 1971 *Wiss. Ber. AEG—Telefunken* **44** 73
Mathur D P, McIntyre R J and Webb P P 1970 *Appl. Optics* **9** 1842
Miller B I and Casey H C 1973 *4th Int. Symp. Gallium Arsenide and Related Compounds* (London and Bristol: Institute of Physics) p231
Reinhart F K 1973 *Appl. Phys. Lett.* **22** 372
Stillman G E, Wolfe C M and Melngailis I 1974 *Appl. Phys. Lett.* **25** 36
Sze S M 1969 *Physics of Semiconductor Devices* (New York: Wiley) pp84—90
Wilson L O and Reinhart F K 1974 *J. Appl. Phys.* **45** 2219

GaAs electroabsorption avalanche photodiode detectors

G E Stillman, C M Wolfe, J A Rossi and J L Ryan

Lincoln Laboratory, Massachusetts Institute of Technology, Lexington, Massachusetts, USA

Abstract. Schottky barrier avalanche photodiodes have been fabricated on n-type high-purity epitaxial GaAs. These devices have their largest response at wavelengths beyond the usual absorption edge for high-purity materials. The absorption mechanism involves the Franz–Keldysh shift of the absorption edge, and the higher response at the longer wavelengths can be explained by a much higher ionization coefficient for holes than for electrons. The results indicate that the ratio of β_p to α_n is even larger than previous measurements have given.

1. Introduction

There has been considerable recent interest in sensitive, wide-bandwidth avalanche photodiodes for the detection of GaAs and AlGaAs room temperature laser emission in fibre-optic communication systems and integrated optical circuits. However, the photo-response of previously reported GaAs avalanche photodiodes (Lindley *et al* 1969) decreases rapidly for wavelengths longer than about $0.86\,\mu m$, so these detectors are not very suitable for detection of room temperature GaAs and low Al-content AlGaAs laser emission with $\lambda \geqslant 0.88\,\mu m$. This is because the detector response depends on band-to-band absorption, whereas the laser emission usually involves a band-to-acceptor transition. Thus, the radiation is emitted at a wavelength significantly longer than the long wavelength threshold of the usual GaAs detector.

Previous results on GaAs (Lindley *et al* 1969) and $In_x Ga_{1-x} As$ (Stillman *et al* 1974a) avalanche photodiodes, which are compatible with integrated optical circuits (Stillman *et al* 1974d) have indicated that wide bandwidth, high gain and low noise performance can be obtained at any wavelength in the 0.80 to $1.06\,\mu m$ range by adjusting the alloy composition to the appropriate value ($0 \leqslant x \leqslant 0.20$). In this paper, however, we describe a new mode of operation of GaAs avalanche photodiodes. The difference between these devices and those previously described is that they have been fabricated on high-purity epitaxial GaAs. This new mode of operation results in a detector response which is nearly ideal for the GaAs and low Al-content AlGaAs room temperature laser and thus avoids the complication of fabricating detectors from $In_x Ga_{1-x} As$ for this application. The absorption beyond the usual band edge in this mode of operation is due to the Franz–Keldysh effect (Franz 1958, Keldysh 1958), and the optimum response for these longer wavelengths is caused by the higher ionization coefficient of holes relative to that of electrons in GaAs (Stillman *et al* 1974b). The results obtained with these electro-absorption avalanche photodiode (EAP) detectors indicate that the difference in the electron and hole ionization coefficients is even larger than previous measurements indicate.

2. Experimental results

The experimental results to be discussed have been obtained on devices fabricated using essentially the same procedure described in the first work on GaAs Schottky barrier avalanche photodiodes (Lindley *et al* 1969). For completeness, the device structure and fabrication procedure will be briefly reviewed here and then the experimental results will be presented and discussed.

2.1. Device structure and fabrication

A cross section of the device structure is shown in figure 1. The device is fabricated on an n-type layer of high-purity GaAs with the desired doping level which has been grown epitaxially on an n⁺ GaAs substrate. An ohmic contact is formed on the

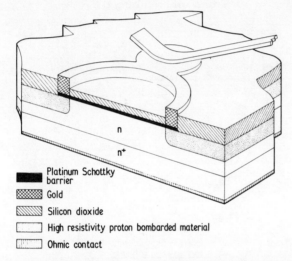

n

n⁺

■ Platinum Schottky barrier

▨ Gold

▨ Silicon dioxide

▯ High resistivity proton bombarded material

▯ Ohmic contact

Figure 1. Pictorial cross section of Schottky barrier avalanche photodiode structure.

n⁺-substrate and the semitransparent (about 100 Å thick) Pt Schottky barrier contact is electroplated on the upper surface. The high resistivity ring shown by the shaded area is formed by first masking the 125 μm diameter active area of the device with a thick layer of Au or photoresist and then irradiating the upper surface of the sample with high energy (300 keV to 3 MeV) protons. The proton bombardment creates a layer of high resistivity GaAs, the thickness of which is determined by the bombardment energy (Foyt *et al* 1969). This guard ring prevents edge breakdown at the perimeter of the planar device. A gold ring and bonding pad are plated around the perimeter of the device for contacting. The semi-transparent Pt Schottky barrier is protected by a previously deposited thin layer of pyrolytic SiO₂. For optical and electrical evaluation the devices are mounted in varactor diode packages.

2.2. Spectral response

In the first GaAs Schottky barrier avalanche photodiodes (Lindley *et al* 1969), it was observed that there was a slight increase in the response at longer wavelengths as the reverse bias voltage was increased from lower to higher values, and this increase was attributed mainly to the Franz–Keldysh shift of the absorption edge. The effect observed was quite small, however, and the photoresponse decreased monotonically for wavelengths longer than about 0·86 μm. By contrast, in the present work, the spectral response of the avalanche photodiodes fabricated in the same way but on high-purity material has a much sharper long wavelength cut-off at low reverse bias voltages, and at high reverse bias voltages, instead of a monotonic decrease in response for wavelengths longer than 0·86 μm, there is a dominant peak in the spectral response in the 0·88–0·91 μm wavelength range.

The variation of the spectral responsivity with bias voltage of an EAP detector fabricated on epitaxial material with $N_D - N_A = 5 \times 10^{14}$ cm^{-3} is shown in figure 2(a)

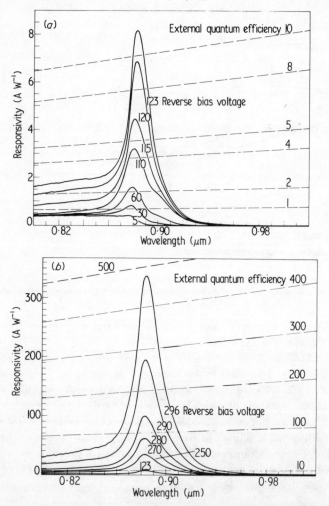

Figure 2. Experimental variation of spectral response with reverse bias voltage for a GaAs Schottky barrier EAP detector for (a) low voltages and (b) high voltages. The epitaxial material was about 20 μm thick and had a net donor concentration of about 5×10^{14} cm^{-3}.

for low reverse bias voltages and in figure 2(*b*) for higher reverse bias voltages. The doping concentration for this particular device was not uniform, but the 20 μm epitaxial layer had a nearly uniform carrier concentration of about $5 \times 10^{14}\,\mathrm{cm}^{-3}$ for the first 15 μm and a high resistivity region approximately 5 μm thick between the uniformly doped layer and the n$^+$ substrate. For this doping profile, the depletion region reaches through to the substrate at about 200 V and the breakdown voltage is 300 V. However, the effect observed is not dependent on the high resistivity region or the reach-through structure. Similar results have been obtained from other devices which were fabricated on uniformly doped epitaxial layers and in which the electric field did not punch through to the substrate before breakdown occurred. Recently, similar results have also been obtained on abrupt p$^+$–n junctions.

The spectral responsivity curve for 5 V reverse bias in figure 2(*a*) shows that the quantum efficiency is essentially constant for wavelengths shorter than 0·87 μm. The long wavelength cut-off starts at a wavelength slightly shorter than 0·88 μm and is quite sharp. As the reverse bias voltage is increased, the long wavelength cut-off shifts to longer wavelengths and a peak develops in the spectral response. The external quantum efficiency at 0·88 μm actually exceeds unity for a bias voltage as low as 30 V, and, because of the reflection losses, this indicates that significant avalanche gain is occurring for this wavelength even at this low voltage. As the reverse bias is increased still further, there is an increase in the gain at all wavelengths, but the increase is larger for the longer wavelengths. Figure 2(*b*) shows the spectral responsivity curves at still higher reverse bias voltages. The peak external quantum efficiency of about 480 for a reverse bias voltage of 296 V corresponds to an avalanche gain of 480 if the quantum efficiency without gain is unity or to a somewhat higher value if the internal quantum efficiency is less than unity. Similar results have been obtained on many different devices fabricated on other high-purity material with $N_D - N_A$ in the 10^{14}–$10^{15}\,\mathrm{cm}^{-3}$ range. However, in devices on some material, the small shoulder which is visible at about 0·90 μm in the spectral response curves of figure 2(*a*) is more pronounced so that the peak response occurs at wavelengths as long as 0·91 μm (Stillman *et al* 1974c). The absorption which is responsible for this shoulder or peak probably involves electroabsorption and some impurity or defect centre. This shoulder is also present in samples fabricated on more heavily doped material.

3. Electroabsorption

The increasing response at the longer wavelengths results from two effects. The absorption for wavelengths longer than the usual band edge involves the Franz–Keldysh effect on band-to-band, exciton, or impurity absorption, or a combination of all three. In this section we will show how the long wavelength cut-off of the GaAs detectors on high-purity material can be shifted significantly to longer wavelengths by just the Franz–Keldysh effect on the band-to-band absorption, and in the following section we will discuss the reasons for the development of the dominant peak in the spectral response.

3.1. Franz–Keldysh effect

The absorption coefficient of a direct bandgap semiconductor in a uniform electric field E, for radiation of wavelength λ, can be expressed in MKS units as (Callaway 1963, 1964; see also Tharmalingham 1963)

$$\alpha(\lambda, E) = \frac{2^{7/3} \mu^{4/3} e^{7/3} |\&.P_{nn'}| E^{1/3}}{\hbar^{8/3} \omega m^2 n \epsilon_0 c} \int_\beta^\infty |Ai(z)|^2 dz. \tag{1}$$

In this equation

$$\beta = \frac{(2\mu)^{1/3}(E_g - \hbar\omega)}{(\hbar e E)^{2/3}},$$

μ is the electron–hole reduced mass, e the electronic charge, $\&$ the radiation vector, $P_{nn'}$ the usual zero field interband matrix element, $\hbar$ is Planck's constant divided by 2π, ω the frequency of the incident radiation of wavelength λ, m the free electron mass, n the index of refraction, ϵ_0 the free space dielectric constant, c the speed of light, and E_g the energy gap at zero field. The Airy functions, $Ai(z)$, are defined by

$$Ai(z) = \frac{1}{\pi} \int_0^\infty \cos\left(sz + \frac{s^3}{3}\right) ds. \tag{2}$$

The interband matrix element can be estimated using either the f-sum rule (Bardeen *et al* 1956) or the $\mathbf{k.p}$ band model (Houghton and Smith 1966). Using the f-sum rule, the interband matrix element can be written as

$$|P_{nn'}|^2 = \tfrac{1}{2} m\hbar\omega f_{nn'} \tag{3}$$

and the sum rule of the f is

$$\Sigma f_{nn'} = 1 + \frac{m}{m_v} \tag{4}$$

where the sum is over direct transitions to all higher bands. Using equations (3) and (4) and $f \simeq 1 + m/m_v$, the Franz–Keldysh absorption coefficient for wavelengths close to the absorption edge can be written as

$$\alpha(\lambda, E) = 1 \cdot 0 \times 10^4 \frac{f}{n} \left(\frac{2\mu}{m}\right)^{4/3} E^{1/3} \int_\beta^\infty |Ai(z)|^2 dz \tag{5}$$

where

$$\beta = 1 \cdot 1 \times 10^5 (E_g - \hbar\omega) \left(\frac{2\mu}{m}\right)^{1/3} E^{-2/3},$$

for α in cm^{-1}, E in V cm^{-1}, and $(E_g - \hbar\omega)$ in eV. In the calculations for GaAs that follow, the parameters that were used were $n = $ constant $= 3 \cdot 63$, $m_v/m = 0 \cdot 087$ (since the light hole band gives the major contribution), and $\mu/m = 0 \cdot 0377$, and the integral of the square of the Airy function was evaluated in terms of the Airy function and its derivative (Callaway 1964) as

$$\int_\beta^\infty |Ai(z)|^2 dz = \left[\left|\left(\frac{dAi(z)}{dz}\right)_\beta\right|^2 + \beta |Ai(\beta)|^2\right]. \tag{6}$$

3.2. Calculated device quantum efficiency

Using equation (5) we can express the absorption coefficient at position x in the depletion region of a GaAs Schottky barrier or p^+–n junction diode as $\alpha(\lambda, x)$ where the variation of electric field with position x, $E(x)$, is determined by the doping profile and applied bias voltage of the device considered. Then the generation rate of electron–hole pairs at position x in the depletion region for a given incident flux of photons of wavelength λ, $\phi(\lambda)$, and sample reflectivity R, can be written as

$$G(\lambda, x) = \phi(\lambda)(1-R)g(\lambda, x) = \phi(\lambda)(1-R)\alpha(\lambda, x)\exp\left[-\int_0^x \alpha(\lambda, x')\,dx'\right] \tag{7}$$

with the assumption that each absorbed photon creates one electron–hole pair. The internal quantum efficiency of the detector for a given wavelength and applied voltage V, without avalanche gain, is then given by

$$\eta_0(\lambda, V) = \frac{\int_0^W G(\lambda, x)\,dx}{\phi(\lambda)(1-R)} = \int_0^W g(\lambda, x)\,dx = \int_0^W \alpha(\lambda, x)\exp\left[-\int_0^x \alpha(\lambda, x')\,dx'\,dx\right] \tag{8}$$

where W is the depletion width corresponding to the applied voltage V.

Equation (8) has been evaluated numerically, using a bandgap energy of $1\cdot42$ eV (Sell and Casey 1974) and the results for two different GaAs Schottky barrier devices with uniform doping profiles are shown in figure 3. The solid curves were calculated for a device on material with $N_D - N_A = 5 \times 10^{14}$ cm^{-3}, and the dashed curves were calculated for a device on material with $N_D - N_A = 2 \times 10^{16}$ cm^{-3}. For both doping levels the

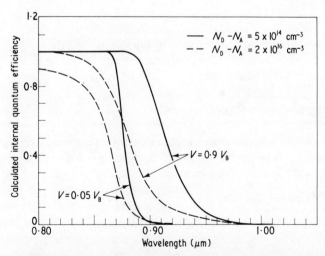

Figure 3. Calculated internal quantum efficiency for two Schottky barrier diodes with $N_D - N_A = 5 \times 10^{14}$ and 2×10^{16} cm^{-3}, respectively. The calculations were done at applied voltages of $0\cdot05$ and $0\cdot9$ of the corresponding breakdown voltages for the case of no multiplication or avalanche gain, but including the Franz–Keldysh electroabsorption.

internal quantum efficiency was evaluated at two different reverse bias voltages: 0·05 and 0·90 of the corresponding breakdown voltage. There is a significant increase in the quantum efficiency of the more heavily doped device at the shorter wavelengths between the two bias levels, and the increase is due to the widening of the depletion region as the bias voltage is increased and the resulting increased collection efficiency for the carriers generated at these wavelengths. At both bias levels, the long wavelength cut-off is more gradual for the device on $2 \times 10^{16} \, \mathrm{cm}^{-3}$ material, and this is due to the higher fields and larger Franz–Keldysh shift of the absorption edge in this material. However, the quantum efficiency is still low because for this doping level the depletion width even at avalanche breakdown is less than $2 \, \mu\mathrm{m}$. The gradual long wavelength cut-off is similar to that observed experimentally on devices of this doping level (Lindley et al 1969). In contrast, at low bias voltages the device on high-purity material has a higher quantum efficiency at the shorter wavelengths and a much sharper long wavelength cut-off due respectively to the wider depletion region and lower electric fields in this device. At high reverse bias voltages, the Franz–Keldysh effect becomes important and, along with the wider depletion region (about $30 \, \mu\mathrm{m}$ at the breakdown voltage), results in a significant extension of the long wavelength cut-off for this device.

4. Avalanche gain

The results presented in the previous section show that the Franz–Keldysh effect in GaAs Schottky barrier detectors on high-purity material can result in an increase of the quantum efficiency to nearly unity at wavelengths which are beyond the long wavelength cut-off at low bias voltages. However, any increase above unit quantum efficiency must be due to avalanche multiplication.

4.1. Quantum efficiency with gain

For a Schottky barrier or p^+–n detector with maximum electric field at $x = 0$ and zero electric field at $x = W$, the electron current increases with increasing x. The differential equation for the electron current $J_n(x)$ can be written as

$$\frac{\mathrm{d}J_n(x)}{\mathrm{d}x} = \alpha_n(x)\,J_n(x) + \beta_p(x)\,J_p(x) + qG(x) \tag{9}$$

where α_n and β_p are the electron and hole ionization coefficients, respectively, $J_p(x)$ is the hole current and $G(x)$ is the generation rate of electron–hole pairs in the depletion region. The total current J is a constant so that

$$J = J_p(x) + J_n(x) = \text{constant}, \tag{10}$$

and the differential equation for the hole current in the same device is

$$\frac{\mathrm{d}J_p(x)}{\mathrm{d}x} = -\alpha_n(x)\,J_n(x) - \beta_p(x)\,J_p(x) - qG(x). \tag{11}$$

Considering only the space charge generated photocurrent, these equations can be solved for the total current with the boundary conditions $J_n(0)$ and $J_p(W) = 0$, and the internal device quantum efficiency with gain can be written as

$$\eta_1(\lambda, V) = \frac{\int_0^W g(\lambda, x) \exp\left[-\int_0^x (\alpha_n - \beta_p)\,dx'\right]\,dx}{\left\{1 - \int_0^W \alpha_n \exp\left[-\int_0^x (\alpha_n - \beta_p)\,dx'\right]\,dx\right\}}$$

$$= \frac{\exp\left[\int_0^W (\alpha_n - \beta_p)\,dx\right].\left\{\int_0^W g(\lambda, x) \exp\left[-\int_0^x (\alpha_n - \beta_p)\,dx'\right]\,dx\right\}}{\left\{1 - \int_0^W \beta_p \exp\left[\int_x^W (\alpha_n - \beta_p)\,dx'\right]\,dx\right\}}$$

(12)

where $g(\lambda, x)$ is defined by equation (7). The device multiplication for a given wavelength and applied voltage is given by $M_0(V) = \eta_1(\lambda, V)/\eta_0(\lambda, V)$ and it can be seen that if $\alpha_n = \beta_p$, $M_0(\lambda, V)$ is independent of wavelength.

Equation (12) can be evaluated numerically if the electron and hole ionization coefficients are known as a function of electric field and if the electric field variation in the device is known.

4.2. Ionization coefficients in GaAs

Until recently, the electron and hole ionization coefficients in GaAs have generally, although not universally, been considered to be equal, based on early photomultiplication measurements on complementary p—n junctions (Logan and Sze 1966). The ionization coefficient data used most frequently are probably those of Hall and Leck (1968) in which they assumed $\alpha_n = \beta_p$, although some workers (for example Salmer *et al* 1973) have used ionization coefficients determined from breakdown voltage measurements, again with the assumption that $\alpha_n = \beta_p$. Recent measurements on GaAs Schottky barrier devices which were fabricated so that electron and hole multiplication could be measured separately in the same device (Stillman *et al* 1974b) have shown conclusively that in fact β_p must be greater than α_n. However, because of various possibilities of contamination of the injected currents, the absolute magnitude and field variation of the ionization coefficients were not well established in that work.

The breakdown voltages and corresponding maximum electric fields calculated for Schottky barriers or abrupt p^+—n junctions for various sets of ionization coefficient data are shown by the broken curves in figure 4, along with some experimental breakdown voltage data for abrupt junctions from the literature. The calculated breakdown voltages were determined using equation (12). The calculated breakdown voltages determined from the data of Hall and Leck (1968) and Stillman *et al* (1974b) are considerably lower than those determined from the data of Logan and Sze (1966). In addition, it is clear that a linear extrapolation of the ionization coefficient data of Stillman *et al* (1974b) to higher fields results in calculated breakdown voltages that are much too low for doping concentrations higher than about $5 \times 10^{15}\,\text{cm}^{-3}$. All of these data give calculated breakdown voltages that are too high when compared to the few experimental

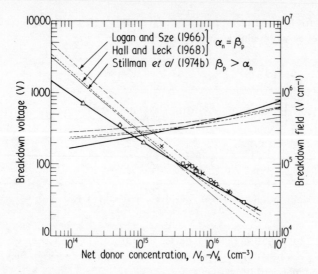

Figure 4. Breakdown voltage (labelled curves) and corresponding breakdown field for abrupt p⁺−n junctions and Schottky barrier diodes on n-type GaAs. The broken curves were calculated using the ionization coefficient data noted and the solid curves were calculated using ionization coefficients which were adjusted to obtain agreement with the breakdown voltages at the lower doping levels. The data points are experimental results from the literature: △ Weinstein and Mlavsky 1963, □ Kuno *et al* 1969, ○ Salmer *et al* 1973, × Miller and Casey 1972, ◇ present work.

breakdown voltage points available for carrier concentrations of about 10^{15} cm⁻³ and lower. This might not be too surprising, since the experimental data are old and the breakdown voltages may have been limited by microplasmas. The experimental point for $N_D - N_A = 5 \times 10^{14}$ cm⁻³, however, was obtained from a Schottky barrier device on material with a uniform doping concentration, verified by $C-V$ measurements, and the multiplication at voltages close to breakdown was quite uniform. The solid curve in figure 4 was obtained by independently adjusting the electron and hole ionization coefficients to give calculated breakdown voltages in agreement with the experimental breakdown voltage data over the lower range of doping concentrations. It was found that when ionization coefficients of the form $\alpha_n = a_n \exp - (b_n/E)^m$ and $\beta_p = a_p \exp - (b_p/E)^m$ were used, the best agreement was obtained with $m = 1$. That is, the curvature of the calculated curve was too great, resulting in even poorer agreement, when $m = 2$ was used. The solid curves in figure 4 were calculated using

$$\alpha_n = 2 \times 10^6 \exp - \left(\frac{2 \times 10^6}{E}\right)^1$$

and

$$\beta_p = 1 \times 10^5 \exp - \left(\frac{5 \times 10^5}{E}\right)^1.$$

The use of ionization coefficients of the form

$$\alpha_n = \alpha_0 \left(\frac{E}{E_0}\right)^m$$

as used by Kuno *et al* (1969) might give better agreement over the entire doping range.

4.3. Calculated results

Equation (12) has been evaluated numerically for several different values of α_n and β_p for a device on material with a net donor concentration of 5×10^{14} cm^{-3}, and the results are given in figure 5.

For figure 5(*a*), the values of the ionization coefficients obtained by Hall and Leck (1968)

$$\alpha_n = \beta_p = 2 \cdot 0 \times 10^5 \exp - \left(\frac{5 \cdot 5 \times 10^5}{E}\right)^2$$

were used, and, as expected for the case where $\alpha_n = \beta_p$, the quantum efficiency is independent of wavelength out to the long wavelength cut-off, and the multiplication is constant for all wavelengths. The results are shown for applied voltages of 0·90, 0·95,

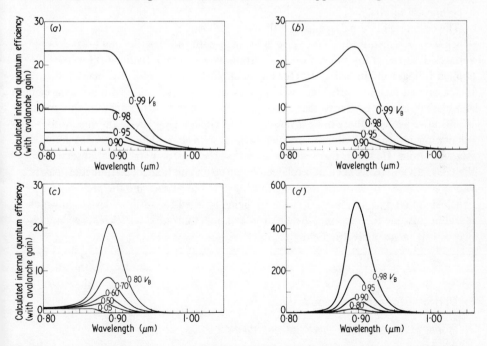

Figure 5. Calculated internal quantum efficiency with gain for a GaAs EAP detector on uniformly doped material with $N_D - N_A = 5 \times 10^{14}$ cm^{-3}, for three different sets of ionization coefficient data, as indicated in the figures. In each case the breakdown voltage determined for the ionization coefficients is shown, and the spectral quantum efficiency curves were calculated for the given fractions of the corresponding breakdown voltages. (*a*) Ionization coefficients of Hall and Leck (1969). (*b*) Ionization coefficient data of Stillman *et al* (1974b). (*c*) and (*d*) Ionization coefficients adjusted to obtain agreement with reported values of breakdown voltage on lightly doped material.

0·98 and 0·99 of the breakdown voltage of 496 V determined using this ionization coefficient data. The value of the multiplication or internal quantum efficiency with gain is quite small and shows no resemblance to the experimental results in figure 2.

Figure 5(*b*) shows the calculated internal quantum efficiency with gain obtained from equation (12) for the case for $\beta_p > \alpha_n$, with

$$\alpha_n = 1\cdot2 \times 10^7 \exp - \left(\frac{2\cdot3 \times 10^6}{E}\right)$$

and

$$\beta_p = 3\cdot6 \times 10^8 \exp - \left(\frac{2\cdot9 \times 10^6}{E}\right)$$

from the data of Stillman *et al* (1974b). The calculated breakdown voltage for this ionization coefficient data was 468 V, and the calculated curves are shown for the same fractions of the breakdown voltage as in figure 5(*a*). These curves show a peak beginning to develop at the absorption edge similar to that observed experimentally, but the multiplication values obtained are still much too low. The quantum efficiency at the peak is too low relative to that at shorter wavelengths when compared with the experimental results.

The calculated internal quantum efficiency with gain for the ionization coefficients which were arbitrarily adjusted to give agreement with the experimental breakdown voltages observed at low net donor concentrations is shown in figure 5(*c*) for low applied voltages and in figure 5(*d*) for higher applied bias voltages. The magnitude of the calculated quantum efficiency with gain as well as the variation with both wavelength and bias voltage are similar to that observed experimentally. Attempts to obtain more quantitative agreement are not warranted because of the accuracy of the absorption calculation and the lack of more detailed breakdown voltage data for low net donor concentrations. However, these results show that the experimental effect observed in the GaAs EAP detectors can be explained by electron and hole ionization coefficients which are consistent with the available breakdown voltage data and the previous conclusion that $\beta_p > \alpha_n$. Work is currently being pursued to fabricate devices in which it will be possible to determine the electron and hole ionization coefficients accurately over a wide range of electric fields.

5. Device performance and applications

5.1. *Speed of response and noise measurements*

The response time of the GaAs EAP detectors has been studied (see Stillman *et al* 1974c) using tunable pulsed room temperature GaAs lasers operating in a grating controlled external cavity (Rossi *et al* 1973). In the wavelength range between about 0·88 and 0·92 μm the measured rise time was less than 1 ns. Since the electroabsorption process is electronic in nature, it is expected to be very fast and should impose no limit on the response times which can be obtained with the EAP detectors. The main limitation to the response time of these detectors will probably result from transit time effects because of the weaker absorption for this process as compared to the usual band-to-band absorption.

Preliminary results of noise measurements with incident radiation of 0·633, 0·800, and 0·880 μm, in which the variation of the noise current with multiplication is essentially the same for all three wavelengths, indicate that the electrons are contributing very little to the multiplication process. Although the problem is complicated by the variation of multiplication with position in the space charge region, it is hoped that a careful analysis of noise measurements can provide an estimate of the effective value of the ratio of the electron and hole ionization coefficients in GaAs.

5.2. Integrated waveguide photodetectors and modulators

Although all of the above discussion has been concerned with discrete EAP detectors, one of the main applications of this device may be as integrated waveguide photodetectors and modulators. High-purity GaAs and AlGaAs waveguides are being studied in many laboratories for use in near-infrared integrated optics, and EAP detectors can be easily incorporated in these waveguides. With zero bias voltage on a detector in such a waveguide, the radiation would pass through the detector with no absorption, but with a high reverse bias the light would be absorbed and detected with avalanche gain. For the power levels that will be used in integrated optical circuits, the Franz–Keldysh absorption mechanism can provide a fast and efficient modulator. The EAP device integrated in a waveguide should perform even better than the discrete devices. In this configuration the light will propagate parallel to the Schottky barrier contact or p–n junction and thus the effective absorption length can be much longer than the depletion layer width and waveguide thickness. Because of the intensity distribution of the light in the waveguide, the generation of electron hole pairs should be more uniform and result in an even higher ratio of hole-to-electron injection and therefore lower noise than is obtained with the discrete device.

Acknowledgments

The authors would like to thank J M Lawless, L Krohn, and T A Lind for assisting with the device fabrication, B J Palm for performing the numerical calculations and C M Penchina for making his computer program for the calculation of the Airy functions available to us. We would also like to thank I Melngailis and J P Donnelly for several helpful discussions. This work was sponsored by the Defense Advanced Research Projects Agency and the Department of the Air Force.

References

Bardeen J, Blatt F J and Hall L H 1956 *Photoconductivity Conference* ed R G Breckenridge, B R Russell and E E Hahn (New York: Wiley) p146
Callaway J 1963 *Phys. Rev.* **130** 549
—— 1964 *Phys. Rev.* **134** A998
Foyt A G, Lindley W T, Wolfe C M and Donnelly J P 1969 *Solid St. Electron.* **12** 209

Franz W 1958 *Z. Naturf.* **13A** 494
Hall R and Leck J H 1968 *Int. J. Electron.* **25** 529
Houghton J and Smith S D 1966 *Infrared Physics* (London: Oxford University Press) p131
Keldysh L V 1958 *Sov. Phys.–JETP* **34** 788
Kuno H J, Collard J R and Gobat A R 1969 *Appl. Phys. Lett.* **14** 343
Logan R A and Sze S M 1966 *J. Phys. Soc. Japan* **21** Suppl 434
Lindley W T, Phelan R J, Wolfe C M and Foyt A G 1969 *Appl. Phys. Lett.* **14** 197
Miller B I and Casey H C 1972 *Proc. 4th Int. Symp. Gallium Arsenide and Related Compounds*
    (London and Bristol: Institute of Physics) p231
Rossi J A, Chinn S R and Heckscher H 1973 *Appl. Phys. Lett.* **23** 25
Salmer G, Pribetich J, Farrayre A and Kramer A 1973 *J. Appl. Phys.* **44** 314
Sell D D and Casey H C 1974 *J. Appl. Phys.* **45** 800
Stillman G E, Wolfe C M, Foyt A G and Lindley W T 1974a *Appl. Phys. Lett.* **24** 8
Stillman G E, Wolfe C M, Rossi J A and Foyt A G 1974b *Appl. Phys. Lett.* **24** 471
Stillman G E, Wolfe C M, Rossi J A and Ryan J L 1974c *Proc. Int. Symp. Optical Acoustical
    Microelectronics* (Polytechnic Institute of Brooklyn)
Stillman G E, Wolfe C M and Melngailis I 1974d *Appl. Phys. Lett.* **25** 36
Tharmalingham K 1963 *Phys. Rev.* **130** 2204
Weinstein M and Mlavsky A I 1963 *Appl. Phys. Lett.* **2** 97

Phase diagram considerations of GaAs hetero-solar cells

D Huber and W Gramann

AEG-Telefunken, Semiconductor Division, Heilbronn, Germany

Abstract. Advanced III–V compound devices require quite often the incorporation of heterostructures. Typically more than one parameter (eg, lattice constant and bandgap of the heterolayer) has to be adjusted to optimize the device properties. From the thermodynamic point of view this means that quaternary mixed III–V compound materials have to be employed. In the paper a simplified method of calculating lattice constants as well as bandgap energies of systems such as GaAlAsP is demonstrated. The data enable one to determine the growth conditions for materials with a given lattice constant and bandgap. Furthermore, experimental results on the fabrication of GaAs hetero-solar cells are presented. In particular the technology for fabricating large area, highly efficient cells with good ohmic contacts is discussed. Solar cells have been realized with areas of several square centimetres and efficiencies of up to 15%. Phase diagram considerations of the Au–AlAs systems lead to the proper alloying conditions for ohmic contacts. Alternatively, an etched contact configuration has been devised.

1. Introduction

GaAs is a solar cell material which has some advantages over Si:

(i) the bandgap is a better match to the solar spectrum;

(ii) the larger bandgap of GaAs produces a higher output voltage;

(iii) the decrease of power output with increasing temperature for GaAs is about half of that for Si cells.

The development of the optoelectronic market has lowered the price of GaAs bulk materials considerably. Today, GaAs costs only a factor two more than high-quality Si for efficient solar cells. Price, therefore, is no longer a determining factor. The main disadvantage of GaAs solar cells so far has been their low efficiency (of the order of 10%). The low efficiency is caused on one hand by the steep absorption edge of GaAs and on the other hand by the high surface recombination velocity. Most of the photon-induced carriers are created within one micrometre of the illuminated surface which requires a junction depth of the same size. Most of the generated carriers recombine via surface recombination centres and are lost for energy conversion.

Very high efficiencies have been obtained (Woodall and Hovel 1972) with solar cells incorporating a GaAs p–n junction and a thin upper layer of p-type $Ga_{1-x}Al_xAs$ grown by liquid phase epitaxy. The close lattice match between the two materials gives a low recombination velocity at the interface. Efficiencies of up to 20% at air mass two (AM2) have been realized with such a structure.

In the first part of the paper results are reported of a large area cell structure featuring a wide bandgap AlAs window material. In the second part ways of technological realization of an improved cell are discussed explicitly.

2. Large area AlAs–GaAs solar cells

The AlAs window layer was grown from a Zn doped melt consisting typically of 10 g Ga, 5 g Al, 14 g GaAs and 1·5 g Zn. The growth temperature was about 950 °C. The melt composition has been calculated (Huber 1971) to give a solid composition of 100% AlAs. Surface photovoltage measurements gave a bandgap energy value of 2·16 eV.

The AlAs layer was doped up to 5×10^{18} cm^3 and was 2 micrometres thick. A junction depth of about 1 micrometre due to the diffusion of Zn into the GaAs was obtained. Ohmic contacts to the front side were fabricated by an evaporated three-layer structure consisting of Au: Zn–Ni–Au. The alloying temperature has been estimated (Huber 1973)'by an ideal solution model to be about 900 °C. The actual alloying temperature was somewhat lower, about 750–800 °C. Figure 1 shows two typical AlAs–GaAs cells with areas of about 1·5 cm^2. The collection efficiency of such a cell is shown in figure 2 and compared to the collection efficiency of a homojunction GaAs cell (Tsaur *et al* 1972). The efficiency at shorter wavelengths is considerably improved

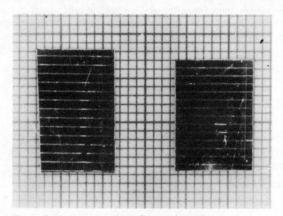

Figure 1. Large area AlAs–GaAs solar cells.

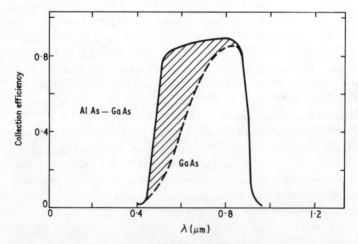

Figure 2. Collection efficiency of an AlAs–GaAs solar cell.

compared to the GaAs cell and drops to zero at a wavelength which corresponds to the bandgap of AlAs. In table 1 state-of-the-art data are gathered for different solar cell technologies: the violet cells of COMSAT, standard silicon solar cells in mass production and two examples of high efficient GaAs solar cells. In comparing the IBM data one has to consider that the area of the IBM cell was only of the order of mm^2.

Table 1. Experimental solar cell data

Material	I_{SC} (mA cm^{-2})	V_{OC} (mV)	FF	η (%)	Measurement conditions	Manufacturer
Si	40	590	0·8	14	AMO$_{spectr}$	COMSAT 1973
Si	39·5	605	0·78	13·2	AMO$_{spectr}$	AEG-TFK 1974
Ga$_{0.3}$Al$_{0.7}$As –GaAs	≤23	≤980	≤0·81	13·4 / 20·3	AMO$_{calc}$ / AM2$_{spectr}$	IBM
AlAs–GaAs	28	975	0·8	15·3	AMO$_{calc}$	AEG-TFK

In table 1 I_{SC} is the short-circuit current, V_{OC} is the open-circuit voltage, FF is the filling factor and η is the collection efficiency.

To improve the efficiency of AlAs–GaAs solar cells several problems must be solved. One has to mention that the chemical stability of AlAs is not the most difficult problem. The stability depends strongly on the liquid phase process. A high Zn concentration (greater than 5 g in the melt) results in deteriorating layers. More often microscopic droplets from the melt, which remain on the epitaxial layer, lead to a fast decomposition of the window material. These droplets consist of the very reactive Ga–Al mixture, which destroys the surface. The most difficult problem is the formation of good ohmic contacts to the top of the solar cell. The very high alloying temperature often leads to a melt-through of the p–n junction. This is a serious problem in making large area GaAs solar cells. To reduce the series resistance and the contact resistance of the cell, the window material is commonly doped with concentrations up to 10^{19} cm^{-3}. Fabrication of the cell by liquid phase epitaxy, however, implies that the diffusion depth is coupled to the Zn concentration in the AlAs layer. A low series resistance is always combined with a short minority lifetime in the highly doped p-GaAs layer which leads to a low collection efficiency. This mechanism makes it impossible to optimize the collection efficiency and the open circuit voltage at the same time. In the next paragraph, ways are evaluated to improve solar cell structures.

3. Improved device structures

Figure 3 gives a survey of the different energy losses in solar cells. Considering AlAs solar cells the following factors can be improved:

(i) absorption in the window;
(ii) voltage factor;
(iii) collection efficiency;
(iv) series resistance.

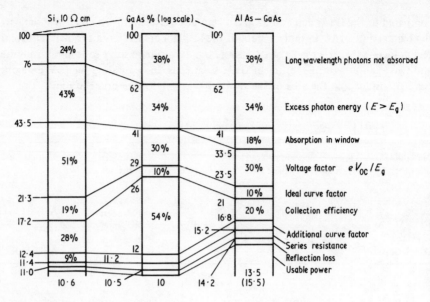

Figure 3. Energy losses in solar cells.

There are two ways to lower the absorption in the window material: (*a*) to make the window thinner, (*b*) to employ an indirect window material. There are some problems in growing a simple heterostructure with a layer thinner than 1 micrometre. The quaternary system $Ga_x Al_{1-x} As_y P_{1-y}$ can serve as a window material with an indirect band larger than the gap of AlAs. Ideally this material should have the same lattice constant as that of the GaAs substrate. The phase rule applied to quaternary III–V mixtures gives one more degree of freedom compared to ternary mixture, provided the number of phases remains unchanged. This degree of freedom is reflected in the fact that in the case of quaternary systems the energy gap may be varied while the lattice constant remains unchanged.

To obtain an optimal heterostructure – ideal lattice match and small absorption loss – one needs to know the dependence of the lattice constant *a* and of the energy gap E_g on the mole fractions *x*, *y* of the quaternary mixture $Ga_x Al_{1-x} As_y P_{1-y}$.

The quaternary system behaves as a strictly regular solution (Ilegems and Panish 1974). Thermodynamic considerations lead to an expression for the lattice constant (Huber 1974) given by $a(x, y) = 5 \cdot 4625 - 0 \cdot 013x + 0 \cdot 1986y - 0 \cdot 0062xy$ (Å).

Based on similar assumptions one can derive the dependence of the energy gap on the mole fraction of a quaternary solution, namely

$$E_{gi}(x, y) = 2 \cdot 5 - 0 \cdot 19xy - 0 \cdot 24x - 0 \cdot 37y \text{ (eV)}.$$

In the calculations the following data of the binary components have been used:

a (GaAs): $5 \cdot 64191$ Å a (AlP): $5 \cdot 4625$ Å

a (GaP): $5 \cdot 4495$ Å a (AlAs): $5 \cdot 6611$ Å for the lattice constants and

E_{gi} (GaAs): 1·75 eV $E_{gi}^{'}$ (AlP): 2·5 eV

E_{gi} (GaP): 2·26 eV E_{gi} (AlAs): 2·15 eV.

This thermodynamic approach to predicting the lattice constants and the energy gaps of III–V mixtures is in excellent agreement with experimental results.

Figure 4 shows some iso-lattice constant lines and iso-energy gap lines of the quaternary GaAlAsP system. The dashed line gives the zero lattice mismatch between a quaternary layer and the GaAs substrate. A mixture of $Ga_{0.65}Al_{0.35}As_{0.98}P_{0.02}$ has an ideal match to the GaAs substrate, which is in good agreement with results on strain-free, DH lasers (Roszgoni and Panish 1973) with low degradation rates. A mixture of 10% AlP and 90% AlP has the highest energy gap among the ideally matching solid solutions.

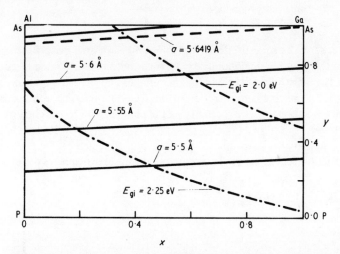

Figure 4. Iso-lattice constant lines and iso-energy gap lines of the quaternary $Ga_xAl_{1-x}As_yP_{1-y}$ system.

A rough estimate of the absorption loss in the ternary AlAsP using published data (Hovel and Woodall 1973, Lorenz *et al* 1970) indicates that a 2 μm thick layer of 20% AlP and 80% AlAs absorbs essentially no light of the solar spectrum, an ideally matched layer only about 1%. Therefore, an ideal GaAs hetero-solar cell structure consists of a 1 micrometre thick $AlP_{0.10}As_{0.90}$ window on the diffused GaAs substrate.

The melt composition of the quaternary liquid Al–Ga–Al–As solution yielding the desired solid solution can be estimated using published results (Ilegems and Panish 1974). A rough estimate yields a phosphorus concentration of about 10^{-4} mol%. Layers grown in such a manner have been examined by x-ray topography and have a perfect match to the substrate.

Considering possible improvements of other parameters mentioned above a new contact structure for AlAsP solar cells is proposed in figure 5. The AlAsP window layer is covered by a 700 Å thick layer of Si_3N_4.

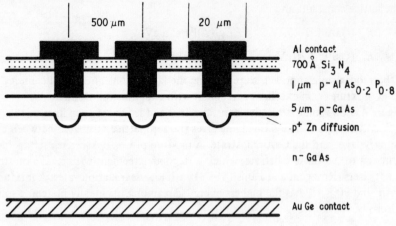

Figure 5. New contact structure for AlAsP solar cells.

This Si_3N_4 layer serves as an antireflection coating and as an etch mask for the AlAsP window. In this way ohmic contacts can be made with Al evaporated directly onto the p-GaAs. The resistivity of the window material has no further influence on the series resistance and on the zinc concentration of the p-GaAs layer. The series resistance may be minimized by optimizing the geometry of the contact structure etched by photoresist techniques.

In conclusion a new solar cell structure has been proposed which can be realized employing known III–V technologies. This structure promises efficiencies of up to 20%, since there is no further absorption loss, and because of the possibility of optimizing the proper contact geometries and concentrations.

Acknowledgments

This work was supported by the Gesellschaft für Weltraumforschung of the Federal Republic of Germany.

References

Hovel H J and Woodall J M 1973 *J. Electrochem. Soc.* **120** 1247
Huber D 1971 *Proc. Int. Conf. Semiconductor Heterojunctions* **1** 195
—— 1973 *Paper presented at 3rd ESSDERC Munich*
—— 1974 to be published
Ilegems and Panish M 1974
Lorenz M R, Chicotka R and Petit G D 1970 *Solid St. Commun.* **8** 693
Rozgonyi G A and Panish M B 1973 *Appl. Phys. Lett.* **23** 533
Tsaur S C, Milnes A G and Feucht D L 1972 *Proc. 4th Int. Symp. Gallium Arsenide and Related Compounds* (London and Bristol: Institute of Physics) p205
Woodall J M and Hovel H J 1972 *Appl. Phys. Lett.* **21** 379

The incorporation of residual impurities in vapour grown GaAs

D J Ashen, P J Dean, D T J Hurle, J B Mullin, A Royle and A M White

Royal Radar Establishment, Malvern, Worcestershire, England

Abstract. A study has been made of the residual impurities and the growth parameters which control their incorporation in VPE GaAs layers. New evidence has been obtained on the role of silicon contamination by comparing layers grown using a BN-lined apparatus with layers grown using an all-silica apparatus. A theoretical model has been derived and used to check the arsenic trichloride pressure dependence of Si and Zn incorporation. Very good agreement can be obtained in predicting the well known molar fraction effect. The presence of Zn acceptors in layers has been established by mass spectrographic and photoluminescence techniques and the presence of at least another acceptor has been inferred from the constancy of the compensation ratio (acceptor to donor ratio). Photoluminescence evidence for an oxygen donor in layers has been noted.

1. Introduction

A knowledge of the identity and behaviour of impurities in GaAs is a vital requirement in the preparation of device-quality material. In this paper we present new evidence both direct and indirect on residual impurities and consider their role in controlling the electrical properties of VPE layers. Although like some previous workers we contrast our results specifically with those obtained in other laboratories it must be considered implicit in such comparisons that the residual impurities found in layers prepared in different laboratories may not always be the same — we are after all considering impurity concentrations in the range 1 part in 10^8–1 part in 10^6. Variations may surely arise as a result of fluctuations in the composition of starting materials let alone the preparation conditions used to grow the layers.

Ideally the problem could perhaps be resolved if one could obtain unequivocal analytical information on the chemical composition of the epitaxial layers and correlate this information unambiguously with other physical properties such as carrier concentration, mobility, trap density, etc. Unfortunately the analytical side is an exceptionally difficult one and fraught with problems some of which will be mentioned later (§2.2). Also, the interpretation of Hall data and photoluminescence spectra although now well developed still fails to give directly unambiguous *quantitative* information on *specific* impurities. One is forced to study the physical phenomena themselves and rely on the interpretation of carefully-devised correlation experiments.

In this paper we have obtained considerable insight on the residual impurity problem from an examination of two specific phenomena; one related to molar fraction effects and the other to compensation behaviour. The former involves the free carrier concentration n and the partial pressure of $AsCl_3$, $p(AsCl_3)$ or its equivalent in related systems.

The molar fraction effect is the dramatic inverse dependence of n on $p(AsCl_3)$. The compensation behaviour can be described by the compensation ratio defined either by K the ratio of acceptors to donors (N_A/N_D) or by R, the ratio of acceptors plus donors to free carriers $(N_D + N_A)/n$. We will use K which is more convenient in chemical interpretations of the acceptor/donor balance. Both these phenomena have been the subject of a number of interesting and valuable studies by other authors. We will discuss first those experiments which relate most directly to our own work and then consider the other pertinent papers within the framework of the discussion.

The most comprehensive study of the molar fraction effect is that of DiLorenzo (1972) following earlier studies (Knight *et al* 1971, DiLorenzo and Moore 1971). The initial reports by Cairns and Fairman (1968, 1970) on the effect do not appear to have been published. DiLorenzo and his colleagues have cogently argued that the molar fraction effect can be explained by a Si contamination model where the activity of Si in the vapour $a(Si)$ and hence in the GaAs layer is controlled by the partial pressure of HCl, $p(HCl)$ through the formation of chlorosilanes $SiCl_{(4-x)}H_x$. This mechanism is in apparent conflict with the view (albeit speculative) of Barry (1971) that arsenic vacancies are involved and also of the work of Hasegawa and Saito (1968) and Furukawa (1967) where they report a basic limitation in the transport and incorporation of Si in GaAs layers. Recently, Wolfe *et al* (1974) report high resolution far infrared spectroscopic evidence for Si as a major residual impurity but additionally they note the presence of another unidentified but dominant donor impurity. This does not conflict with Si being primarily responsible for the molar fraction effect since their residual impurities were found under conditions where $p(HCl)$ would effect considerable suppression of the Si uptake in layers.

Silicon has also been implicated in its amphoteric role (Nakagawa and Ikoma 1971, Ikoma and Nakagawa 1972, Nakanisi and Kasiwagi 1974, DiLorenzo 1972) in explaining the relative constancy of the compensation ratio K for GaAs layers. This view is implicit in many other papers although Okamoto *et al* (1973) argue that the acceptor is not Si but a Ga vacancy.

The following study was therefore undertaken in an effort to assess quantitatively and qualitatively the function of Si and other impurities on the molar fraction effect and on the compensation parameter K. A feature of the study was the strategy of preventing Si contamination of the reacting media through the use of a BN system in order to delineate the role of Si by controlled doping experiments. Other residual impurities have been identified by mass spectrographic (MS) and photoluminescence (PL) techniques. Their role in compensation is discussed.

2. Experimental details

2.1. Preparation and characterization of layers

Epitaxial layers of GaAs were grown by the $AsCl_3:Ga:H_2$ system using fairly conventional procedures. The silica reactor tube, however, could be lined internally with a pyrolytic BN tube some 18 inches long and 1 inch in diameter. The Ga source boat and substrate holder together with the $AsCl_3/H_2$ mixture feed-in were also made of pyrolytic BN. Transport of epi-GaAs could thus be accomplished in a non-silicon containing environment. The $AsCl_3$ partial pressures were set by passing Ag/Pd-purified H_2 (at approximately 100 ml min^{-1}) through a thermostatically controlled bubbler unit. All the runs including the initial saturation of the Ga were carried out in an acid-etched and carefully dried reactor which was conditioned prior to each run with H_2 and $AsCl_3$. It was intended that this would limit the build-up of trace impurities in the reactor and set a constant background level. As a further precaution the upstream connectors and taps were protected against atmospheric contamination by a blanket stream of purified H_2. The epitaxial layers were deposited on Cr-doped semi-insulating substrates cut from LEC crystals. The substrates were orientated about 2° off the (100) face, chemically polished in Br_2/CH_3OH and etched in 3:1:1 ($H_2SO_4:H_2O_2:H_2O$). The source and substrate temperatures were respectively 800 and 750 °C during growth.

Carrier concentration and Hall mobility measurements were made in a field of 2 kG using the van der Pauw technique on clover-shaped specimens.

The PL measurements were made using apparatus and techniques reported elsewhere (Dean 1974, Ashen *et al* 1975). The bound exciton spectra were obtained by illuminating the specimen surface at liquid He temperatures with radiation from a 488 nm argon ion laser and recording the high-resolution spectra photographically. It was also considered essential for identification purposes to record both the pair spectra and the free-to-bound spectra at low levels of illumination and at various temperatures from liquid He up to about 20 K.

Direct mass spectrographic analysis of the epitaxial layers was carried out by J B Clegg at Mullard, Redhill, using the rotating disc electrode technique (Clegg *et al* 1970). Whilst the layer rotated a counter electrode was gradually moved radially in order to scan the whole surface. Standard spark source MS was carried out on the Ga samples by G W Blackmore at AML, Poole.

2.2. Chemical and photoluminescence evidence

The results of the MS and PL analyses are shown in table 1. The MS results on four epitaxial layers grown in the silica reactor show significant surface contamination in spite of very careful handling and cleaning. Further, only when about 5 μm has been removed from the surface are the results of the impurity content of the layer significant. This arises because each spark creates a 0·4–1 μm crater finely controlled by the narrow electrode gap (25–50 μm). Even so the surface removal cannot be supposed to be completely uniform.

Silicon is clearly present in the layers at concentrations (4–10 × 10^{15} atom/cm^3) sufficient to produce the donor concentrations found. Zinc is also present at the detection limit (3 × 10^{15} atom/cm^3) in one layer. The identification of Ca and K in the layers is probably associated with surface contamination and the difficulty of removing these elements completely by surface sparking. The presence of Cu (layer A) and Zn (layer B) may account for them being p-type. The presence of substantial boron in the Si-doped substrates is noteworthy. Also the Si content of a Ga source (after use) is greater than the Si content of the starting material or the Si content of a source run in BN.

These MS analyses support the findings of DiLorenzo (1971) and Kim *et al* (1973) who both used ion beam mass spectrometers.

Table 1. Summary of analytical results on epi—GaAs

Mass spectrographic ($\times 10^{14}$ atom/cm^3)

Layer	B	Al	Si	S	Cl	K	Cu	Zn	P	Ca	Cr	Layer
Surface A	100	100	400	100	≤40	80	40	40	≤40	≤80	≤40	A = SN14R45 35 μm p_{293} $4{\cdot}8\times10^{15}$ cm^{-3} μ_{293} 376 cm^2 V^{-1} s^{-1}
Surface B	100	100	<80	400	1200	80	200	200	30	200	20	B = SN14R46 28 μm p_{293} $1{\cdot}5\times10^{15}$ cm^{-3} μ_{293} 383 cm^2 V^{-1} s^{-1}
Surface C	20	100	400	800	800	300	300	100	≤40	300	≤40	C = S19R1A 18 μm n_{77} $1{\cdot}5\times10^{14}$ cm^{-3} μ_{77} 60 100 cm^2 V^{-1} s^{-1}
Surface D	10	200	1600	300	300	300	≤40	80	20	200	≤40	D = SN26R3A 37 μm n_{77} $1{\cdot}3\times10^{15}$ cm^{-3} μ_{77} 51 700 cm^2 V^{-1} s^{-1}
Substrate												Substrate
Layer A	80	100	100	≤40	≤40	80	80	≤20	≤40	≤80	≤40	A = A267/Si-doped/(100); 1×10^{18} carriers cm^{-3}
Layer B	<10	<10	≤80	≤40	<80	10	≤20	30	<10	30	20	B = A327/Cr-doped/(100); $1{\cdot}8\times10^{8}$ Ω cm
Layer C	<20	20	40	<80	100	20	≤40	≤40	≤40	80	≤40	C = A343/Cr-doped/(100); $3{\cdot}5\times10^{8}$ Ω cm
Layer D	<10	10	80	≤10	80	30	≤40	<30	<10	80	≤40	D = A365/Cr-doped/(100); $4{\cdot}0\times10^{8}$ Ω cm
Substrate A	16400	60	14400	4	<10	20	100	80	20	≤80	≤40	
Substrate B	40	<20	14400	80	<10	20	20	20	20	<10	2400	
Gallium 1	300	70	300	<100	<500	10	<7	<10	90	<200	<10	Gallium 1: Typical JMC Grade Al
Gallium 2	100	100	900	400	≤1000	20	<30	<20	10	<50	<20	Gallium 2: From SiO$_2$ Boat
Gallium 3	2000	100	300	500	≤300	100	<10	<10	20	30	<30	Gallium 3: From BN Boat

Photoluminescence

Layer type	C_A	Si_A	Ge_A	Sn_A	Zn_A	O_D
Vapour Epi†	§	§	=		=	=
Liquid Epi‡	=	=	=	¶	=	=

† Typical of 95% of 230 + layers.
‡ Typical of 95% of 60 + layers.
§ Seen under exceptional circumstances.
¶ Seen if dopant.
‖ Identified by excitons and F → B spectra.

A very detailed photoluminescence (PL) study of the identification of acceptors in epi-GaAs has been carried out by the authors (Ashen *et al* 1975). A brief summary of the results is given in table 1. The PL study was made on over 230 VPE and 60 LPE layers of which at least 30 layers were from other laboratories throughout the world. The results are significant in three respects: (i) the typical VPE layer contained Zn and no other acceptor within say about 1% of the Zn composition. This was true of all the S16, 17, 20, 21, 22 and 23 layers. Si acceptors were found in about four VPE layers from other laboratories and were introduced in RRE layers only under exceptional conditions; (ii) the LPE layers differed markedly from the VPE layers in the impurities found; they generally contained C, Si and a trace of Ge; (iii) the VPE layers often contained a donor line (White *et al* 1974) which is attributed to oxygen.

The significance of these findings particularly in relationship to compensation will be discussed in §4.

2.3. *Experiments in the BN apparatus*

A comparison was made between layers grown using undoped Ga sources (S16, 17, 20, 21, 22 and 23) and a Si-doped source (S23 R6+) using the BN system described. In each series a study was made of the relationship between the carrier concentration found in the epitaxial layer and the pressure of $AsCl_3$ used in the run. Six different sources were examined to produce the non-intentionally doped layers in order to obtain statistically reliable information. The experimental data are reproduced in table 2. The pressure relationship is examined in figure 1.

There are two distinctive features in the results of the non-intentionally doped layers:

(i) They are all p-type with only one exception – the very first run which has been omitted. Six p-type results have also been omitted because of difficulty in contacting and measurement.
(ii) There is no systematic relationship between lg (hole concentration h) and lg $p(AsCl_3)$. The S22 series of runs is somewhat suspect since the layer-substrate interfaces were poor (which was a feature of the particular crystal used for the substrates) and the hole concentration in these layers was fairly high.

The specific effect of the silica apparatus itself was also investigated in two further series of runs. These runs followed on from the two previous series just discussed which produced p-type layers. In the further series the BN liner was removed but the BN source boat and BN substrate holder were retained. The layers produced immediately went n-type (a single exception was caused by growth at a high partial pressure of $AsCl_3$; the layers reverted to n-type on subsequent growth at low pressures of $AsCl_3$). In a further series of runs, again following an established pattern of p-type layers, the effect of placing a slice of Si between the source and substrate whilst retaining the BN liner was examined. This series also produced heavy n-type doping.

One concludes from these experiments that:

(i) the silica apparatus is a major source of contaminating donors;
(ii) the BN system eliminates silica contamination.

The possibility that BN could give rise to p-type impurities cannot be ruled out. On the

Table 2. Electrical properties of non-intentionally doped layers grown in BN apparatus

Source	Run	$AsCl_3$ (atmosphere $\times 10^3$)	Thickness (μm)	$n \times 10^{14}$ cm^{-3}		μ (cm^2 V^{-1} s^{-1})	
				n_{293}	n_{77}	μ_{293}	μ_{77}
16	2	8·5	12	5·8		400	
	5	8·7	20	1·7	0·17	344	1500
17	1	9·9	22	6·2	3·0	364	7850
	2	9·4	18	0·59		422	
	3	8·6	19	0·14		356	
	4	7·5	22	0·12		291	
20	2	5·2	20	0·45		421	
	3	4·7	23	8·6		382	
	6	7·5	15	18		351	
	7	8·4	10	0·95		407	
21	1	4·3	17	0·21		404	
	2	4·4	32	1·3		368	
	3	4·3	32	1·8		446	
	6	4·2	5	0·25	0·12	405	2310
22	1	4·5	12	6·4	2·0	360	5280
	2	4·2	4	46	23	430	4160
	3	5·8	21	11	5·9	370	6400
	4	5·6	23	25		310	
	6	6·8	21	3·6	2·1	420	6920
	7	6·8	25	6·5	4·0	440	6820
	8	7·1	18	1·2		390	
23	2	5·0	5	5·3	2·4	365	4680
	3	5·3	5	1·5		292	
	5	5·0	8	0·48		403	

other hand the acceptor concentration ($\simeq h$) is no greater or less on average than the acceptor concentration found in n-type layers grown in silica (§2.4).

A detailed study of the PL spectra of the p-type layers again revealed the presence of Zn acceptors in all the layers. There was no evidence of Si acceptors.

The BN system was then used to grow layers from a Si-doped source — again one which had produced p-type layers when undoped. The dependence of n on $p(AsCl_3)$ was examined and the absence of a memory effect was noted by following a high $p(AsCl_3)$ run with a low one. Some traces of silica were noted on the source after the third run but these did not build up subsequently. The layers produced were all heavily n-type (table 3) and they also revealed a molar fraction effect which is discussed in the next section.

2.4. Experiments in silica apparatus

An initial study on the molar fraction effect was carried out in an all-silica apparatus in order to compare the results with published work and also with the results using a Si-doped source in the BN apparatus. The results of this experiment are reproduced in table 4. They are compared with related work in figure 2.

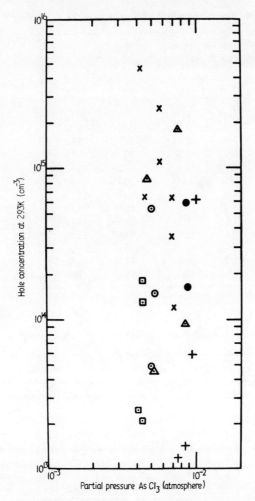

Figure 1. Hole concentration plotted as a function of AsCl₃ partial pressure for non-intentionally doped layers grown in the BN apparatus (S16, ●; S17, +; S20, △; S21, ▢; S22, ×; S23, ⊙).

The basis of the comparisons in figure 2 is the similarity in the slopes of the $\lg n(77\,\mathrm{K})-\lg p$ relationships: p represents $p(\mathrm{AsCl_3})$ in all the lines except that of Kennedy *et al* (1974) where it represents $p(\mathrm{GaCl})$. Only the slope of Aoki and Yamaguchi's line (1972) is dissimilar. In fact it may not be strictly comparable since the line is drawn from 'arbitrarily' selected data which vary due to memory effects at the three selected AsCl₃ pressures.

The displacement of the S23 Si-doped line upwards of the S13 line is a consequence of higher Si-doping. Kennedy *et al*'s (1974) result involves $p\,(\mathrm{GaCl})$ which is proportional to $p\,(\mathrm{HCl})$. A quantitative comparison of the pressure dependence obtained by different authors has been made by fitting the data to a relationship of the form $n = Cp^m$ and comparing values obtained for the index m. The individual donor and

Table 3. Electrical properties of Si-doped layers grown in the BN apparatus

Run	AsCl$_3$ (atmosphere × 10^3)	Thickness (μm)	Carrier parameter (× 10^{17} cm^{-3})		Mobility (cm^2 V^{-1} s^{-1})	
			n_{293}	n_{77}	μ_{293}	μ_{77}
6	4·9	28	12·4	12·4	2450	2440
7	4·9	21	3·8	3·7	3040	2800
8	4·9	27	5·0	5·1	2780	2680
9	6·8	27	2·7	2·6	3700	3650
10	6·8	26	3·6	3·5	3520	3380
11	6·8	28	11	11	2800	2770
12	9·2	26	2·0	1·9	4050	4210
13	9·2	22	3·3	3·1	3810	3830
14	9·2	27	2·5	2·3	3900	3830
15	13	24	1·1	0·96	4480	5300
16	13	30	0·88	0·74	4550	5450
17	13	27	1·2	1·1	4430	4500
18	4·5	27	37	38	1670	1760

Table 4. Electrical properties of non-intentionally doped layers grown in the silica apparatus

Run	AsCl$_3$ (atmosphere × 10^3)	Thickness (μm)	Carrier parameter (× 10^{14} cm^{-3})					Mobility (cm^2 V^{-1} s^{-1})	
			n_{293}	n_{77}	N_D	N_A	$N_D + N_A$	μ_{293}	μ_{77}
3	8·6	10	0·26	0·5	0·3	4·8	10	6900	73100
4	7·7	12	0·46	0·72	8·3	7·6	16	6080	54100
5	6·8	13	4·7	3·7	11	7·3	18	5020	57300
6	6·0	15	1·3	2·4	9·4	7·1	17	6360	59200
7†	8·6	37	8·3	7·1	20	13	33	6940	42900
8	6·8	25	3·4	3·2	10	7	17	7290	58000
9	6·8	16	5·1	2·5	16	13	29	5800	41000
10	6·0	22	21	7·4	21	13	34	5000	39200
11	5·3	20	27	10	23	13	36	5200	40900
12†	9·2	25	3·0	3·0	8·1	5·2	13	8100	71000
13†	9·8	52	0·94	0·91	5·2	4·3	9·6	8250	80000

† Upside down.

acceptor concentrations in the undoped layers were obtained from Wolfe *et al*'s (1970) data. It is implicitly assumed that the shallow donor and acceptor impurities are fully ionized and uninfluenced by deep levels.

The results of this analysis are shown in table 5. It enables a direct comparison to be made between the donor impurities in the undoped layers and the Si in the doped layers. In this comparison we believe that Si$_{As}$ is not a significant acceptor in our undoped layers. Correlations with $N_D + N_A$ are 'fortuitous' in that there is an apparently constant

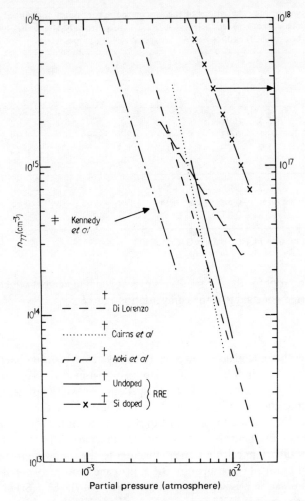

Figure 2. Free electron concentration plotted as a function of the AsCl$_3$† (or GaCl‡) partial pressure for (*a*) non-intentionally doped layers grown in the silica apparatus or (*b*) Si doped layers grown in the BN apparatus.

relationship between N_D and N_A. For each series the index $\bar{m}$ was obtained by calculating a least squares fit of y on x and x on y: from this we deduced the correlation coefficient r which is a measure of the quality of fit to a straight line and also the mean slope $\bar{m}$ which is the slope equally inclined to the regression lines.

Values of $\bar{m}$ for the plots of $n(77\,\text{K})$ have a spread of values from $-1\cdot7$ to $-6\cdot7$. The relatively large negative value for $\bar{m}$ in the S13 series could be taken as evidence for acceptor contamination, hence it would seem more consistent to examine the slope parameter for N_D which we believe is principally due to Si. The good agreement and correlation between $\bar{m}$ for N_D in the comparable undoped sources with $\bar{m}$ for $n(77\,\text{K})$ (which is equivalent to N_D where $N_D \gg N_A$) obtained from results using the S23 Si-doped source is entirely consistent with Si being the main impurity in the molar

Table 5. Comparison of the calculations of the average slope parameter $\bar{m}$ deduced from the results of various workers. $\bar{m}$ was calculated from regression analysis of $\lg n = \lg p^m$ where $p = p(\text{AsCl}_3)$ or $p(\text{GaCl})$ and $n = n_{77}, N_D, N_A$ or $N_D + N_A$ respectively

Impurity parameter	n_{77}		N_D		N_A		$N_D + N_A$		No. of data pts
	$\bar{m}$	r	$\bar{m}$	r	$\bar{m}$	r	$\bar{m}$	r	
Cairns and Fairman (1968)†	−5·4	−0·94							10
DiLorenzo (1971)† (100) only	−3·6	−0·92	−2·8	−0·99	−2·7	−0·99	−2·7	−0·99	8
Aoki and Yamaguchi (1972)†	−1·7	−0·99	−1·5	−1·00	−1·5	−0·95	−1·9	−0·78	3
Kennedy *et al* (1974)‡	−3·4	−0·97	−3·1	−0·99	−2·6	−0·88	−3·0	−0·99	8
S13 identical conditions†	−6·7	−0·91	−3·3	−0·91	−2·3	−0·78	−2·8	−0·80	8
S13 all results† RRE	−4·4	−0·58	−2·5	−0·65	−2·0	−0·63	−2·3	−0·65	11
S23 Si-doped in BN†	−2·8	−0·85							13

n, N_D etc proportional to $p(\text{AsCl}_3)$† or $p(\text{GaCl})$‡.

fraction effect. The values of $\bar{m}$ for the N_A results also show good agreement which is consistent with the reported constancy of the compensation ratio K.

3. Thermochemistry

A full treatment of the thermochemistry involved in this system has been undertaken but is too long to reproduce here in full. Instead a simplified treatment is given together with the essential results of full calculations. The principal objective of the present treatment has been to explicitly deduce the value of the coefficient $\bar{m}$ in the n–$p(\text{AsCl}_3)^m$ relationship for Si and Zn in an effort to throw light on the source and mechanism of contamination by these elements. It must be understood that the limitations and advantages of equilibrium calculations are inherent to the following treatment.

In a similar detailed and valuable treatment by DiLorenzo and Moore (1971) of the same system the Si contamination was considered as the activity of the Si, $a(\text{Si})$ in the deposition region as a function of the HCl concentration $p(\text{HCl})$. The form of $\lg a(\text{Si})$–$\lg p(\text{HCl})$ was a gentle 'parabola' peaking near a value of $p(\text{HCl})$ of 7×10^{-3}. However, in the range of interest of $p(\text{HCl}) = 3$–20×10^{-3} a linear approximation shows a slope greater than -2 but certainly indicative of the mechanism claimed. The significant changes reported for the enthalpy of SiClH_3 by Hunt and Sirtl (1970) from -48 ± 15 kcal to -34 kcal may be important in this connection. We now consider the fate of impurity contaminants during transport and during incorporation in the crystal lattice. We consider the latter process first.

3.1. Incorporation of Si in the crystal

Si can be incorporated on either or both the V(Ga) or V(As) sub-lattice sites. In a previous treatment (Ashen *et al* 1974) it was shown that the value of the ratio $[\text{Si}_{\text{Ga}}]/[\text{Si}_{\text{As}}]$ was $>10^3$ for typical VPE growth but about 0·25 for LPE growth (at about 750 °C). We therefore need to consider only the incorporation of Si on the V_{Ga} site. (An equivalent argument can be advanced for the other site.)

$$\mathrm{Si_V} + \mathrm{V_{Ga}} \overset{K_{sg}}{\rightleftharpoons} \mathrm{Si^+_{Ga}} + e^- \tag{1}$$

where K_{sg} is the equilibrium constant for the reaction. Under conditions of dilute doping (less than approximately 10^{17} atoms/cm^3) growth will take place under intrinsic conditions and $[e^-] = K_i^{1/2}$. Thus

$$[\mathrm{Si^+_{Ga}}] = K_{sg} K_i^{-1/2} [\mathrm{V_{Ga}}] p(\mathrm{Si}). \tag{2}$$

The arsenic vapour pressure controls the gallium vacancy concentration through the following reaction

$$\tfrac{1}{2} \mathrm{As_{2(V)}} \overset{K_{\mathrm{As_2V}}}{\rightleftharpoons} \mathrm{As_{As}} + \mathrm{V_{Ga}} \tag{3}$$

that is

$$K_{\mathrm{As_2V}} = [\mathrm{V_{Ga}}] p(\mathrm{As_2})^{-1/2} \tag{4}$$

since the activity of As atoms on As sites can be taken to be unity. The decomposition of the AsCl$_3$ to elemental arsenic can be assumed to be complete at the growth temperatures involved and the arsenic balance can be written

$$p^0(\mathrm{AsCl_3}) = 2p(\mathrm{As_2}) + 4p(\mathrm{As_4}) \tag{5}$$

where $p^0(\mathrm{AsCl_3})$ is the input pressure of AsCl$_3$. Evaluation of the equilibrium ratio between As$_2$ and As$_4$:

$$\mathrm{As_{4(V)}} \overset{K_{\mathrm{As}}}{\rightleftharpoons} 2\,\mathrm{As_{2(V)}} \tag{6}$$

with

$$K_{\mathrm{As}} = p(\mathrm{As_2})^2 p(\mathrm{As_4})^{-1} \tag{7}$$

shows that for the commonly used range of values of $p^0(\mathrm{AsCl_3})$, As$_4$ is the dominant species and from (5)

$$p(\mathrm{As_4}) \sim \tfrac{1}{4} p^0(\mathrm{AsCl_3})$$

hence from (7)

$$p(\mathrm{As_2}) = K_{\mathrm{As}}^{1/2} p(\mathrm{As_4})^{1/2} \cong \tfrac{1}{2} K_{\mathrm{As}}^{1/2} p^0(\mathrm{AsCl_3})^{1/2}. \tag{8}$$

Hence the incorporation of Si donors in the crystal can be deduced from:

$$[\mathrm{Si^+_{Ga}}] = K_{sg} K_i^{-1/2} K_{\mathrm{As_2V}}\, K_{\mathrm{As}}^{1/4} 2^{-1/2} p^0(\mathrm{AsCl_3})^{1/4} p(\mathrm{Si}). \tag{9}$$

Thus from a knowledge of the equilibrium constants (Logan and Hurle 1971, Boucher and Hollan 1970) and $p(\mathrm{Si})$, which we will now consider, we can predict the Si donor concentration in the crystal.

3.2. Transport processes and their control of silicon activity

We consider first the potential attack on the silica tube and the 'pick-up' reactions involving HCl. They are:

$$(4-x)\mathrm{HCl} + \mathrm{SiO_2} + (x)\mathrm{H_2} \rightleftharpoons \mathrm{SiCl_{4-x}H_x} + 2\mathrm{H_2O} \tag{10}$$

where $x = 0, 1, 2, 3$ corresponding to the chlorosilanes $SiCl_4$, $SiHCl_3$, SiH_2Cl_2 and SiH_3Cl. An equivalent series of deposition reactions in the cooler substrate region can take place as illustrated by the following reactions:

$$SiCl_{4-x}H_x + (2-x)H_2 \rightleftharpoons Si + (4-x)HCl. \tag{11}$$

The relative importance of the various chlorosilanes on the reaction depends critically, as noted earlier, on the values of their enthalpies of formation and on the pressure of HCl. This will be considered later but to illustrate the pressure dependence we need to relate $p(HCl)$ to the value of $p^0(AsCl_3)$. It is controlled by the dominant transport reaction for GaAs:

$$GaCl + \tfrac{1}{4}As_4 + \tfrac{1}{2}H_2 \overset{K_{GA}}{\rightleftharpoons} GaAs + HCl \tag{12}$$

where

$$K_{GA} = p(HCl)\, p(GaCl)^{-1}\, p(As_4)^{-1/4}. \tag{13}$$

The chlorine conservation is:

$$3p^0(AsCl_3) = p(HCl) + p(GaCl) \tag{14}$$

so that from (13) and (14)

$$p(HCl) = 3p^0(AsCl_3)\,[1 + K_{GA}^{-1}\,p(As_4)^{-1/4}]^{-1} \tag{15}$$

where for a liquid Ga source As conservation demands that

$$p(As_4) = \tfrac{1}{4}p^0(AsCl_3). \tag{16}$$

At 1093 K using a value of $K_{GA}(1093)$ of 2·8 (Boucher and Hollan 1970) we calculate $p(HCl)(1093)$; it ranges from $2·8 \times 10^{-3}$ to $1·7 \times 10^{-2}$ for the $p^0(AsCl_3)$ range of interest namely 3×10^{-3} to $1·5 \times 10^{-2}$. This determines the dominant chlorosilanes involved. For illustrative purposes we will consider $SiCl_4$ only.

We note from (10) that the presence of residual water vapour in the system strongly suppresses the pick-up of chlorosilanes but assume for now that the system is perfectly dry. In this case from (10) putting $x = 0$ oxygen conservation demands that:

$$2p(SiCl_4) = p(H_2O). \tag{17}$$

Thus from (10) where the equilibrium constant for the reaction with $x = 0$ is K_1 we find:

$$p(SiCl_4) = \left(\frac{K_1}{4}\, p(HCl)^4\right)^{1/3}. \tag{18}$$

If K_5 is the equilibrium constant for the deposition reaction (11) with $x = 0$ then:

$$p(Si) = K_5 p(HCl)^{-4} \left(\frac{K_1}{4}\, p(HCl)^4\right)^{1/3} \tag{19}$$

$$= K_5 \left(\frac{K_1}{4}\right)^{1/3} p(HCl)^{-8/3} \tag{20}$$

or using (15) and (16)

$$p(\text{Si}) = K_5\left(\frac{K_1}{4}\right)^{1/3}\{3p^0(\text{AsCl}_3)[1 + 2^{1/2}K_{\text{GA}}^{-1}p^0(\text{AsCl}_3)^{-1/4}]^{-1}\}^{-8/3}. \qquad (21)$$

Substitution of (21) in (9) predicts the donor concentration in the crystal. We note that $[\text{Si}_{\text{Ga}}^+]$ is proportional to

$$f(p^0)^{-8/3}[p^0(\text{AsCl}_3)]^m$$

where $m = -2\cdot42$ and where

$$f(p^0) = [1 + 2^{1/2}K_{\text{GA}}^{-1}(p^0)^{-1/4}]^{-1}.$$

At low $p(\text{HCl})$, where SiHCl_3 becomes important, m should be smaller; if SiHCl_3 dominates then $m = -1\cdot75$. If we approximate the function $f(p^0)^{-8/3}$ by the form $\alpha(p^0)^q$ then over the range $3\times10^{-3} < p^0 < 1\cdot5\times10^{-2}$ we obtain $q \sim -0\cdot42$. Hence we expect the slope of $\lg N_\text{D}$–$\lg p^0(\text{AsCl}_3)$ to be $-2\cdot8$ when SiCl_4 dominates and $-2\cdot2$ when SiHCl_3 dominates. The experimental situation is intermediate with SiCl_4 tending to be more important. The prediction is close to the observed range of values found, namely $-2\cdot5$ to $-3\cdot3$.

3.3. *Transport processes and their control of Zn activity*

A full thermochemical treatment of the quantitative aspects of Zn contamination from potential sources such as the AsCl_3 and the Ga has been undertaken, but only a summary of the results can be reproduced here. Zinc diffusion from the substrate is another possibility which needs to be assessed experimentally. On general considerations of vapour pressure it would seem unlikely that the Zn could arise from ZnCl_2 in the AsCl_3 assuming the AsCl_3 is 'five or six nines' pure. However, even minute traces of Zn in the Ga (less than 10^{14} atoms/cm^3) could be transported. Although Zn was not detected in the Ga samples analysed in table 1, Zn has been found in equivalent samples at the 1–2×10^{15} atoms/cm^3 concentration. Assuming that Zn is present in the Ga the most important question is whether the amount transported and incorporated would show an AsCl_3 pressure dependence similar to that for Si. If the pick-up and deposition occurred through the reaction:

$$\text{Zn} + 2\text{HCl} \rightleftharpoons \text{ZnCl}_2 + \text{H}_2$$

then $p(\text{Zn}_\text{v})$ would be to a first-order independent of

$$p(\text{HCl}) \text{ [ie, } p(\text{Zn}_\text{v}) \propto p(\text{HCl})^2 p(\text{HCl})^{-2}].$$

But, if there was a diffusion limitation to the rate of pick-up then $[\text{Zn}_{\text{Ga}}]$ would depend on $p^0(\text{AsCl}_3)$ only through the concentration of $[\text{V}_{\text{Ga}}]$; $[\text{Zn}_{\text{Ga}}] \propto p(\text{AsCl}_3)^{1/4}$. However, the development of a Zn depletion boundary layer in the Ga is to be expected and this would tend to produce a constant value of $p(\text{ZnCl}_2)$ irrespective of the AsCl_3 pressure. In this situation it can be shown that $[\text{Zn}_{\text{Ga}}]$ varies approximately as $p^0(\text{AsCl}_3)^{-2\cdot1}$. If the Zn is picked up from the crust of GaAs on the Ga, $[\text{Zn}_{\text{Ga}}]$ would vary as $p^0(\text{AsCl}_3)^{-0\cdot9}$. All of these values of the modulus of the index m are smaller than that obtained for silicon.

4. Discussion

The residual impurities found in VPE layers depend on the conditions of growth in addition to the composition of the starting materials. The present work along with work discussed in §1 clearly shows that Si can be a dominant donor; indeed the quantitative results on the molar fraction effect support a Si contamination model. Where does the Si come from? Is it from the silica walls as the analysis in §3.2 or from the Ga source perhaps through its reaction with the silica boat? A detailed treatment of the former model taking into account all the chlorosilanes results in a mean slope $\bar{m}$ for the molar fraction effect of $-2 \cdot 5$ which is fully consistent with the experimental results. Silicon picked up directly from the Ga of the Ga/GaAs source produces a model where $\bar{m} = \frac{1}{4}$. However, if the pick-up of Si from the source is diffusion limited by transport in the liquid as discussed above for Zn then $m = -4 \cdot 4$ or if it is picked up from the GaAs crust then $m = -3 \cdot 2$. The precision of the experimental results is not sufficient for us definitely to decide between the mechanisms of attack of the SiO_2 tube and pick-up from the GaAs crust, but the BN results suggest that it is attack of the SiO_2 which is the dominant source.

The chlorosilane model was tested recently by Kennedy *et al* (1974) by introducing excess HCl into the deposition zone during GaAs growth. They claim not to find the inverse HCl pressure dependence of n. However, it can be argued that the added HCl pressure should only show an effect when it becomes comparable to the input HCl pressure. In fact it is at this pressure of about 10^{-2} atoms that the results reported by Kennedy *et al* (1974) appear to show the inverse pressure dependence.

The model analysed in §3 assumes that the partial pressure of non-intentionally added water vapour was insignificant. The effect of water vapour has been analysed by DiLorenzo and Moore (1971), Rai-Choudhury (1971) and Weiner (1972) and it is evident that pressures as low or lower than 10^{-7} atmospheres could be significant. Theoretically the equilibrium model would indicate that the chlorosilane formation will be suppressed when the amount of water becomes comparable with the amount formed by reaction (10). It will be more significant the lower the pressure of HCl. In fact it is because of reaction (10) that one gets a molar fraction effect which should be completely suppressed by an excess of water vapour.

Silicon is not necessarily the dominant residual donor. We cited the work of Wolfe *et al* (1974) earlier. Ihara *et al* (1974) by growing in N_2 atmosphere claim to have suppressed the Si, Zn and Ge which were found in layers grown in H_2, and they were left with C donors. DiLorenzo (1972) suggested that the major residual impurity found in layers grown on (311)B and (211)B surfaces was a group VI element although this may not be entirely consistent with the pressure dependence of n found on these faces. Se and Te did not show such an effect at least on (100) surfaces. More work using definitive identification techniques such as far infrared spectroscopic techniques, eg Stradling *et al* (1973), would greatly help our understanding of the residual donors. Techniques which could routinely and readily identify residual donors would be a boon.

The identity and role of residual acceptors is especially interesting in view of the close compensation which is found between acceptors and donors. In fact it is just this constancy of the ratio K which invites the view that the acceptors and donors are amphoterically situated Si atoms (Nakagawa *et al* 1971, Ikoma and Nakagawa 1972, DiLorenzo 1972, Nakanisi and Kasiwagi 1974). Nakanisi and Kasiwagi (1974) claim an ionization energy for Si of 31 meV which is at variance with our own findings (White *et al* 1973, 34·9 meV, and that of Ozeki *et al* 1973, 35·1 meV). We identify Zn with an ionization energy of 31·1 eV. The use of D/A pair spectra is in our experience very difficult for impurity identification. We have already cited Okamoto's view that the acceptor is probably a Ga vacancy. The apparent elimination of Si in N_2 grown layers (Ihara *et al* 1974) reduced K from between 0·4—0·8 found by DiLorenzo (1972) and ourselves to an apparently constant 0·2.

Experimentally Si is not found as an acceptor in our standard VPE layers. However, we have found evidence of Si acceptors in layers grown elsewhere. In a detailed theoretical study (Ashen *et al* 1974) we reported that under our standard VPE growth conditions the ratio of Si donors to Si acceptors was greater than 1000 to 1. The experimental PL evidence supported this result. Can we explain our residual acceptors then as being due to residual Zn, the only acceptors found in standard VPE PL spectra? The constancy of K is incompatible with this model. Even if the $AsCl_3$ pressure dependence of Si and Zn doping incorporation were the same it is unthinkable that the balance of Si and Zn would be so finely and reproducibly compensated in so many layers produced in so many different laboratories. One is forced to the conclusion that there is another unidentified acceptor present. This would probably be true even if we accept that the mobility scattering model is too naïve in its interpretation of N_D and N_A and that a further scattering mode is present.

We are not in a position to speculate convincingly on the role of O_2 in the VPE layers at the moment. All the evidence of a recent identification (White *et al* 1974) suggests that O_2 is present in our VPE layers as a donor. More work is required here.

It is evident even with our knowledge of residual impurities in VPE layers that we are still some way from fully explaining their electrical properties; in particular the identity of the dominant compensating acceptor in n-type material remains to be established.

Acknowledgments

It is a pleasure to acknowledge the excellent analytical support given by Mr J B Clegg of Mullard, Redhill, and Mr G W Blackmore of the Admiralty Materials Laboratory. The help and skilled assistance of Mr C A Jones, Mr S Benn, Mr D K Wright, Mrs J Cooke and Miss S G Stephenson is very much appreciated. Contributed by the permission of the Director of RRE. Copyright Controller HMSO.

References

Aoki T and Yamaguchi N 1972 *J. Appl. Phys.* **11** 1775
Ashen D J, Dean P J, Hurle D T J, Mullin J B and White A M 1975 *J. Phys. Chem. Solids* to be published
Barry B E 1971 *3rd Int. Symp. Gallium Arsenide and Related Compounds* (London: Institute of Physics) p172
Boucher A and Hollan L 1970 *J. Electrochem. Soc.* **117** 932
Cairns B and Fairman R 1968 *J. Electrochem. Soc.* **115** 327c
—— 1970 *J. Electrochem. Soc.* **117** 197c
Clegg J B, Millett E J and Roberts J A 1970 *Anal. Chem.* **42** 713

Dean P J 1974 *J. Phys. (Paris)* **35** suppl C-3 127
DiLorenzo J V 1971 *J. Electrochem. Soc.* **118** 1645
—— 1972 *J. Cryst. Growth* **17** 189
DiLorenzo J V and Moore G E 1971 *J. Electrochem. Soc.* **118** 1824
Furukawa Y 1967 *Japan. J. Appl. Phys.* **6** 1344
Hasegawa F and Saito T 1968 *Japan. J. Appl. Phys.* **7** 1342
Hunt L P and Sirtl E 1970 *J. Electrochem. Soc.* **117** 3
Ihara M, Dazai K and Ryuzan O 1974 *J. Appl. Phys.* **45** 528
Ikoma H and Nakagawa M 1972 *Japan. J. Appl. Phys.* **11** 338
Kennedy J K, Potter W D and Davies D E 1974 *J. Cryst. Growth* **24/25** 233
Kim H B, Barrett D L and Sweeney G G 1973 *4th Int. Symp. Gallium Arsenide and Related
    Compounds* (London: Institute of Physics) p88
Knight S, Dawson L R, DiLorenzo J V and Johnson W A 1971 *3rd Int. Symp. Gallium Arsenide
    and Related Compounds* (London: Institute of Physics) p108
Logan R M and Hurle D T J 1971 *J. Phys. Chem. Solids* **32** 1739
Nakagawa M and Ikoma H 1971 *Japan. J. Appl. Phys.* **10** 1345
Nakanisi T and Kasiwagi M 1974 *Japan. J. Appl. Phys.* **13** 484
Okamato H, Sakata S and Sakai K 1973 *Japan. J. Appl. Phys.* **44** 1316
Ozeki M, Nakai K, Dazai K and Ryuzan O 1973 *Japan. J. Appl. Phys.* **12** 478
Rai-Choudhury P 1971 *J. Cryst. Growth* **11** 113
Stradling R A, Eaves L, Hoult R A, Muira N, Simmonds F E and Bradley C S 1973 *4th Int. Symp.
    Gallium Arsenide and Related Compounds* (London: Institute of Physics) p65
Weiner M E 1972 *J. Electrochem. Soc.* **119** 496
White A M, Dean P J, Ashen D J, Mullin J B and Day B 1974 *12th Int. Conf. Phys. of Semicond.,
    Stuttgart*
White A M, Dean P J, Ashen D J, Mullin J B, Webb M, Day B and Greene P D 1973 *J. Phys. C:
    Solid St. Phys.* **6** L243
Wolfe C M, Korn D M and Stillman G E 1974 *Appl. Phys. Lett.* **28** 78
Wolfe C M, Stillman G E and Dimmock J O 1970 *J. Appl. Phys.* **41** 504

Diffusion length studies in n gallium arsenide

A M Sekela, D L Feucht and A G Milnes

Carnegie-Mellon University, Pittsburgh, Pennsylvania, USA

Abstract. Hole diffusion lengths have been studied in n gallium arsenide from the current response to a scanning electron microscope beam applied to a Schottky barrier junction region. The validity of the procedure has been examined by a model that includes geometric and surface recombination effects.

Material from many sources was studied with net electron concentrations in the range 7×10^{14} to $2 \times 10^{18} \, cm^{-3}$. The diffusion lengths in ingot material (Czochralski or horizontal Bridgman) were typically in the range 0·3 to 1·5 μm, but occasional specimens of tellurium-doped horizontal Bridgman ingot material showed significantly longer values (3 μm or more). This did not correlate directly with crystal perfection as judged by etch pit counts or mobility.

Similarly, large variations were found in epitaxially-grown material. In vapour-phase-grown specimens at the low end of the doping range considered L_p varied from 11 μm to less than 1 μm depending on the specimen and the supplier. In the preparation of liquid-phase epitaxial layers a change from 0·65 to 7·8 μm could be produced by baking of the gallium melt (1000 °C, 15 h in hydrogen) prior to growth.

It is concluded that the hole diffusion lengths in gallium arsenide are being affected by a lifetime controlling recombination centre that still has to be identified.

1. Introduction

Hole diffusion lengths in n gallium arsenide have been measured using the voltage dependence of cathodoluminescence (Wittry and Kyser 1966), the short-circuit current of diffused diodes under 20 meV irradiation (Aukerman *et al* 1967) and grown hetero-structure laser diodes with light injection (Casey *et al* 1973, Young and Rowland 1973), measurements of photoluminescence and optical absorption in samples of tellurium-doped ingots (Hwang 1969), and the short-circuit current of diffused diodes (Wittry and Kyser 1965) and Schottky barriers (Ryan and Eberhardt 1972, van Opdorp *et al* 1974) with scanning electron microscope beam injection. Similar techniques have been reported for electron diffusion length in p gallium arsenide by Ashley *et al* (1973), Ettenberg *et al* (1973), Acket *et al* (1974), and others reviewed by Casey *et al* (1973).

Reported values of hole diffusion length in ingot gallium arsenide range from 0·3 μm for a net electron concentration $n = N_D - N_A = 2 \times 10^{17}$ electrons/cm³ (dopant not given, Aukerman *et al* 1967) to 4 μm for $n = 5·1 \times 10^{16} \, cm^{-3}$ germanium-doped material (Wittry and Kyser 1965). Liquid epitaxial tin-doped layers grown by Casey *et al* (1973) showed diffusion lengths from 4·5 μm at $n = 1 \times 10^{17} \, cm^{-3}$ to 0·35 μm at $n = 6 \times 10^{18} \, cm^{-3}$, while Ryan and Eberhardt (1972) report a value of 200 μm in high purity $(n < 10^{14} \, cm^{-3})$ liquid epitaxial material. Hwang (1969) and Casey *et al* (1973) relate measured values of hole diffusion length to net electron concentration n: Hwang obtained nearly constant values around 2 μm for $n < 1 \times 10^{18} \, cm^{-3}$ and decreasing values for $n > 2 \times 10^{18}\,\mathrm{cm}^{-3}$, while the data of Casey *et al* show similar behaviour with more scatter for $n < 1 \times 10^{18}\,\mathrm{cm}^{-3}$.

This paper reports an investigation into hole diffusion length in n gallium arsenide produced by vapour epitaxy, liquid epitaxy, and Czochralski and horizontal Bridgman growth methods, from many different sources. The range of net electron concentrations is from $7 \cdot 2 \times 10^{14}$ to $2 \times 10^{18}\,\mathrm{cm}^{-3}$, and dopants include tin, germanium, silicon, selenium and tellurium, with silicon and tellurium the principal ingot dopants. An SEM beam-induced current technique was used with Schottky barrier collectors to measure hole diffusion length, and, within the limitations reported below, gave reliable results. Fabrication of samples did not require temperatures above 200 °C and is less likely than a technique using p–n junctions to have changed their carrier lifetimes.

This work developed from an interest in the hole diffusion lengths in specimens of n gallium arsenide with net electron concentrations in the range 10^{16} to $5 \times 10^{17}\,\mathrm{cm}^{-3}$ and used in our work on gallium arsenide and aluminium–gallium arsenide heteroface solar cells.

2. Experimental details

2.1. Sample preparation

Schottky barriers are formed on gallium arsenide by evaporating $0 \cdot 6\,\mathrm{mm}$ aluminium dots through a metal mask. Ingot samples with as-sawn surfaces are first polished by etching for 30 min in $1:1:5\ \mathrm{H_2O:H_2O_2:H_2SO_4}$ in a rotating beaker to give damage-free surfaces. All samples are boiled in concentrated aqueous HCl and dried with a blast of hydrocarbon-free nitrogen immediately before loading into the evaporator.

A low index sample orientation, such as (100), allows a flat surface perpendicular to a barrier to be formed by cleaving. Indium contacts are made to the barrier and substrate and connections are brought out through the header leads. The leads are also used to support the sample with the cleaved surface up.

2.2. Measurement technique

A JEOL JSM-2 scanning electron microscope is used to inject carriers and measure the point of injection relative to the barrier. Induced sample current is measured point-by-point with an electrometer in the current mode while the sample bias is kept slightly negative by typically $0 \cdot 001\,\mathrm{V}$ and monitored to be constant with another electrometer in the voltage mode (input resistance $10^{14}\,\Omega$), as shown schematically in figure 1. The sweep circuits of the microscope are disabled and the beam position is changed with the sweep bias controls and recorded by exposing a spot of light representing the beam on to an undeveloped photograph of the sample and its barrier made previously by a scan in the secondary mode.

A measurement of diffusion length typically consists of 20 current measurements made on a line perpendicular to the barrier. The length of the line is determined by changing the magnification of the microscope until, with the barrier at one side of the CRT image, a stationary beam at the other side induces a current just detectable above the noise (induced background) current of the diode. The maximum induced current

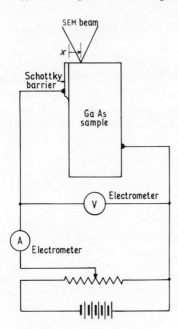

Figure 1. Schematic representation of SEM beam-induced current method of measuring minority carrier diffusion length.

with the beam at the barrier is about 10^{-6} A for a beam of 5×10^{-11} A accelerated by 25 kV. The noise current with the beam on the sample but very far from the barrier is typically 10^{-9} A. It is a function of beam current; a lower beam current would give a smaller induced background current but in this system results in a poorer signal-to-noise ratio. The choice of 25 kV for the beam acceleration voltage was governed also by signal-to-noise considerations. The size of the injection region under the beam, of the order of 1μm, is a function of acceleration voltage (Hackett 1972, Bresse and Lafeuille 1971), and limits the smallest diffusion length that can be measured. With 25 kV, 0.35μm was the smallest diffusion length observed.

3. Modelling and interpretation of the technique

The logarithm of induced current is plotted against beam position on the sample measured from the barrier. A straight line shows an exponential dependence, $I \propto \exp(-x/x_0)$. Most experimental plots have a curved section between the barrier and straight line region. Samples with diffusion lengths around 1μm give plots such as figure 2, where the response curves down as the beam approaches the barrier; those with diffusion lengths of several micrometres show increasing slope near the barrier, as in figure 3. Curvature down can be explained by the finite size of the injection region (Hackett *et al* 1972). Upward curvature has been attributed to carrier recombination taking place at the surface under the SEM beam (Hackett *et al* 1972, Ryan and Eberhardt 1972). However, modelling suggests that the curvature is also a function of the geometry of the system.

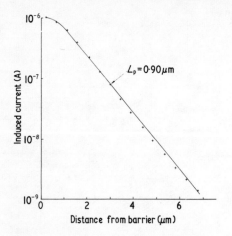

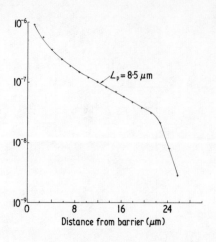

Figure 2. Typical plot of beam-induced current against distance for a gallium arsenide sample (64) with low diffusion length. The sample is from a horizontal Bridgman ingot doped with selenium with a net electron concentration of $1 \times 10^{17}\,\mathrm{cm}^{-3}$.

Figure 3. Response for an epitaxial gallium arsenide sample (5) with long diffusion length. The sample is a vapour epitaxial layer doped with tin with a net electron concentration $5 \times 10^{15}\,\mathrm{cm}^{-3}$ on an n^{+} substrate. The abrupt change of diffusion length at the layer-substrate interface is visible.

The relationship between diffusion length and the shape of the induced current plot is shown by solving a model of a point source of excess minority carriers near a surface with finite recombination velocity (van Roosebroeck 1955) for the current collected by a barrier (or the depletion edge), modelled as a surface with infinite recombination velocity and replaced with the negative image of the source model (Bresse and Lafeuille 1971), as shown in figure 4. The total collected current for the source at a given distance

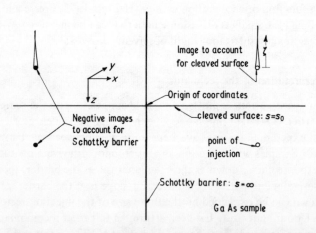

Figure 4. Point source model used to determine the effect of surface recombination velocity s_0 on the measurement of diffusion length. The finite value of s_0 causes the upper image sources to have negative source tails (van Roosebroeck 1955).

from the barrier x_1 is found by numerically integrating the current density at the barrier,

$$I = -qD_\rho \frac{\partial \Delta p}{\partial x} = -qD_p \frac{R_0}{2\pi D_0} (x_1) \left[\frac{R_1 + L_p}{L_p R_1^3} \exp\left(-\frac{R_1}{L_p}\right) + \frac{R_2 + L_p}{L_p R_2^3} \right.$$

$$\left. \times \exp\left(-\frac{R_2}{L_p}\right) - \frac{2 s_0 \tau_p}{L_p^2} \int_0^\infty \frac{R_3 + L_p}{L_p R_3^3} \exp\left(\frac{-s_0 \tau_p \zeta - R_3 L_p}{L_p^2}\right) d\zeta \right]$$

$$R_1 = [(x - x_1)^2 + (y - y_1)^2 + (z - z_1)^2]^{1/2}$$

$$R_2 = [(x - x_1)^2 + (y - y_1)^2 + (z + z_1)^2]^{1/2}$$

$$R_3 = [(x - x_1)^2 + (y - y_1)^2 + (z + z_1 + \zeta)^2]^{1/2}$$

over the barrier half-plane. In this equation, I is the current density at the point (x, y, z), (x_1, y_1, z_1) are the coordinates of the point of injection, s_0 is the recombination velocity at the injection surface, R_0 is the rate of injection, ζ is a variable of integration used in calculating the contribution of the image tail, and the rest of the symbols are standard as in van Roosebroeck (1955).

The computed results show curvature that depends on diffusion length and surface recombination velocity, as in figures 5(a) and 5(b). The apparent slope of the straight line portion, x_0^{-1}, within three decades of the maximum, depends primarily on the diffusion length, but also slightly on the surface recombination velocity: x_0 changes from 0·70 to 0·67 μm as s_0 is changed from 0 to 10^6 cm s^{-1} in the modelled response for a sample with diffusion length 0·70 μm, figure 5(a). A model plot for a sample with diffusion length 8·0 μm, figure 5(b), shows x_0 changing from 8·0 to 6·5 μm as s_0 increases from 0 to 10^6 cm s^{-1}. This is a worst-case error of 19%, as all but one of the samples in the experimental study showed x_0 below 8·6 μm. The surface recombination velocity of cleaved gallium arsenide is reported by Wittry and Kyser (1966) to increase with increasing carrier concentration to a value of $2·9 \times 10^6$ cm s^{-1} at $n = 3 \times 10^{18}$ elec-

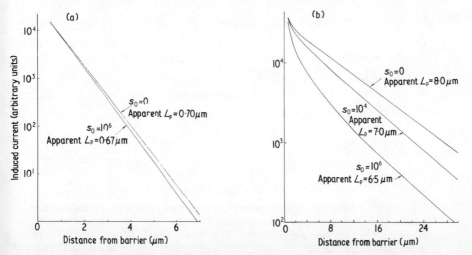

Figure 5. (a) Computed response plot for a diffusion length of 0·7 μm and s_0 of 0 and 10^6 cm s^{-1}. (b) Computed response plot for a diffusion length of 8 μm and s_0 of 0, 10^4 and 10^6 cm s^{-1}.

trons/cm^3. The reported value for a sample with a net carrier concentration of $5 \cdot 1 \times 10^{16} cm^{-3}$ is $6 \cdot 2 \times 10^5 cm s^{-1}$. No attempt was made to correct for the surface effect in our experimental measurements; x_0 is taken directly as the diffusion length and thus is slightly lower than the true value.

The modelled response for a diffusion length over $100 \mu m$ shows that within a diffusion length of the barrier the induced current drops more than three decades, even assuming $s_0 = 10^4 cm s^{-1}$. The curvature in this case appears to be caused primarily by the form of the radial dependence of current density in equation (1). Estimation of long diffusion lengths with this geometry is therefore not practical.

For easily interpretable measurements with epitaxial layers, the layer thickness should be several diffusion lengths so that the upward curvature present within a diffusion length of the barrier is not dominant.

The lowest diffusion length we have observed is $0 \cdot 35 \mu m$. We believe this to be approaching the lower limit of detectability because of the finite size of the injection region (approximately $1 \mu m$). The finite size of the injection region does not weaken the conclusions derived from the point model because in principle the model can be extended to account for the injection region by replacing the single point source with several sources. The result would be the superposition of similar response curves.

4. Experimental results and discussion

Hole diffusion lengths for a number of n gallium arsenide samples are plotted in figure 6 against net electron concentration, $N_D - N_A$, as measured from the slope of

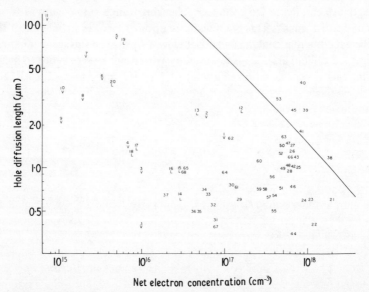

Figure 6. Hole diffusion length against net electron concentration for n gallium arsenide samples. Samples 1 to 11 are of vapour epitaxial material, 12 to 20 of liquid epitaxial material, 21 to 37 of silicon-doped ingots, 38 to 62 of tellurium-doped ingots, 63 to 65 of selenium-doped ingots, 66 to 67 are of tin-doped ingots and 68 is of a germanium-doped ingot. The epitaxial samples, except for 12 which is tellurium-doped, are either tin-doped or undoped.

reverse bias-capacitance plots (Boonton 72A capacitance meter at 1 MHz). The line drawn in figure 6 represents the maximum possible diffusion length computed from the probability of direct hole—electron recombination across the bandgap (Ryan and Eberhardt 1972), assuming that the diffusion constant for minority holes at any electron concentration is the same as the diffusion constant of majority holes of the same concentration as determined by mobility (Wittry and Kyser 1966). The line shows how hole diffusion length would depend on electron concentration in gallium arsenide modelled in this way, and is not intended as an absolute benchmark.

Specimens with the same net electron concentration are seen to vary considerably in diffusion length. Attempts to reconcile this scatter by relating diffusion length to total impurity concentration were made. Inferring the total impurity density $N_D + N_A$ by measuring the electron mobility below 100 K with the van der Pauw (1958) technique was inconclusive because most of the samples have total impurity densities above the 3×10^{17} cm^{-3} limit for this measurement (Wolfe *et al* 1970). However, considerable compensation is suggested by the low mobilities measured.

For vapour epitaxial layers with net electron concentrations in the range 7×10^{14} to 5×10^{15} cm^{-3} diffusion lengths between 11 and 5 μm are seen. However, we have also seen for this concentration range low diffusion lengths for vapour epitaxial material with good crystal structure as judged by electron mobility and etch pit density. In the liquid epitaxial samples, all but one of which were grown in our laboratory, diffusion length is seen to increase, and net electron concentration decrease, with the amount of time the melt (gallium metal and gallium arsenide source crystal) is baked under pure hydrogen at 1000 °C (the seed crystal is not present during baking). Baking at 800 °C is not nearly as effective in removing the implied lifetime killer. One impurity that might be removed by baking is oxygen. Experiments are in progress to determine the effect on diffusion length of deliberate addition of gallium oxide to the baked melt before growth.

No patterns are apparent in the diffusion lengths of ingot samples except for a small group of tellurium-doped horizontal Bridgman crystals whose diffusion lengths lie on or near the theoretical maximum. Similar samples from the same manufacturers, even from the opposite ends of the same ingots, have low diffusion lengths. Dislocation densities in all samples are between 10^3 and 10^6 cm^{-2}, but no clear correlation with diffusion length can be found. Effort has not been available to interpret lifetime in terms of specific recombination processes such as band to band, band to impurity or Auger state transitions.

The tellurium-doped samples represented as having unusually long diffusion lengths give induced current plots similar to those for samples 39 and 40 shown in figure 7. More curvature than is expected from the model or seen in plots for other samples with high and low diffusion lengths makes interpretation of these plots hazardous. The plot for sample 39 is nearly normal: the slope yields 2·5 μm for the diffusion length from about 3 μm from the barrier to the point where the current is one decade below its peak. The plot for sample 40, however, cannot be construed as approaching a straight line. This curvature implies another mechanism for carrier transport, such as internal emission and reabsorption of photons (Dumke 1957). Casey *et al* (1973) found a possibility of internal re-excitation in their tin-doped liquid epitaxial layers with net electron concentrations between 1 and 4×10^{18} cm^{-3} by measuring the relative photoluminescence.

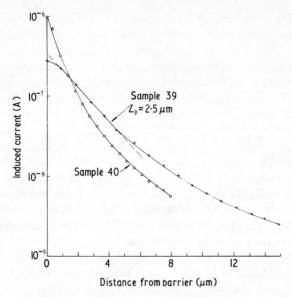

Figure 7. Response plots for two tellurium-doped ingot samples. The net electron concentration of 40 is $8.5 \times 10^{17}\,\mathrm{cm^{-3}}$, while that for 39 is $9.2 \times 10^{17}\,\mathrm{cm^{-3}}$. The line is the theoretical diffusion length for recombination only across the bandgap.

Mass spectrometry shows the impurity concentrations in sample 40, with net electron concentration $8.5 \times 10^{17}\,\mathrm{cm^{-3}}$, to be unusually large: tellurium is present at $6 \times 10^{18}\,\mathrm{cm^{-3}}$, silicon at $7.8 \times 10^{17}\,\mathrm{cm^{-3}}$ and oxygen at $6 \times 10^{17}\,\mathrm{cm^{-3}}$, while in sample 47 ($L_p \simeq 1.5\,\mu\mathrm{m}$), with a net electron density of 6.1×10^{17} and which gave a normal induced current plot, tellurium is present at $6.1 \times 10^{17}\,\mathrm{cm^{-3}}$, silicon at $1.0 \times 10^{16}\,\mathrm{cm^{-3}}$ and oxygen at $2.4 \times 10^{17}\,\mathrm{cm^{-3}}$. Kressel *et al* (1968) reported segregation of tellurium in excess of $2 \times 10^{18}\,\mathrm{cm^{-3}}$ in liquid epitaxial gallium arsenide, but related this to lower radiative efficiency. The oxygen values above are probably unreasonably high because very low-temperature cryo-pumping was not used.

The excess curvature is not seen in the other samples represented in figure 6; these gave response plots with adequate straight line regions for interpretation of diffusion length.

The technique results in well-behaved response plots when applied to n gallium arsenide—phosphide epitaxial layers on gallium arsenide substrates. Values of hole diffusion length of 1.0 and $1.5\,\mu\mathrm{m}$ were obtained for material with 40 and 24% gallium phosphide and net electron densities of $1.6 \times 10^{16}\,\mathrm{cm^{-3}}$ from one source, and 2.7 to $3.6\,\mu\mathrm{m}$ with 30% gallium phosphide and net electron densities from 5.9×10^{15} to $1.2 \times 10^{16}\,\mathrm{cm^{-3}}$ from another laboratory. As with gallium arsenide, diffusion length is determined in part by who makes the material.

5. Conclusions

The SEM beam-induced current measurement of hole diffusion length in n gallium arsenide described in this paper is a good method within the limitations imposed by sample geometry and signal-to-noise ratio of the instruments. It should be easy to use the method to monitor changes in diffusion length resulting from changes in growth technology. Reduction of the levels of silicon, oxygen, carbon, or deep impurities in the finished ingots or epitaxial layers might improve the hole diffusion lengths. Although improvement of crystalline quality of GaAs as judged by increased electron mobility and reduced etch pit density is a desirable goal, it does not appear to correlate directly with improved diffusion length.

Acknowledgments

This paper reports work supported by the National Aeronautics and Space Administration, Langley, Virginia. The authors are grateful to Mr G Walker for helpful discussions and Mr D Wang for preparing some of the epitaxial samples. The following persons and organizations also aided this work through helpful discussions or assistance in obtaining samples: D J Ashen, K L Ashley, H Beneking, C J Hwang, H B Kim, R Mattis, J B Mullin, R Stirn, J Woodall, Bell and Howell, Electronic Materials Corporation, Laser Diode Laboratories, Litton Industries, MCP Electronics, Monsanto, Motorola, Photo-Electronic Materials Corporation, Raytheon, Texas Materials Laboratories, and Wacker Chemitronic.

References

Acket G A, Nijman W and Lam H't 1974 *J. Appl. Phys.* **45** 3033
Ashley K L, Carr D L and Romano-Moran R 1973 *Appl. Phys. Lett.* **22** 23
Aukerman L W, Millea M F and McColl M 1967 *J. Appl. Phys.* **38** 685
Bresse J F and Lafeuille D 1971 *Electron Microscopy and Analysis* (London: Institute of Physics) p220
Casey H C, Miller B I and Pinkas E 1973 *J. Appl. Phys.* **44** 1281
Dumke W P 1957 *Phys. Rev.* **105** 139
Ettenberg M, Kressel H and Gilbert S L 1973 *J. Appl. Phys.* **44** 827
Hackett W H, Saul R H, Dixon R W and Kamlott G W 1972 *J. Appl. Phys.* **43** 2857
Hwang C J 1969 *J. Appl. Phys.* **40** 3731
Kressel H, Hawrylo F Z, Abrahams M S and Buiocchi C J 1968 *J. Appl. Phys.* **39** 5139
van Opdorp C, Peters R C and Klerk M 1974 *Appl. Phys. Lett.* **24** 125
van der Pauw L J 1958 *Phillips Res. Rep.* **13** 1
van Roosebroeck W 1955 *J. Appl. Phys.* **26** 380
Ryan R D and Eberhardt J E 1972 *Solid St. Electron.* **15** 865
Wittry D B and Kyser D F 1965 *J. Appl. Phys.* **36** 1387
—— 1966 *J. Phys. Soc. Japan* **21** suppl. 312
Wolfe C M, Stillman G E and Dimmock J O 1970 *J. Appl. Phys.* **41** 504
Young M L and Rowland M C 1973 *Phys. Stat. Solidi* A **16** 603

On the oxygen donor concentration in GaP grown from a solution of GaP in Ga

R C Peters and A T Vink

Philips Research Laboratories, Eindhoven, The Netherlands

Abstract. The results of measurements of the oxygen concentration in n- and p-type GaP are presented. The measurements were performed on LPE layers and on crystals grown from unseeded solutions of GaP in Ga. High temperature Hall effect measurements reveal a deep donor level in O-doped n-type GaP. Evidence is presented that this level is due to O_P. The maximum concentration of oxygen in these crystals is found to be of the order of 6×10^{16} cm^{-3}. In p-type crystals, the oxygen concentration was determined from the degree of compensation of the Zn acceptors. In addition the results of photo-capacitance measurements are presented.

The data support the interpretation of the oxygen incorporation in terms of an equilibrium model for the case that excess Ga_2O_3 is added to the growth solution.

1. Introduction

Oxygen substituted on P-sites in GaP acts as a deep donor with an ionization energy of the donor electron of 896 meV and plays an important role in the luminescence properties of this compound (Dean 1969, Bhargava 1971). Its presence is a prerequisite for the formation of radiative $Zn_{Ga}O_P$ complexes. On the other hand, complexes like $Si_{Ga}O_P$ may be formed, which are believed to be efficient non-radiative centres (Bhargava 1972, Bachrach *et al* 1972), and some workers consider isolated O_P as the main non-radiative centre in p-type Zn, O doped GaP (Jayson *et al* 1972). It is thus of importance to have reliable data on the substitutional oxygen concentration $[O_P]$. Published data on the total oxygen concentration and on the oxygen concentration as substitutional donors, ZnO complexes and interstitials are in poor agreement, however. Unfortunately, it is rather difficult to obtain such data by simple chemical analysis.

Direct measurements of the *total* oxygen concentration were reported by Kim (1969, 1971), using ^{3}He activation analysis and by Lightowlers *et al* (1973) using proton activation analysis of ^{18}O doped GaP. Kim observed total oxygen concentrations of about 10^{19} cm^{-3} but the presence of inclusions of Ga_2O_3 was suspected. Taking care to avoid Ga_2O_3 inclusions, Lightowlers determined a total oxygen concentration of approximately 7×10^{16} cm^{-3} in p-type GaP. Various determinations of the *substitutional* oxygen concentration, $[O_P]$, have been published. Foster and Scardefield (1969) used the degree of compensation, as found from Hall effect data and the measurement of $[Zn_{Ga}]$ using radioactive Zn, and found $[O_P] \simeq 7 \times 10^{16}$ cm^{-3} in p-type GaP crystals grown from solution starting at 1144 °C. Saul and Hackett (1970) also determined $[O_P]$ in p-type GaP from the degree of compensation now obtained from capacitance measurements. They report $[O_P] \simeq 3 \times 10^{17}$ cm^{-3} as the solubility limit at 1025 °C in crystals doped with 6×10^{17} Zn/cm^3. This value has been corrected upwards for the influence of O on the solubility of Zn in GaP. Such a correction was not applied by Foster and Scardefield, since they adopted a higher value for n_i, the intrinsic carrier concentration. From spectroscopic measurements (Dishman *et al* 1970) a value of $[O_P] \simeq 10^{17}$ cm^{-3} was deduced in p-type crystals.

A more direct determination of $[O_P]$ in n- and p-type material is found in photo-capacitance measurements (Kukimoto *et al* 1972, 1973). From measurements on p–n junctions they found $[O_P] \simeq 1.5 \times 10^{16}$ cm^{-3} in p-type GaP and from measurements on Schottky barrier diodes $[O_P] \simeq 2.8 \times 10^{16}$ cm^{-3} in p-type GaP, doped with Zn and O, and $[O_P] \simeq 8 \times 10^{15}$ cm^{-3} in n-type GaP, doped with Te and O. To obtain the total O_P concentration in p-type material, these values should be corrected upwards for the Zn_{Ga}–O_P complex concentration; this is believed to be an addition of about 30% at most, however (Wiley 1971).

Two remarks can be made here. First, the highest and lowest reported values of $[O_P]$ in p-type GaP differ by about a factor of ten. Second, $[O_P]$ is rather low in n-type GaP (Kukimoto *et al* 1973), which is relevant when red LED are made by diffusion of Zn into n-type O-doped GaP. In this paper we will present another fairly direct way of determining $[O_P]$ in O-doped n-type GaP, namely by high-temperature Hall effect measurements. In O-doped GaP that is weakly n-type at 300 K (eg, n = 10^{16} cm^{-3}) the Fermi level E_F will be, at low temperatures, on one of the shallow donors, generally S_P and Si_{Ga}. The oxygen donor is thus uncompensated. At sufficiently high temperatures the ionization of the oxygen donor will contribute to the carrier concentration, but since the level is very deep, no complete ionization can be achieved at the highest practical temperatures, which are around 1000 K. $[O_P]$ can be calculated from the data, however, using the relevant Hall formula.

In the following, the sample preparation, the experimental data, the calculation of $[O_P]$ from the data and the results will be presented in turn. The results of some photo-capacitance measurements, and of the determination of $[O_P]$ from the degree of compensation in p-type crystals, will also be given. Finally, the results will be discussed and compared with data from the literature.

2. Crystal preparation

The initial measurements were performed on O-doped GaP, grown by the slow cooling of a Ga melt saturated with GaP and Ga$_2$O$_3$. The evacuated quartz ampoules containing the melt were cooled from about 1100 to about 800 °C at 5 °C h^{-1}. Most of the measurements, however, were done on LPE layers grown on the (111)-P side of GaP substrates (LEC) in a vertical open tube system (Peters 1973). This system was made of quartz, but a pyrolytic BN crucible was used to contain the melt of Ga† saturated with GaP‡ and Ga$_2$O$_3$†.

The GaP used was spectrochemically pure; in addition specific sulphur analyses (Gijsbers *et al* 1974) showed $[S_P]$ to be less than or equal to 0.5 ppm by weight.

---

† Ga and Ga$_2$O$_3$ were Alusuisse 6N quality.
‡ GaP, obtained from Philips Maarheeze, was synthesized by the method of Grimmeiss *et al* (1965), but with additional purification of PH$_3$.

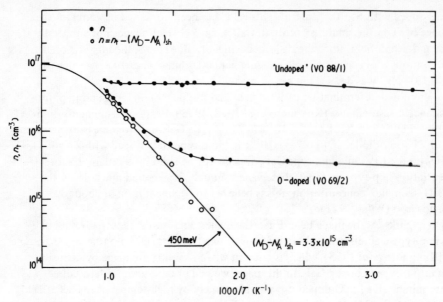

Figure 1. Results of Hall effect measurements using a Hall factor of 1, on an O-doped crystal (VO69/2) and an 'undoped' crystal (VO88/1).

The substrates were dipped into the melt at about 1050 °C and the whole was cooled at 1 °C min^{-1} to about 850 °C, resulting in layers that are thick enough to remove the substrate for Hall effect measurements.

3. High-temperature Hall-effect measurements

Diamond-shaped Hall samples were cut from the epiwafers and the substrate was removed. For the measurements, using the Van der Pauw technique, small Sn contacts were alloyed on the four edges and the sample was mounted on an Al_2O_3 holder. It was kept in an argon flow and heated by a bifilary wound furnace. In calculating the total free carrier concentration, n_t, from the data, a Hall factor of 1 was used. The results representative for O-doped layers (VO69/2) and 'undoped' layers (VO88/1) are shown in figure 1. We will first consider VO69/2. At room temperature $n_t = 3 \cdot 0 \times 10^{15}$ cm^{-3}. With increasing temperature n_t first slightly increases as a result of further ionization of the shallow donor levels. Above about 600 K n_t increases strongly, showing the presence of a deep donor level. To obtain the contribution from the deep donor n to the total free carrier concentration n_t the concentration of uncompensated shallow donors $(N_D - N_A)_{sh}$ must be subtracted. Additional Hall effect measurements from 77 K to room temperature show that S_P is the residual shallow donor level with $(N_D - N_A)_{sh} = 3 \cdot 3 \times 10^{15}$ cm^{-3}. The results obtained by subtracting $(N_D - N_A)_{sh} = 3 \cdot 3 \times 10^{15}$ cm^{-3} are also shown in figure 1. At 1000 K the contribution of the deep donor is $n = 3 \cdot 4 \times 10^{16}$ cm^{-3}. The activation energy of about 450 meV derived from figure 1 corresponds to an ionization energy of about 840 meV (see §4) which is in good agreement with expectations for the oxygen donor. As expected for such a deep level, no complete ionization is achieved at 1000 K. Further evidence that oxygen is involved is twofold:

Firstly, similar observations were made on all O-doped samples examined, grown both from unseeded Ga-solutions and by LPE, although for these experiments different GaP starting material and growth conditions were used.

Secondly, in LPE layers that were grown taking care to avoid oxygen contamination, no ionization of a deep level is seen in the Hall measurements at high temperature. The results of measurements on such a sample, VO88/1, are also shown in figure 1. The value of $n_t = 4 \cdot 3 \times 10^{16}$ cm^{-3} at 300 K is considerably higher than that for samples with O-doping, but the additional steep rise near 1000 K is now nearly absent. Careful measurements on this sample showed that at 1000 K n due to the deep level is less than or approximately equal to 5×10^{15} cm^{-3}. The calculation of [O$_P$] from these measurements is the subject of the next section.

4. Calculation of the oxygen donor concentration

From photocapacitance studies strong evidence was found (Kukimoto *et al* 1972) that in n-type GaP, like that used in our experiments, the O$_P$-centre can bind two electrons. Very recently the two levels of O$_P$ were predicted theoretically by applying the pseudo-impurity theory to GaP : O$_P$ (Pantelides 1974). A Hall formula applicable in the situation of an uncompensated centre that can bind two electrons, has been derived by Champness (1956), assuming g to be 2 for the first captured particle and $g = 1$ for the second one† (spins paired). The result,

$$n^3 + 2n^2 N_c \exp\left(-\frac{E_1}{kT}\right) + n\left\{N_c^2 \exp\left(-\frac{E_1 + E_2}{kT}\right) - 2N_c N_D \exp\left(-\frac{E_1}{kT}\right)\right\}$$

$$- 2N_D N_c^2 \exp\left(-\frac{E_1 + E_2}{kT}\right) = 0 \qquad (1)$$

leads to

$$N_D = \frac{n^3 + 2n^2 N_c \exp\left(-\dfrac{E_1}{kT}\right) + nN_c^2 \exp\left(-\dfrac{E_1 + E_2}{kT}\right)}{2nN_c \exp\left(-\dfrac{E_1}{kT}\right) + 2N_c^2 \exp\left(-\dfrac{E_1 + E_2}{kT}\right)}, \qquad (2)$$

in which $N_c = N_c' . T^{3/2} = 8 \times 10^{15} T^{3/2}$ (Vink *et al* 1972) is the effective density of states of the conduction band, E_1 is the energy for thermal release of the first of the two bound electrons and E_2 that for the second one. N_D denotes the concentration of the centre, in this case oxygen. We have now first to determine E_1 and E_2. Kukimoto *et al* (1973) have concluded that the activation energies for thermal emission of the two states are nearly equal, 0·78 and 0·76 eV respectively. For simplicity we use $E_1 = E_2$; the best fit to the data in figure 1 is then obtained with $E_1 = E_2 = 840$ meV. With these values substituted in equation (2), the relation between n at 1000 K and N_D was calculated. The result is shown in figure 2. A value of n at 1000 K of $3 \cdot 4 \times 10^{16}$ cm^{-3}, as

† A larger value of g leads to an increase in N_D (see §7).

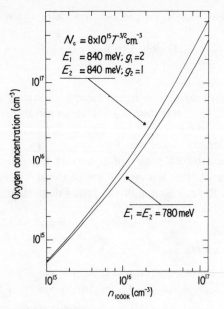

Figure 2. Concentration of substitutional oxygen $[O_P]$ as a function of n at 1000 K from equation (2).

found for sample VO69/2, corresponds to $N_D = 5 \cdot 5 \times 10^{16} \, \text{cm}^{-3}$. Table 1 gives a survey of data on several O-doped crystals. Recently an interpretation of the ionization energy of the second electron E_2, differing from that of Kukimoto *et al* (1973), was presented by Grimmeiss *et al* (1974). In this interpretation, E_2 is 650 meV. In view of our data, indicating a lower limit of both the ionization energies E_1 and E_2 of about 800 meV, this interpretation cannot be correct, however.

5. Oxygen concentration determined from photocapacitance measurements

In addition to the Hall-effect measurements, photocapacitance measurements were performed on two of our crystals by C H Henry†. A value for $[O_P] = 8 \times 10^{15} \, \text{cm}^{-3}$ has been derived from these measurements on a Schottky barrier diode on the surface of the LPE layer VO69/2. In this case the Schottky barrier diode was on the last grown surface of the LPE layer, which means that the value of $[O_P]$ refers to GaP grown around 850 °C. A similar measurement was performed on a Te- and O-doped GaP LPE layer (VO61) with a net carrier concentration of $n = 5 \cdot 5 \times 10^{17} \, \text{cm}^{-3}$ at room temperature. In this layer, grown at 1020 °C, $[O_P] \cong 8 \times 10^{15} \, \text{cm}^{-3}$ was found.

6. Oxygen concentration in p-type ZnO doped GaP, determined from the degree of compensation

The concentration of compensating substitutional oxygen donors in p-type Zn and O doped GaP has been estimated from the degree of compensation found from room

† C H Henry, Bell Laboratories, Murray Hill, NJ, USA.

Table 1. Survey of the data obtained from high temperature Hall effect measurements, using the ionization energies $E_1 = E_2 = 840$ meV for the two electrons oxygen

Sample No.	Growth method	Growth temp. (°C)	Ga_2O_3 addition to the melt	Mobility $(cm^2V^{-1}s^{-1})$		n_t at 300 K (cm^{-3})	$(N_D - N_A)$sh. (cm^{-3})	n_{deep} at 1000 K (cm^{-3})	[O_P] (2 el model) (cm^{-3})
				300 K	1000 K				
SV11	Solution growth	1100–800	Excess	135	12	2.6×10^{16}	3.2×10^{16}	2.6×10^{16}	$\sim 3.6 \times 10^{16}$
VO53/1H$_1$	LPE	1050–850	Excess	91	12.5	1.2×10^{16}	1.6×10^{16}	3.9×10^{16}	$\sim 7.4 \times 10^{16}$†
VO53/1H$_2$	LPE	1050–850	Excess	142	14.1	1.5×10^{16}	1.8×10^{16}	2.6×10^{16}	$\sim 3.6 \times 10^{16}$
VO68	LPE	1050–850	No	124	8.7	1.0×10^{16}	1.3×10^{16}	$\leqslant 1.4 \times 10^{16}$	$\leqslant 1.3 \times 10^{16}$
VO69/2	LPE	1050–850	Excess	90	10.2	2.9×10^{15}	3.3×10^{15}	3.3×10^{16}	5.5×10^{16} (a)‡ 5×10^{16} (b)
VO88/1	LPE	1050–850	No; care was taken to avoid O contamination	148	12.0	4.0×10^{16}	5.0×10^{16}	$< 5 \times 10^{15}$	$< 3.3 \times 10^{15}$

(a) Calculated from n_{deep} at 1000 K.
(b) Obtained from the fit in figure 1.

† Using $E_1 = E_2 = 780$ meV results in [O_P] = 4.6×10^{16} cm^{-3}.
‡ Using $E_1 = E_2 = 780$ meV results in [O_P] = 3.4×10^{16} cm^{-3}.

temperature Hall-effect data. The method is similar to the one used by Foster and Scardefield (1969) and Saul and Hackett (1970). Our measurements were performed on p-type (Zn doped) LPE layers grown between 1050 and 1030 °C from melts containing excess Ga_2O_3 (type I) and melts nearly saturated with Ga_2O_3 (type II). The degree of compensation was estimated from the comparison with p-type layers grown from melts without Ga_2O_3 addition. The results are summarized in figure 3.

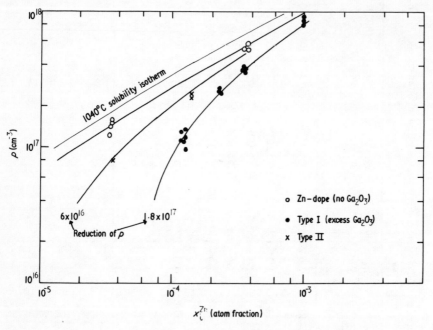

Figure 3. Room temperature Hall effect data on Zn-doped, Zn- and O-doped (type I) and Zn- and O-doped (type II) LPE layers grown from 1050 to 1030 °C.

The departure of the Zn-doped crystals from the solubility isotherm at 1040 °C is due to the loss of Zn from the solution prior to growth, which was very reproducible and agrees with the analytically determined Zn loss.

From the data for type I samples, $[O_P] \simeq 3 \cdot 5 \times 10^{17} cm^{-3}$ is derived, confirming the results of Saul and Hackett (1970). For type II samples, $[O_P] \simeq 1 \cdot 2 \times 10^{17} cm^{-3}$ is found. In both cases, these concentrations are corrected for the influence of oxygen doping on the incorporation of zinc, using $n_i = 1 \cdot 6 \times 10^{17} cm^{-3}$ at the growth temperature.

7. Discussion of the results

A comparison of the room temperature free electron concentration of undoped crystals, and those doped with oxygen, indicates that generally n_t is significantly reduced by the addition of oxygen to the melt (table 1). This can be understood from the fact that O_P can bind two electrons; in binding the second electron, O_P acts as a compensating acceptor, thus lowering E_F.

In the presence of oxygen, the incorporation of Si_{Ga} donors is generally suppressed due to the oxidation of the Si present. This could also account for the reduction of n_t observed in oxygen doped crystals.

We therefore made a detailed analysis of the shallow donors in sample VO88/1, using the Hall effect data of figure 1, photoluminescence studies at $1 \cdot 6$ K, and optical absorption measurements on bound excitons, as described earlier (Vink *et al* 1972), in order to determine $[S_P]$. From these measurements, VO88/1 can approximately be characterized as follows: $[S_P] \simeq 2 \times 10^{16} \, \mathrm{cm}^{-3}$; $[Si_{Ga}] \simeq 10^{17} \, \mathrm{cm}^{-3}$; $[C_P] \simeq 7 \times 10^{16} \, \mathrm{cm}^{-3}$.

Since we, and also Bachrach *et al* (1972), found that the addition of Ga_2O_3 to the melt strongly reduces the incorporation of Si in GaP, the position of the Fermi level in samples like VO53/1 and VO69/2 can only be accounted for if the incorporation of compensating carbon acceptors, C_P, is also suppressed by the addition of Ga_2O_3.

The results of the determination of $[O_P]$ in different samples, presented in this paper, are summarized in table 2, together with the available data from the literature. With respect to the conditions of oxygen doping, two groups of crystals can be distinguished:

Type I crystals, grown from a melt with excess Ga_2O_3 added. According to Kowalchik *et al* (1972), this leads to co-precipitation of Ga_2O_3 in the grown GaP crystal.

Type II crystals, grown from a melt nearly saturated with Ga_2O_3. In this case, the absence of excess Ga_2O_3 prevents its co-precipitation.

In these two cases, the actual oxygen concentration in the melt is assumed to be nearly the same. Based on the incorporation of Te in GaP, Jordan *et al* (1973) suggested that the two different modes of oxygen doping (type I and type II) lead to two different incorporation mechanisms. In type I crystals the large number of Ga_2O_3 inclusions allows the oxygen to reach the equilibrium concentration. This is referred to as equilibrium incorporation. In type II crystals the oxygen incorporation is supposed to be surface controlled, referred to as non-equilibrium incorporation. The simplest reaction to describe the equilibrium incorporation of oxygen is given by $O_1 + V_P \rightleftharpoons O_P^+ + e^-$. Since the concentration of electrons at the growth temperature depends on the position of the Fermi level, one expects the oxygen concentration to depend on whether the material is intrinsic, extrinsic p-type, or extrinsic n-type. A comparison of $[O_P]$ in p-type samples, $[O_P] \simeq 3 \cdot 5 \times 10^{17} \, \mathrm{cm}^{-3}$, in intrinsic n-type GaP (VO69/2), $[O_P] \simeq 6 \times 10^{16} \, \mathrm{cm}^{-3}$ and in extrinsic n-type GaP (VO61) (Te doped, $n \simeq 5 \cdot 5 \times 10^{17} \, \mathrm{cm}^{-3}$) $[O_P] = 8 \times 10^{15} \, \mathrm{cm}^{-3}$, confirms the idea of equilibrium incorporation. All these crystals were grown around the same temperature and from solutions with excess Ga_2O_3 added. A strong indication to equilibrium incorporation of oxygen in these crystals is found from a study of the dope dependence of the free electron recombination by Van der Does de Bye (1974). From the dope dependence of the minority carrier lifetime τ_{min} in our p-type ZnO doped LPE layers, the concentration of ZnO complexes and total substitutional oxygen concentration have been calculated. The oxygen concentration thus determined in type I crystals, having a net carrier concentration of $p = 4 \times 10^{17} \, \mathrm{cm}^{-3}$, has a value of $[O_P] \simeq 2 \times 10^{17} \, \mathrm{cm}^{-3}$. Furthermore, the oxygen concentration, $[O_P]$, turns out to be a function of the Zn concentration in the solid. For $[Zn_{Ga}] \gtrsim n_i$, the oxygen concentration increases with increasing $[Zn_{Ga}]$. This is just what is expected for equilibrium incorporation of O.

The dependence of $[O_P]$ on $[Zn_{Ga}]$ is also expected to be seen in the results of

Table 2. Survey of the results of different determinations of the oxygen concentration in GaP

Reference	Sample No.	Mode of O-doping (see text)	Type of crystal at growth temp.	Temperature of growth (°C)	Measuring method	Oxygen concentration (cm^{-3})	Oxygen in crystal measured
a	VO53	Type I	Intrinsic (n)	1050–850	High T Hall effect	$7\cdot4 \times 10^{16}$	O$_P$
a	VO53	I	Intrinsic (n)	1050–850	High T Hall effect	$3\cdot6 \times 10^{16}$	O$_P$
a	VO69/2	I	Intrinsic (n)	1050–850	High T Hall effect	$5\cdot5 \times 10^{16}$	O$_P$
a	SV11	I	Intrinsic (n)	1100–800	High T Hall effect	$3\cdot6 \times 10^{16}$	O$_P$
a	VO88/1	No O-dope	Intrinsic (n)	1050–850	High T Hall effect	$< 3\cdot3 \times 10^{15}$	O$_P$
a	VO69/2	Type I	Intrinsic (n)	850	Photocapacitance	8×10^{15}	O$_P$
a	VO61	I	Extrinsic (n)	1040	Photocapacitance	8×10^{15}	O$_P$
a		I	Extrinsic (p)	1040	Degree of compensation	$3\cdot5 \times 10^{17}$	O$_P$
a		II	Extrinsic (p)	1040	Degree of compensation	$1\cdot2 \times 10^{17}$	O$_P$
b		I	Extrinsic (p)	1050–1030	τ_{min}	2×10^{17}	O$_P$
c		II	Extrinsic (p)	1144–850	Degree of compensation	7×10^{16}	O$_P$
d		I	Extrinsic (p)	1025	Degree of compensation	$2\cdot6 \times 10^{17}$	O$_P$
e		I	Extrinsic (p)	1100	Spectroscopic	$\sim 1\cdot5 \times 10^{17}$	O$_P$
f		II	Extrinsic (p)	1025	Photocapacitance on p–n junction	$1\cdot5 \times 10^{16}$	O$_P$
g		II	Extrinsic (p)	1025	Photocapacitance on Schottky diode	$2\cdot8 \times 10^{16}$	O$_P$
g		II	Extrinsic (n)	1025	Photocapacitance on Schottky diode	$\sim 8 \times 10^{15}$	O$_P$
h		I	?	?	^{3}He activation	10^{19}	O$_{total}$
i		II	Extrinsic (p)	1050	Proton activation	$5\cdot5 \times 10^{16}$	O$_{total}$

a This paper, b Van der Does de Bye (1974), c Foster and Scardefield (1969), d Saul and Hackett (1970), e Jordan et al (1972), f Kukimoto et al (1972), g Kuki.oto et al (1973), h Kim (1971), i Lightowlers et al (1973).

[O_P] from the degree of compensation (figure 3). These measurements are not accurate enough, however, to reveal such dependence. The solubility of oxygen in GaP, determined in this paper, and the available literature data, as summarized in table 2, are shown as a function of reciprocal temperature in figure 4. The equilibrium incorporation in p-type and intrinsic GaP, calculated by Jordan *et al* (1973) is also shown in figure 4. All results on type I crystals, including the photocapacitance measurement on VO69/2, can be satisfactorily described by the equilibrium incorporation model.

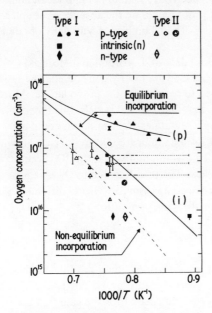

Figure 4. Data of O_P as a function of the reciprocal temperatue of growth

▲ ♤ Saul and Hackett (1970)
● ○ This paper (compensation measurement)
x   Van der Does de Bye (1974)
■   This paper (Hall effect at high temperature). The broken curve indicates temperature region where most of the layer thickness has been grown.
▮ ♦ This paper (photocapacitance)
⏀ Lightowlers *et al* (1973)
⊛ ⊘ Kukimoto *et al* (1973)
⋋ Foster and Scardefield (1969)
△ Jordan *et al* (1973).

For type II crystals no effect on the oxygen incorporation is expected, regardless of whether they are n- or p-type. This is not confirmed, however, by the data of Kukimoto *et al* (1973). In their type II crystals, [O_P] = 6–10 × 10^{15} cm^{-3} for n-type and [O_P] = 2·2–3·4 × 10^{16} cm^{-3} for p-type crystals. The result on p-type crystals fits well, however, with the results of Lightowlers *et al* (1973) on their type II crystals. Furthermore, for p-type crystals grown from a melt nearly saturated with Ga_2O_3 (type II) a lower oxygen concentration is expected than for type I p-type crystals, since the surface controlled incorporation results in lower oxygen concentrations (Jordan *et al* 1973).

As mentioned in §6, this difference between type I and type II crystals is indeed found. As can be seen from figure 4, the reduction of $[O_P]$ is less than expected, however. This could be attributable to the inaccuracy of this determination. A more direct proof for equilibrium or surface controlled incorporation could be provided by a study of the effect of substrate orientation, for example (111) A and (111) B faces, on $[O_P]$. For n-type, O-doped crystals such experiments are feasible, using the high temperature Hall effect technique. Under similar growth conditions we found for Te-donors an orientation effect of nearly one order of magnitude between (111) A and (111) B faces, confirming a non-equilibrium incorporation. Another proof would be the absence of coupled incorporation of Zn and O in type II crystals. The technique of Van der Does de Bye is very suitable for such a study.

Finally, the electron spin resonance (ESR) data of Toyotomi and Morigaki (1970) obtained on solution-grown, n-type, O-doped GaP are briefly considered. The broad (150 Oe) resonance line observed is ascribed to oxygen donors in the lattice. Assuming the identification to be correct, it is likely that in practical crystals, this broad line can be detected only when $[O_P]$ is greater than a few times 10^{16} cm^{-3}; this could agree with our data. An interesting point is that the ESR data indicate that in the case of two electrons on O_P, the spins are not paired. This would imply that the $[O_P]$ values calculated with equation (2) are somewhat too low. The ESR line due to O_P is reported to saturate very easily. This makes a quantitative comparison of $[O_P]$ in different crystals difficult and ESR probably less suitable for studying orientation effects on $[O_P]$.

8. Conclusion

High temperature Hall effect measurements on a deep donor level in O-doped, n-type GaP are reported. Evidence is presented to show that this level is due to O_P. The maximum concentration of oxygen in these crystals, calculated for the case that O_P can bind two electrons, is found to be $\simeq 6 \times 10^{16}$ cm^{-3}. Comparison of the data obtained on intrinsic n-type crystals (Hall effect measurements), on p-type crystals (degree of compensation), and on extrinsic n-type crystals (photocapacitance), supports the interpretation of the oxygen doping with an equilibrium incorporation model for the case that crystals are grown from melts with excess Ga_2O_3 added.

Experiments that could provide further evidence for the equilibrium model for type I crystals and for the non-equilibrium model for type II crystals are suggested.

Acknowledgments

The authors are indebted to Dr C H Henry, Bell Laboratories, for performing photocapacitance measurements on VO69 and VO61. We wish to thank Mr A de Vos for his assistance in growing the crystals, Miss H Mulder for the sample preparation and Hall effect measurements and Dr T N Morgan for a reprint of the paper presented at the Semiconductor Physics Conference (Stuttgart).

References

Bachrach R Z, Lorimor O G, Dawson L R and Wolfstirn K B 1972 *J. Appl. Phys.* **43** 5098
Bhargava R N 1971 *Philips Techn. Rev.* **32** 261

Champness C H 1956 *Proc. Phys. Soc.* B **69** 1335

Dean P J 1969 *Appl. Solid St. Sci.* vol I ed R Wolfe (New York: Academic Press)

Dishman J M, DiDomenico M and Caruso R 1970 *Phys. Rev.* B **2** 1988

van der Does de Bye J A W 1974 *Electrochem. Soc. Spring Meeting San Francisco*

Foster L M and Scardefield J 1969 *J. Electrochem. Soc.* **116** 494

Gijsbers K, Bastings L and van de Leest R 1974 *Analyst* **99** 376

Grimmeiss H G, Kishio W and Scholz H 1965 *Philips Techn. Rev.* **26** 136

Grimmeiss H G, Ledebo L A, Ovren C and Morgan T N 1974 *Paper 16A04 12th Int. Conf. Physics of Semiconductors Stuttgart.*

Jayson J S, Bachrach R Z, Dapkus P D and Schumaker N E *Phys. Rev.* B **6** 2357

Jordan A S, Derick L, Caruso R and Kowalchik M 1972 *J. Electrochem. Soc.* **119** 1585

Jordan A S, Trumbore F A, Wolfstirn K B, Kowalchik M and Roccasecca D D 1973 *J. Electrochem. Soc.* **120** 791

Kim C K 1969 *Radiochem. Radioanal. Lett.* **2** 53

—— 1971 *Anal. Chim. Acta* **54** 407

Kowalchik M, Jordan A S and Read M H 1972 *J. Electrochem. Soc.* **119** 756

Kukimoto H, Henry C H and Merritt F R 1973 *Phys. Rev.* B **7** 2486

Kukimoto H, Henry C H and Miller G L 1972 *Appl. Phys. Lett.* **21** 251

Lightowlers E C, North J C, Jordan A S, Derick L and Merz J L 1973 *J. Appl. Phys.* **44** 4758

Pantelides S T 1974 *Solid St. Commun.* **14** 1255

Peters R C 1973 *Proc. 4th Int. Symp. GaAs and Related Compounds* (London and Bristol: Institute of Physics) p55

Saul R H and Hackett W H 1970 *J. Electrochem. Soc.* **117** 921

Toyotomi S and Morigaki K 1970 *J. Phys. Soc. Japan* **29** 800

Vink A T, Bosman A J, van der Does de Bye J A W and Peters R C 1972 *J. Luminesc.* **5** 57

Wiley J D 1971 *J. Phys. Chem. Solids* **32** 2053

Photoemission and secondary emission of gallium phosphide epitaxial layers

C Piaget, P Guittard, J P André and P Saget

Laboratoires d'Electronique et de Physique Appliquée, 3 avenue Descartes 94450 Limeil-Brévannes, France

Abstract. Photoemission and secondary emission have been studied on Zn doped gallium phosphide vapour phase homoepitaxial layers. Activation to negative electron affinity is obtained by heat cleaning followed by Cs or CsO absorption.

A careful analysis of photoemission spectral response has been carried out; it includes optical, electron transport and surface emission properties of both X and Γ minima of the conduction band. From this analysis electron diffusion lengths, as well as emission probabilities, for electrons thermalized in both minima may be extracted. Diffusion lengths in the X minimum range from 0.7 to $3\,\mu$m, depending on doping level, while diffusion lengths in the Γ minimum are independent of it, typically $0.07\,\mu$m. The emission probability of electrons thermalized in the Γ minimum is high, close to 0.5, but, in opposition to usual statements, the emission probability of electrons from the X minimum is always very low; about 0.05 with Cs activation and no more than 0.17 for CsO activation. The emission process is discussed in terms of band structure.

A comparative study of secondary emission and photoemission shows that reflection secondary emission in the $0-4$ keV primary energy range cannot be described as a thermalized electron emission, even when negative electron affinity is achieved: it is shown that the energy distribution of secondary electrons is far broader than that of photoelectrons; moreover, the escape depth of secondary electrons is shown to vary continuously from 0.01 to $0.2\,\mu$m as the work function is lowered from 5.5 to 1.3 eV. The usually quoted diffusion length value of $0.2\,\mu$m, deduced from secondary emission measurements, is consequently interpreted as the escape depth of hot electrons induced in the GaP crystal by primary electrons.

1. Introduction

Negative electron affinity emitters have been studied extensively during the last ten years (review papers by Williams and Tietjen 1971, Bell 1973). This activity has been mainly devoted to $1-1.4$ eV band gap semiconductors for near infrared detection, while a rather restricted research activity has been dealing with GaP emitters.

The largest part of the available information regarding these emitters was already given in 1969 by Simon and Williams: photoemission (Williams and Simon 1967) and secondary emission (Simon and Williams 1968) are shown to be explained by a three-step process:

electron excitation by photons or primary electrons; followed by a rapid thermalization to the bottom of conduction band via phonon interactions
thermal diffusion (with diffusion length L)
emission into vacuum through the surface region (with emission probability P).

Typical values for $10^{19}\,\text{cm}^{-3}$ p-type GaP appear to be $0.2\,\mu\text{m}$ for diffusion length and 0·5 for emission probability as confirmed from photoemission (Garbe 1969b) or secondary emission (Simon *et al* 1968) measurements.

Since 1969, photomultipliers have been made including GaP–Cs dynodes, allowing good pulse height resolution owing to their high secondary emission coefficient.

In 1970 Coates showed that secondary emission measured on a photomultiplier GaP dynode was best described by a much smaller escape depth L than previously stated. On the other hand Aphonina *et al* (1971) point out a non linear increase in δ with primary energy in the 0·1 to 2·5 keV range, and more recently (Aphonina and Stuchinskii 1974) estimate that secondary emission from GaAs as well as GaP is mainly dependent upon electron scattering in the solid. Although these questions may seem to be purely academic they are of great interest as far as present and future devices are concerned. Our work on polycrystalline GaP dynodes for photomultipliers as well as the interest of transmission secondary emitters and cold cathodes has led us to make a comparative study of photoemission and secondary emission properties of GaP–Cs emitters.

2. Experimental conditions

Samples are 30 to $50\,\mu\text{m}$ thick vapour phase epitaxial layers, grown by the hydride method on 100 (few degrees off in the 110 direction) GaP substrates; p type is achieved by Zn doping in the 10^{16} to $3 \times 10^{19}\,\text{cm}^{-3}$ range. Samples are introduced after growth in the ultra high vacuum chamber; a pressure lower than $10^{-10}\,\text{Torr}$ is reached and maintained during the whole experiment. The activation procedure is similar in every respect to that used for low band gap (Ga, In)As samples (Piaget *et al* 1974); heat cleaning is processed here between 650 and 680 °C by Joule or electron bombardment heating; Cs is evaporated from a chromate channel source. When, occasionally, oxygen is used, it is admitted through a leak valve. Photoemission measurements are performed using a high resolution grating monochromator, great care is taken for flux calibration, specially for relative spectral dependence (with thermopile), absolute calibration is achieved with a silicon photodiode standard. Secondary emission is measured in steady conditions, at low beam current, in the 0·5 to 4 keV primary energy range.

3. Photoemission

3.1. Analysis of experimental results

A typical spectral response curve, obtained for optimum activated samples, is plotted in figure 1. Two energy ranges can be distinguished (2·25 to 2·6 and 2·6 to 3·5 eV), the second one being related to the increase of the optical absorption coefficient α for $h\nu > 2.5$ eV. This increase is due to the onset of direct transitions ($\Gamma_{15}^{\text{v}} \rightarrow \Gamma_{1}^{\text{c}}$) which become dominant above 2·7 eV (see Zallen and Paul 1964).

Clearly, a diffusion model involving a single band would be inadequate to analyse our data over the whole energy range. In the case of GaAs, James and Moll (1962) were able to fit their experimental results with a two band diffusion model, where they mainly assume that carriers of the higher conduction minimum tend to be transferred

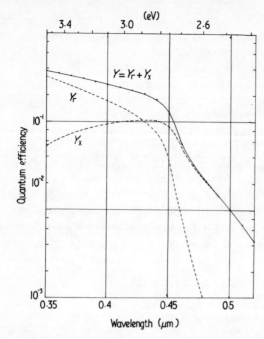

Figure 1. Typical spectral response for optimum activated samples. Experimental points for a $1 \cdot 2 \times 10^{18}$ cm^{-3} doped layer, theoretical curve with $P_\Gamma = 0 \cdot 483$, $L_\Gamma = 0 \cdot 079$ μm, $P_x = 0 \cdot 175$, $L_x = 1 \cdot 1$ μm. Broken curves: $Y_x Y_\Gamma$, full curve $Y = Y_x + Y_\Gamma$.

to the lower conduction minimum rather than to recombine directly. The same assumption seems reasonable for GaP too, reminiscent of the high electron–phonon interaction probabilities as well as the large density of states in the X minima. Solving the two equations describing the diffusion at both conduction band minima one obtains their respective photoemissive quantum efficiencies:

$$Y_\Gamma = \frac{P_\Gamma F_\Gamma}{1 + 1/\alpha L_\Gamma} \tag{1}$$

$$Y_X = \frac{P_X F_X}{1 + 1/\alpha L_X} \left(1 + \frac{F_\Gamma/F_X}{(1 + L_\Gamma/L_X)(1 + \alpha L_\Gamma)}\right) \tag{2}$$

where Γ and X subscripts refer to corresponding conduction band minima, P are emission probabilities, L diffusion lengths, and F_Γ and F_X the probabilities for electrons to be excited into the corresponding minimum at photon absorption. These probabilities may be computed, as James and Moll did, from the band structure, they may be extracted easily for GaP from optical absorption coefficients: up to $2 \cdot 5$ eV available α values (Spitzer 1959, Zallen and Paul 1964) have a very good fit to the following relation

$$\alpha(\text{cm}^{-1}) = 7 \cdot 6 \times 10^3 \, (h\nu \, (\text{eV}) - 2 \cdot 215)^2 \tag{3}$$

Now, if we assume that, above $2 \cdot 5$ eV, relation (3) is still valid to evaluate the transition probability $F_X \alpha$ from Γ_{15}^v to X_1^c, both F_X and F_Γ can be deduced over the whole energy range, since $F_X + F_\Gamma = 1$, using values of α reported by Spitzer (1959), Zallen and Paul (1964) and Morgan (1974), see figure 2. Practically, L_X and P_X are first extracted from the linear part (corresponding to the 2·4 to 2·6 eV range) of the usual Y^{-1} (α^{-1}) graph, P_Γ and L_Γ are then iteratively computed using relations (1) and (2). Using this procedure, we are able to fit our experimental data within 10% accuracy (see figure 1).

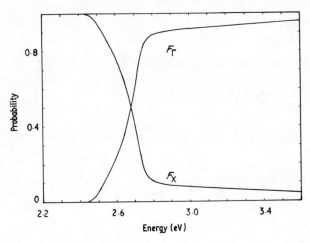

Figure 2. Probability for electrons to be excited into X_1^c and Γ_1^c minima at photon absorption.

3.2. Comments

L and P values obtained in this way are gathered together in table 1. Measured L_X values are much higher than previously reported for bulk samples (Williams and Simon 1967, Garbe 1969b), but in better agreement with those obtained on epitaxial layers by other methods than photoemission (Bachrach and Lorimer 1972, Gershenson and Mikulyak 1966, Hackett 1972). On general grounds these values seem to be more reasonable for an indirect gap semiconductor.

A still more surprising result concerns emission probabilities. As P_Γ is always of the order of 0·5, P_X is always less or equal to 0·05 after Cs activation; it may be increased to 0·1—0·17 by CsO activation but remains still much smaller than P_Γ although the vacuum

Table 1. Electron transport and emission parameters of GaP—Cs emitters for different doping levels. L are diffusion lengths, P emission probabilities; subscripts Γ and X refer to the corresponding conduction band minima, SE to secondary electrons.

$N_A - N_D$ (cm^{-3})	L_X (μm)	P_X	L_Γ (μm)	P_Γ	L_{SE} (μm)	P_{SE}
2×10^{19}	0·7	0·04	0·035	0·5		
1×10^{18}	1·2	0·05—0·17†	0·07	0·59	0·2	0·55
3×10^{17}	3·0	0·025	0·06	0·50	0·2	0·42

† Cs-O activation

energy level is about 1 eV lower than the conduction band in the bulk. The most likely reason for this result seems to be related to the 100 **k** vector orientation for electrons thermalized in the X minima (Bell 1973).

Indeed, LEED experiments reveal that heating the (100) surface for a good desorption of residual surface impurities gives rise to (110) oriented facets. Then, we write that, during the electron emission process across the surface region

(i) the total energy is conserved
(ii) the transverse wave vector is conserved

If no phonon scattering takes place in the surface region, we obtain:

$$(E_{vac} - E_F) + \frac{\hbar^2}{2m} (k_T^2 + k_N^2) = E_X - E_F \tag{4}$$

$$k_T = k_T \text{ (in the solid at X point)}$$

$$\cong \frac{2\pi}{a} \cos ([100], [110])$$

$$= \frac{\pi\sqrt{2}}{a} \tag{5}$$

In (4) and (5) k_N and k_T denote respectively the normal and transverse components of **k** in vacuum, E_X is the bulk value of the minimum of the conduction band, E_F the Fermi level, E_{vac} the potential energy of electrons in vacuum and a the lattice periodicity.

Electrons can only be emitted if $k_N \geqslant 0$ or if

$$E_X - E_{vac} \geqslant \frac{\hbar^2 \pi^2}{ma^2} \tag{6}$$

which is nearly equal to 2·5 eV for GaP. This high value implies that no emission is possible in actual conditions from the X minimum, without a collision with a phonon at the surface, which is a low probability event.

If electron–phonon interactions are taken into account in the bent band region, the electron energies are somewhat lowered, but their wave vectors are scattered all around X points in the conduction band so that some electrons may reach the surface with $k_T \ll \pi\sqrt{2}/a$ and be emitted. If the band bending were larger than about 0·8 eV (which in our opinion, it is not) electrons could be 'heated' to Γ (or K) points in the Brillouin zone and then have a high emission probability. Because of the band structure this last phenomenon has a high probability in GaAs, even the absence of any field, this explains the high P_X values measured for this semiconductor (James and Moll 1969, Garbe 1969b).

4. Secondary emission

The variation of secondary emission with primary energy has been measured in the same activation states as photoemission. Escape depth L_{SE} and emission probability

P_{SE} are calculated using the three step model with the following assumptions:

(*a*) Constant energy loss according to Young's (1957) measurements on Al_2O_3 for low energy electrons, the mean energy for electron–hole pair creation is taken equal to 6·54 eV (Kobayashi 1972).

(*b*) One dimension diffusion equation in semi infinite sample with a mean diffusion length L_{SE}.

(*c*) Mean emission probability for all secondaries P_{SE}.

The results of these measurements are given in table 1 (columns 6 and 7). They call for some remarks:

L_{SE} and P_{SE} are independent of doping level.

They agree with previous published values as well as with our own measurements on polycrystalline layers.

L_{SE} corresponds to neither L_X nor L_Γ; a model including diffusion at both minima (equivalent to James and Moll's for photoemission) does not give good results, either.

These remarks suggest that secondary electrons may be emitted before thermalization.

Energy distribution of secondary electrons has been measured to check this assumption. A four grid retarding field Auger spectrometer has been used for this purpose with a small 10 mV analysing voltage. Figure 3 gives energy distributions of secondary electrons for 200 and 1000 eV primary energies, that of photoelectrons (for white light illumination) is also given for comparison and as a check for resolution.

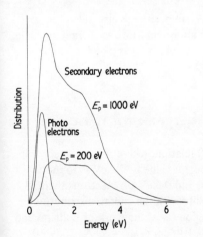

Figure 3. Energy distribution for GaP–Cs emitters ($N_A = 2 \times 10^{19}$ cm^{-3}), secondary electrons (200 and 1000 eV primary energies), photoelectrons (white light illumination).

The secondary electron distribution is about 2·5 eV (FWHM) broad while photoelectrons are emitted from the conduction band edge with an 0·4 eV broad distribution. Only few secondary electrons may be expected to come from conduction band edge, even for 1000 eV primary energy.

A large number of electrons have energies up to 1·8 eV, more than thermalized electrons this figure may be compared to a kinetic energy of 1·75 eV for hot electrons, as calculated for GaP on the basis of Zulliger's (1971) calculations for Si, and with $\epsilon = 6 \cdot 54$ eV (Kobayashi 1972).

A second experiment has been performed to go further into details: the caesium deposition on a clean surface has been processed in 15 steps; after each step, spectral photoemission yield $Y(h\nu)$ was measured (giving the work function ϕ and when NEA is achieved L_Γ, P_Γ and L_X, P_X) as well as the energy dependence of the secondary emission yield $\delta(E_p)$ (from which L_{SE} and P_{SE} are calculated). Emission probabilities P_Γ and P_{SE} are shown in figure 4 as a function of ϕ, the relative dependence of photo-

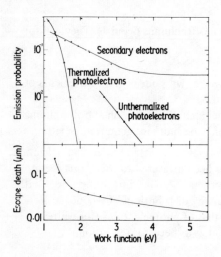

Figure 4. Emission probabilities and escape depth as a function of work function for secondary electrons, unthermalized photoelectrons (relative) and thermalized in the Γ minimum photoelectrons. $N_A = 4 \times 10^{17}$ cm^{-3}.

emission for 3·9 eV photons when electron affinity is still positive is also given for comparison. These curves may be considered, at first approximation, as an energy distribution analysis of emitted electrons (the vacuum level acting as a high-pass energy filter):

$$Y(\phi) \quad \text{or} \quad \delta(\phi) = KP(\phi) = \int_\phi^\infty N(E) \, dE \qquad (7)$$

where K depends on electron excitation and transport properties, $N(E)$ is the energy distribution of electrons after emission.

These curves follow approximately an $\exp(-\phi/A)$ relation, where A may be compared to an 'equivalent temperature' of the emitted electrons. A value of 5·8 eV is found for secondaries, 2·4 eV for unthermalized photoelectrons and 0·6 eV for electrons thermalized in the Γ minimum. As this last value fits reasonably with the energy distribution resulting from phonon scattering in the bent band region (as in the work of Williams and Simon 1967, Escher and Schade 1973) with $N_A = 4 \times 10^{17}$ cm^{-3}, $l_{ph} = 39$ Å, $\epsilon_{ph} = 50$ meV), we must conclude, from this experiment too, that the major part of secondary electrons are not thermalized before emission.

The lower part of figure 4 shows the corresponding dependence of L_{SE} on the work function. A continuous increase in L_{SE} from 0·015 to 0·2 μm does not indicate, either, the onset of a new type of diffusion process (at band edge) for electrons when negative electron affinity is achieved.

This last result explains why 'diffusion lengths' extracted from secondary emission on well activated GaP have always led to values of the order of 0·2 μm, whatever the sample was.

5. Discussion

The present results have a straightforward interest for device problems.

5.1. Reflection secondary emitters

Photomultiplier dynodes work generally at low primary energies, typically from 250 to 1000 eV. As secondaries are mainly hot electrons at emission, diffusion length, and thereby crystal quality, have nearly no effect on secondary emission. This explains why polycrystalline dynodes show as good secondary emission as single crystals or epitaxial layers. We have already observed that GaP layers with grain size much smaller than 1 μm give as good results as large size ones. As a consequence, secondary emission of GaP dynodes is only dependent of activation quality.

Secondary electron emission probability has been shown to vary slowly with work function. P_{SE} decreases only by less than 10% for an increase in work function of 0·1 eV (see figure 4), this is the reason why the secondary emission of photomultiplier dynodes is very stable and, moreover, why statistical spatial fluctuations on δ are so low for such an heterogeneous surface as that of polycrystalline GaP.

A last point concerns the design of electron optics for multiplier chains. Secondary electrons from GaP–Cs emitters must not be considered as electrons with nearly zero kinetic energy; their energy distribution (figure 3) must be taken into account.

5.2. Transmission secondary emitters

For such devices, long escape lengths are necessary, of the order of technologically achievable sample thicknesses. Negative electron affinity GaP–Cs transmission emitters appear now to be possible as L_X diffusion lengths are of the order of a few micrometres. Unfortunately we have also shown that the emission probability, for those electrons which are thermalized in the X-minima, is very low (about 5×10^{-2}). If this is actually due, as we think, to faceting effects of the (100) surface, it may be solved if unfaceting cleaning techniques are found (this would also be of great interest for image tubes III–V photocathodes). If not, only direct bandgap semiconductors are to be considered for transmission mode secondary emitters.

6. Conclusion

Photoemission and secondary emission measurements have been compared: p type GaP epitaxial layers may have electron diffusion length at X_1 minima of the order of a few microns. When negative electron affinity is achieved electrons may be emitted from the Γ_1^c minimum with a high emission probability (0·5 or more) but those from the X_1 minima have a poor one (0·05 for GaP–Cs, 0·17 for GaP–CsO). Reflection secondary electrons have been shown to be mainly unthermalized for primary energies up to 1 keV, escape depth may increase up to 0·2 μm depending on activation conditions.

Acknowledgments

This work was supported by the Délégation Générale pour la Recherche Scientifique under contract number 737 1066 00 221 75 01.

The authors thank D Poulain for his constant help, they are greatly indebted to J P Hurault for discussions on theoretical aspects and comments on manuscript.

References

Aphonina L Ph *et al* 1971 *C R Acad. Sci. URSS* **35** 1046
Aphonina L Ph and Stuchinskii G B 1974 *Sov. Phys.–Solid St.* **15** 1448
Bachrach R Z and Lorimer O G 1972 *J. Appl. Phys.* **43** 500
Bell R L 1973 *Negative Electron Affinity Devices* (Oxford: Clarendon Press)
Coates P B 1970 *J. Phys. D: Appl. Phys.* **3** L25
Escher J S and Schade H 1973 *J. Appl. Phys.* **44** 5309
Garbe S 1969a *Solid St. Electron* **12** 893
—— 1969b *Phys. Stat. Solidi* **33** K87
Gershenson M and Mikulyak R M 1966 *Appl. Phys. Lett.* **8** 245
Hackett W H *et al* 1972 *J. Appl. Phys.* **43** 2857
James L W and Moll J L 1969 *Phys. Rev.* **183** 740
Kobayashi T 1972 *Appl. Phys. Rev.* **21** 150
Morgan A E 1974 *Surface Sci.* **43** 150
Piaget C, Polaert R and Richard J C 1974 *6th Symp. Photoelectronic Image Devices*
Simon R E, Sommer A H, Tietjen J J and Williams B F 1968 *Appl. Phys. Lett.* **13** 355
Simon R E and Williams B F 1968 *IEEE Proc.* **NS15** 167
Spitzer S W G 1959 *J. Phys. Chem. Solids* **11** 339
Williams B F and Simon R E 1967 *Phys. Rev. Lett.* **18** 485
Williams B F and Tietjen J J 1971 *IEEE Proc.* **59** 1489
Young J R 1957 *J. Appl. Phys.* **28** 524
Zallen R and Paul W 1964 *Phys. Rev.* **134** A1628
Zulliger H R 1971 *J. Appl. Phys.* **42** 5570

Precision lattice parameter measurements on doped gallium arsenide

C M H Driscoll, A F W Willoughby
Engineering Materials Laboratories, The University, Southampton, UK

J B Mullin and B W Straughan
Royal Radar Establishment, Malvern, UK

Abstract. Precision lattice parameter measurements have been used to study doping effects in gallium arsenide in order to provide a basis for the understanding of strain-controlled degradation of semiconductor devices. A comprehensive characterization of the change in lattice parameter of gallium arsenide produced by a number of group IV and group VI dopants over a wide concentration range has been made. This study shows that for substitutional impurities, such as tin and tellurium, with larger atomic radii than the matrix atoms involved in the replacement, a monotonic increase in lattice parameter results as the doping level is increased above 10^{18} cm^{-3}, and above these levels deviations from Vegard's law were found. At doping concentration exceeding 10^{19} cm^{-3}, changes greater than 100 parts per million in the lattice parameter are observed, with accompanying broadening of the diffraction profile, indicating possible inhomogeneous distribution of impurities at these higher doping levels. The lattice mismatch, due to differences in doping level, between a high purity gallium arsenide layer and a heavily doped substrate has been shown by lattice parameter measurement to produce significant interfacial strain in the layer of a typical device structure. This study will be of value in assessing the likely dopants responsible for interfacial strain between layers and substrates as an aid to gallium arsenide device processing.

1. Introduction

Recent reports (Hasegawa and Ito 1972, Reinhart and Logan 1973, Petroff and Hartmann 1973) have indicated that strain introduced into gallium arsenide during semiconductor processing may be a controlling factor in device performance. One source of strain in a device structure is that arising from lattice mismatch in the growth of epitaxial layers of gallium arsenide, where the substrate and layer are doped to different levels (Kishino and Iida 1972). In view of this and to provide a basis for the understanding of strain-controlled degradation of semiconductor devices, precision lattice parameter measurements have been used to study doping effects in gallium arsenide. This x-ray technique is a convenient and non-destructive means of assessing strain since the fractional change in lattice parameter $\Delta a/a$ gives a direct measure of lattice deformation.

The change in lattice parameter produced by some common dopants from groups IV and VI of the periodic table over a wide concentration range has been made both on melt-grown and epitaxial gallium arsenide. These results have been used to gauge the impurities likely to give rise to significant strain levels in gallium arsenide, and have been related to lattice parameter measurements made on a device section consisting of a high purity epitaxial layer of gallium arsenide on a heavily doped substrate. Theoretical models, assuming elastic isotropy, have been used to examine the experimental behaviour.

2. Experimental technique

2.1. Crystal growth

Gallium arsenide single crystals grown by a variety of techniques and possessing a range of doping properties, as shown in table 1, have been used in this study. Undoped melt-grown and epitaxial gallium arsenide were required to obtain a lattice parameter for strain-free material to be compared with that for heavily doped gallium arsenide. The melt-grown material used in the doping studies was pulled as single crystals by the Liquid Encapsulation Technique (Mullin *et al* 1965) in a pressure puller with high purity boric oxide as the encapsulant using an overpressure of 100 psi nitrogen. Chromium and group IV doped crystals were produced by adding the element to the melt, while group VI dopants were added to the melt as the gallium compound (eg gallium sulphide). Crystals grown with a deliberately high dislocation density were seeded onto [100] seeds which had been plastically deformed on a four-point bending apparatus. Low dislocation density gallium arsenide was grown by suitably programming the shape of the crystal growth interface. LPE and VPE gallium arsenide had thicknesses generally greater than 20 micrometres, the x-ray penetration depth. The layers were grown either on semi-insulating or n^+ tellurium-doped gallium arsenide substrates.

2.2. Sample preparation and measurement

The automatic precision x-ray goniometer, used in this study, is based on the Bond lattice parameter technique (Bond 1960, Baker *et al* 1968, 1969), in which eccentricity, absorption and zero errors are eliminated by measuring the angle between two reflecting positions of a crystal instead of that between the diffracted beam positions (Driscoll and Willoughby 1972). Using the (800) Cu $K\beta$ reflection at a high Bragg diffracting angle of $80°$, whilst maintaining the temperature constant at 27 ± 0.1 °C, the precision in lattice parameter was better than 1 part per million. In general terms, this means that lattice deformation produced by elastic stress of magnitude greater than 10^6 dynes cm^{-2} ($=10^2$ kN m^{-2}) or impurities in concentrations approaching or greater than 10^{18} cm^{-3} can be monitored by this single crystal x-ray technique.

Melt-grown samples were sectioned for lattice parameter measurement parallel to a (100) plane by means of a diamond saw and were then ground and chemically polished (Willoughby *et al* 1971). Specimen surfaces were cleaned with trichloroethylene before the prepared sample was mounted on the x-ray goniometer using silicone grease to obtain a strain-free mounting. Lattice parameter scans, similar to those shown in figure 1, were made across all specimens in order to obtain a mean lattice parameter characteristic of the material and a standard deviation from the mean, giving a measure of the uniformity of the specimen, as given in table 1. Doping concentrations were assessed electrically by means of the Hall effect and doping profiles through epitaxial layers were obtained from differential capacitance measurements. Oxygen doping levels were obtained by gamma photon activation and liquid helium cryopumped mass spectrometry (Blackmore *et al*)†.

† G W Blackmore, B Clegg, L G Harvey, G A Heath, J F Hislop and J B Mullin to be published.

Table 1. Summary of lattice parameter measurements on gallium arsenide as a function of growth and doping properties

Sample data		X-ray data	
Doping Carrier concentration (cm^{-3}) Dislocation density ρ (cm^{-2})	Growth† layer thickness	Lattice parameter (Å) at 27 °C	Broadening (¾ breadth) (min of arc)
A. Undoped GaAs			
low ρ	Melt (a, b)	5·65325 ± 0·00002	10
	Epitaxy (c, d, e) 20–140 μm	5·65325 ± 0·00002	9·75
high $\rho \sim 10^7$	Melt (a)	5·65327 ± 0·00004	12·25
B. Group IV doped GaAs			
Si 1 × 10^{17}	Melt (a)	5·65325 ± 0·00001	9·75
3 × 10^{18}	Melt (f)	5·65323 ± 0·00001	9·75
Ge 8 × 10^{17}	Melt (a)	5·65331 ± 0·00020	13
2 × 10^{18}	Melt (a)	5·65334 ± 0·00001	12
Sn 1 × 10^{17}	Melt (b)	5·65325 ± 0·00001	10·5
3 × 10^{18}	Epitaxy (e) 14 μm	5·65355 ± 0·00002	10
1 × 10^{19}	Epitaxy (g) 23 μm	5·65420 ± 0·00005	11
Pb 4 × 10^{17}	Epitaxy (d) 91 μm	5·65324 ± 0·00003	10
2 × 10^{18}	Epitaxy (e) 6 μm	5·65330 ± 0·00003	9·75
4 × 10^{18}	Epitaxy (e) 12 μm	5·65329 ± 0·00003	10
C. Group VI doped GaAs			
O 10^{17}	Melt (a)	5·65318 ± 0·00005	10
S 2 × 10^{18}	Melt (a)	5·65323 ± 0·00001	11·75
4 × 10^{18}	Melt (a)	5·65329 ± 0·00002	12·25
Se 2 × 10^{18} low ρ	Melt (a)	5·65319 ± 0·00002	11·75
$\rho \sim 10^7$	Melt (a)	5·65325 ± 0·00010	12·5
Te 1 × 10^{18}	Melt (a)	5·65325 ± 0·00003	10·25
5 × 10^{18}	Melt (a)	5·65341 ± 0·00008	11·25
6 × 10^{18}	Melt (a)	5·65351 ± 0·00005	10
7 × 10^{18}	Melt (a)	5·65360 ± 0·00005	10
1 × 10^{19}	Melt (a)	5·65391 ± 0·00020	11·5
1 × 10^{19}	Epitaxy (g) 100 μm	5·65432 ± 0·00002	12·25
Cr 10^{17}	Melt (a)	5·65315 ± 0·00007	11·25

† Growth methods and establishments
(a) Czochralski Liquid Encapsulation   RRE (Malvern)
(b) Horizontal Bridgman   Mullard (Redhill)
(c) Vapour epitaxy   RRE & Plessey (Caswell)
(d) Liquid epitaxy from Ga solution   Mullard & STL (Harlow)
(e) Liquid epitaxy from Pb solution   STL (Harlow)
(f) Zero dislocation melt grown   Monsanto
(g) Liquid epitaxy   SERL (Baldock)

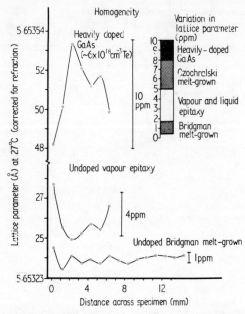

Figure 1. Lattice parameter scans across three gallium arsenide samples with different growth and doping properties, and the general variation in lattice parameter, giving a measure of the homogeneity, across different gallium arsenide specimens.

3. Results

3.1. Lattice parameter measurements on undoped gallium arsenide

Lattice parameter measurements were made on a wide range of undoped gallium arsenide in order to characterize effectively melt-grown, VPE and LPE material with doping levels below $10^{17}\,cm^{-3}$ and dislocation densities less than $10^4\,cm^{-2}$, assessed by etch-pit and x-ray topography measurements. The results are shown in the histograms of figure 2 for melt-grown and epitaxial material. The initial results on these two classes of material showed differences in lattice parameter, but, by referring the results to a common standard at 27 °C, it was found that both melt-grown and epitaxial material had the same mean lattice parameter, despite the wide differences in their growth temperatures. The mean lattice parameter for both types of undoped gallium arsenide was $5·65325 \pm 0·00002$ Å at 27 °C, and both types of material showed a similar gaussian distribution about the mean.

From studies of the variation in lattice parameter across undoped gallium arsenide samples, as shown in figure 1, horizontal Bridgman single crystals grown close to the stoichiometric composition were found to be the most uniform with a variation of less than 2 parts per million. Variations in lattice parameter across VPE and LPE gallium arsenide were generally in the range of 2 to 5 parts per million, being most uniform for samples grown from gallium or lead-rich solution on semi-insulating substrates. LEC gallium arsenide was generally less uniform than the horizontally grown gallium arsenide, and variations of lattice parameter across these former samples were in the range 4 to 8 parts per million, being independent of growth axis.

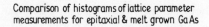

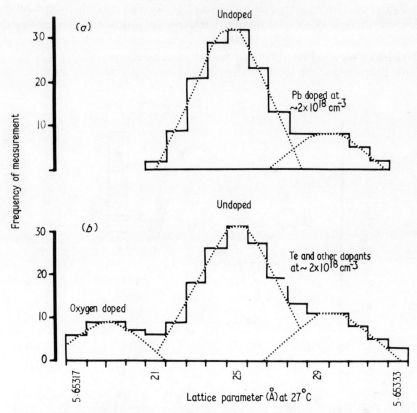

Figure 2. Comparison of histograms of lattice parameter measurements for (*a*) 20 samples of melt- and vapour-grown epitaxial gallium arsenide; (*b*) 25 samples of melt-grown gallium arsenide.

The breadth of the x-ray diffraction profile, which can provide information concerning non-uniform strain, was found to be $10'4'' \pm 13''$ for the undoped melt-grown gallium arsenide. The value for undoped epitaxial material was generally lower than that for melt gallium arsenide, being $9'48'' \pm 15''$. This small difference in the breadth between melt and epitaxial material may not be significant, however, since the major part of the x-ray diffraction profile, amounting to about $8'30''$, is due to instrumental broadening, primarily related to the effects of the diffraction geometry using an x-ray beam of 1 mm diameter at the sample surface.

3.2. The effect of dislocations and mechanical surface damage on the x-ray diffraction profile

The effect of dislocations on the x-ray diffraction profile was studied in undoped melt-grown gallium arsenide with dislocation densities between 10^5 and 10^7 cm^{-2}. It was found that, although no significant change in the mean lattice parameter resulted, the breadth of the x-ray diffraction profile was increased with increasing dislocation density, the extra broadening being over 2 minutes of arc for material with 10^7 dislocations/cm^2, as shown in figure 3. In an approximate analysis, this broadening, produced by dislocations, is in agreement with Kurtz's theory (Kurtz *et al* 1956) and, according to simple broadening theory (Friedel 1964), represents a non-uniform strain of amplitude ± 50 parts per million at the 10^7 dislocation level.

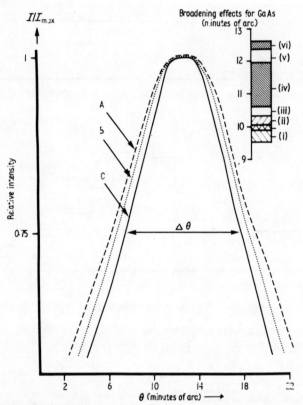

Figure 3. The peak of a typical line profile obtained in this work – (800) reflection, Cu Kβ radiation – and broadening effects for gallium arsenide. A: GaAs with 10^7 cm^{-2} dislocations, B: GaAs heavily doped with 6×10^{18} cm^{-3} Te atoms, C: Undoped GaAs with low dislocation density; (i) undoped epitaxy, (ii) undoped melt-grown, (iii) 10^5 cm^{-2} dislocations, (iv) heavily doped melt-grown and epitaxy, (v) 10^7 cm^{-2} dislocations, $\Delta\theta \simeq 3bN_\rho^{1/2}$, (vi) diamond saw damage.

The effect of mechanical surface damage on the x-ray diffraction profile was studied on a melt-grown gallium arsenide face, cut with a diamond saw without any subsequent chemical treatment. This was observed to produce a lattice parameter increase of $10^\cdot$ parts per million and an increase in the profile breadth of 2·5 minutes of arc, as compared with the values for undoped material, as shown in figure 3. This strain was completely eliminated by surface grinding and chemical polishing (Willoughby *et al* 1971) which removed about 100 micrometres from the damaged surface.

These results indicate the importance of removing surface damage and other potential sources of strain during device fabrication.

3.3. Lattice parameter measurements on doped gallium arsenide

The results of the study of the effects of dopants on the lattice parameter are summarized in table 1. The measurements were made on a range of melt and epitaxial gallium arsenide containing one of a number of doping elements. The doped specimens covered a wide impurity concentration range and included the group VI donor impurities tellurium, selenium and sulphur, which might be expected to replace arsenic lattice atoms; the amphoteric impurities silicon, germanium, tin and lead, which might prefer a particular lattice site depending upon the impurity concentration and vacancy equilibrium; and chromium and oxygen whose location in the gallium arsenide lattice is uncertain, although they might be expected to be a deep acceptor and donor, respectively (Madelung 1964, Allen 1968, Spitzer and Panish 1969).

A consideration of the Pauling atomic radii of these impurities (Pauling 1960), as compared with those of the lattice atoms which they are considered to replace, indicates that oxygen, sulphur and selenium are smaller than the matrix atoms, while tellurium, tin and lead are thought to be larger. Silicon and germanium give a reasonable fit to the matrix atoms, although silicon is smaller than the matrix gallium atom.

Presentation of the results within these three categories, which will be discussed in section 4, shows some interesting and generally consistent effects of heavy doping on the lattice parameter of gallium arsenide.

3.3.1. Dopants with smaller atomic radii than the matrix atoms which they are considered to replace.

Lattice parameter measurements on gallium arsenide doped with oxygen, which has a smaller atomic radius than either of the matrix atoms involved in a possible replacement, show a general decrease in lattice spacing as compared with that for undoped material, as shown in the histograms of figure 2. For gallium arsenide doped with oxygen at the 10^{17} cm^{-3} concentration level, this decrease in lattice parameter is about 10 parts per million.

Doping with selenium or sulphur at a concentration of 2×10^{18} cm^{-3} in low dislocation density gallium arsenide was observed to produce a small decrease in lattice parameter. This decrease was barely significant at 3 parts per million for sulphur-doped gallium arsenide, while for selenium-doped material the decrease was larger at 10 parts per million. However, these results were not confirmed on other selenium- or sulphur-doped gallium arsenide samples. Another sample doped with selenium to the same level of 2×10^{18} cm^{-3}, but with a high dislocation density of 10^7 cm^{-2}, showed no decrease in mean lattice parameter, although the lattice parameter variation across the sample was large at ±18 parts per million. Lattice parameter studies on undoped gallium arsenide showed that dislocations at concentrations of 10^7 cm^{-2} produce no significant change in the mean lattice parameter, although complicating factors may arise in high dislocation density doped material owing to the possible interaction of impurities with dislocations (Bullough and Newman 1963). Another sample doped with sulphur to a higher concentration of 4×10^{18} cm^{-3} showed a slight *increase* in lattice parameter of 7 parts per million. The barely significant result on the sample with the lower doping level may indicate that sulphur in sufficient concentration expands the gallium arsenide lattice, and that, in such a case, the sulphur atom may be associated with an atomic volume greater than that inferred from its covalent radius (King 1966). However, in view of this lack of reproducibility for both selenium and sulphur doped gallium arsenide, no definite conclusions can be made at present as to the effect of these impurities on the gallium arsenide lattice.

Although the site of chromium in gallium arsenide is uncertain, measurements on chromium-doped material have shown a decrease in lattice parameter, as compared with that for undoped material.

3.3.2. Dopants with larger atomic radii than the matrix atoms which they are considered to replace. Lattice parameter measurements on gallium arsenide doped with impurities, such as tellurium, tin and lead, with larger atomic radii than the matrix atoms involved in the replacement, show an increase in lattice parameter as compared with that for undoped material, as shown in table 1. This increase in lattice parameter is observed at doping levels greater than 10^{18} cm^{-3} and was observed, for tellurium and tin doped samples, in both melt-grown and epitaxial gallium arsenide. For these samples, large changes in lattice parameter of greater than 100 parts per million were measured at doping concentrations exceeding 10^{19} cm^{-3} (assessed electrically), with an accompanying broadening of the x-ray diffraction profile by up to 2 minutes of arc.

Lattice parameter measurements on liquid epitaxial gallium arsenide grown either from lead-rich solution or doped with lead showed a small increase of about 7 parts per million for concentrations of a few parts times 10^{18} cm^{-3}.

3.3.3. Amphoteric impurities with atomic radii of comparable size to the matrix atoms. Lattice parameter measurements on gallium arsenide doped with the amphoteric impurity silicon up to a concentration of 3×10^{18} cm^{-3} show a barely significant decrease in lattice parameter of about 3 parts per million at the high doping level. At silicon concentrations of the order of 10^{17} cm^{-3}, this impurity is considered to act as a donor replacing gallium atoms in the gallium arsenide lattice (Madelung 1964), and the silicon atom is smaller than the matrix gallium atom. This might be consistent with the lattice parameter results, assuming that silicon behaves in a similar way to impurities, such as oxygen, with small atomic radii, in decreasing the lattice parameter. However, from reported radioactive studies on the incorporation of silicon in gallium arsenide (Whelan *et al* 1961), as the concentration of silicon increases above 10^{18} cm^{-3} the ratio of the silicon donor to the total silicon concentration decreases due to an increasing number of compensating silicon acceptors.

Lattice parameter measurements on gallium arsenide doped with germanium at concentrations up to 2×10^{18} cm^{-3} (assessed electrically) showed an increase of about 15 parts per million. This change in lattice parameter could not be readily explained in terms of differences between the atomic radii of the impurity and matrix atoms, since the germanium atomic radius is in reasonable agreement with those of the host atoms on the Pauling model. The nonuniformity of the lattice parameter variations for one particular germanium-doped crystal was very large, and strain effects as monitored both by the change and variation in lattice parameter, may be contributory to reported inhomogeneous electrical and optical properties of such material.† The present measurements have also shown that a significant degree of broadening of over 2 minutes of arc occurs in the diffraction profiles from these samples.

3.4. *Lattice parameter measurements on a step-etched gallium arsenide epitaxial layer grown on a heavily doped substrate*

In view of the large changes in lattice parameter that have been observed in heavily doped gallium arsenide, particularly in the tellurium and tin doped material, strain induced in a high purity layer of gallium arsenide grown epitaxially on a heavily doped substrate has been investigated by the precision lattice parameter technique. The doping concentration throughout the n-type vapour epitaxial layer was measured by the differential capacitance technique to be in the range 10^{13} to 10^{14} cm^{-3}. This layer was grown on a heavily tellurium-doped substrate of thickness $430\,\mu$m with a doping concentration of 5×10^{18} cm^{-3}. At this doping level, the lattice parameter of the substrate was about 30 parts per million higher than that of undoped gallium arsenide. Thus, a potential source of strain existed from the lattice mismatch, due to differences in doping level, between the high purity layer and the heavily doped substrate. This structure is similar to that used in making gallium arsenide microwave devices. In order to reveal strain in the layer, the specimen was step-etched, as shown in figure 4, with $H_3PO_4 : H_2SO_4 : H_2O : H_2O_2$ in the ratio $20:10:10:3$. The method involved making lattice parameter measurements on a layer etched in steps, so that an assessment of strain as a function of depth through the layer could be made.

At layer depths greater than about 40 micrometres from the layer–substrate interface, the lattice parameter of the layer was typical of that for undoped gallium arsenide at about 5·65325 Å and remained uniform through to its top surface. However, the lattice parameter of the layer showed a large increase to 5·65338 Å on the step 27 micrometres from the interface. Since the x-ray penetration depth is about 20 micrometres and the diffraction profile will be dominated by reflections from the upper diffracting planes, this means that the growth of a high purity gallium arsenide layer on a heavily doped substrate can produce strain in the layer extending up to 20 micrometres from the interface. A possible explanation of the results in terms of a high doping concentration at the beginning of the growth of the layer is ruled out by the differential capacitance measurements. In the case investigated, the tensile strain in the layer for diffracting planes parallel to the layer–substrate interface is 23 parts per million. The x-ray diffraction profile breadth for reflections from the layer increased slightly with decreasing depth toward the layer–substrate interface, and for reflections from the 27 micrometre step was about 10 minutes of arc, which is just below the higher limit for that expected from undoped epitaxial gallium arsenide. The diffraction profiles for the tellurium-doped substrate, which were obtained from its lower face, showed an increased broadening of nearly 11 minutes of arc, which is consistent with that obtained from heavily tellurium-doped gallium arsenide.

---

† E V Solovjova, V V Karateev, M G Milvidskii and A V Govorkov 1974 unpublished.

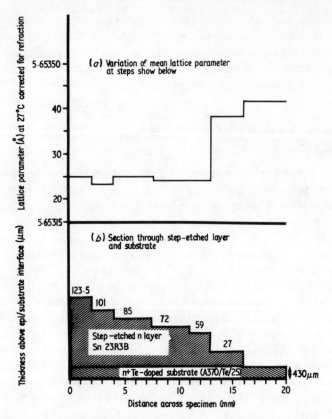

Figure 4. Study of a step-etched undoped epitaxial gallium arsenide layer on a tellurium-doped gallium arsenide substrate.

4. Discussion

4.1. The lattice parameter of undoped gallium arsenide layers and substrates

An interesting outcome of the present study is the agreement between the lattice parameters of undoped melt-grown and epitaxial gallium arsenide. The calculations of Logan and Hurle (1971), using standard methods of chemical thermodynamics, of point defect concentrations in gallium arsenide suggest that experimental vacancy concentrations of the order of 10^{18} to $10^{19}\,\mathrm{cm^{-3}}$ at the melting point of the compound are theoretically feasible. On this model, one would also expect vacancy concentrations of about $10^{15}\,\mathrm{cm^{-3}}$ at 750 °C, a temperature typical for epitaxial growth. This difference in the vacancy concentration between melt and epitaxial gallium arsenide might be reflected in differences in their lattice parameters if residual point defects were present. The observation that there is no difference between the lattice parameters of melt and epitaxial material indicates that there are no residual vacancy effects above the 1 part per million level. This confirms the results of high temperature annealing and quenching studies on gallium arsenide (Driscoll and Willoughby 1972, Potts and Pearson 1966) concerning the migration and possible aggregation of defects after cooling from growth temperatures. The undoped material contains no more than 2 parts per million impurity atoms which, assuming a 10% difference between lattice and doping atom radii, one would expect an effect on the lattice parameter of about 0·2 parts per million (ie, below the experimental detection limit). The presence of significant vacancy—impurity complexes above the 2 parts per million level, therefore, seems unlikely in view of these results.

4.2. Doping effects on the lattice parameter of gallium arsenide

The changes in lattice parameter observed with group IV and VI dopants at approximately the same concentration are shown in table 2. Although some results do not conform to the misfit expected, the general trend of these changes with position in the periodic table are in agreement with the sign of the atomic mismatch, as predicted by differences between the covalent radii of atoms involved in a possible replacement. Thus, for both the group IV and VI dopants at a concentration of a few parts times $10^{18} \, cm^{-3}$, a reversal in the sign of the fractional change in lattice parameter is observed within each group as the atomic number is increased. For the group IV impurities, silicon is observed to slightly decrease the lattice spacing, while germanium, tin and lead give rise to an increase. Although the results on sulphur and selenium doped gallium arsenide are inconclusive, a similar trend is indicated with the group VI dopants. Oxygen is found to decrease the lattice spacing, while doping with tellurium is observed to cause an increase.

The principle anomalies, however, can be seen by comparing the results with the simple predictions of Vegard's law, also shown in table 2. While the effects on silicon

Table 2. Comparisons of the atomic mismatch and change in lattice parameter for group IV and VI dopants in gallium arsenide
(Covalent radii for As and Ga are 1·18 and 1·26 Å, respectively)

Impurity	Carrier concentration (cm^{-3})	Covalent radius (Å)	Possible location	Atomic mismatch (Å)	Fractional change in lattice parameter (ppm)	Change predicted on Vegard's law (ppm)
Group IV						
Si	3×10^{18}	1·17	Si_{Ga}	−0·09	− 3	− 5
Ge	2×10^{18}	1·22	Ge_{Ga}	−0·04		
			Ge_{As}	+0·04	+16	+1·5
Sn	3×10^{18}	1·40	Sn_{Ga}	+0·14	+53	+8
Pb	2×10^{18}	1·46	Pb_{Ga}	+0·20	+ 8	+7·5
			Pb_{As}	+0·28		+10
Group VI .						
O	10^{17}	0·66			−12	
S	2×10^{18}	1·04	S_{As}	−0·14	——	−5
Se	2×10^{18}	1·14	Se_{As}	−0·04	——	−1·5
Te	5×10^{18}	1·32	Te_{As}	+0·14	+28	+13
Cr		1·17			−18	

and lead are close to that expected, tin and tellurium (discussed below) show a change larger than expected, while germanium, expected to be a good fit in the gallium arsenide lattice, shows a very large deviation from the expected change. A possible explanation for the differences between the results on silicon and germanium doped gallium arsenide may arise from differences between the effects of electrically inactive dopant in these samples. Whelan *et al* (1961) have found that in silicon doped gallium arsenide with electron concentrations of 3×10^{18} cm^{-3} the major part of the total silicon concentration, corresponding to about 75%, is associated with the silicon donor. However, the same authors (Whelan *et al* 1960) have found that in germanium doped gallium arsenide with electron concentrations of $1 \cdot 5 \times 10^{18}$ cm^{-3} only 20% of the total germanium concentration is associated with the germanium donor. Since the remaining parts of the total germanium and silicon concentrations in these samples are likely to be primarily in the form of neutral nearest neighbour pairs (Madelung 1964), their effect on the lattice parameter might be quite different to the simple substitutional donors.

Dopants which have been shown to produce large increases in the lattice parameter of gallium arsenide have been tellurium and tin at doping concentrations above 10^{18} cm^{-3}, as shown in figure 5. Previous reports have confirmed this observation (Black and Lublin

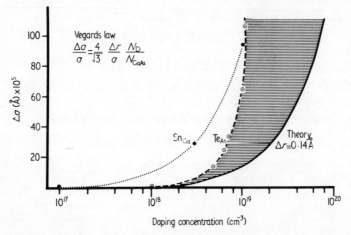

Figure 5. The effect of heavy doping on the lattice parameter of gallium arsenide: comparison of experimental variations with theoretical predictions using Vegard's law.

1964, Pierron and McNeely 1968, Willoughby *et al* 1971 and Kuznetsov *et al* 1973). The experimental points of figure 5 show a smooth monotonic increase in parameter with doping level, suggesting that the lattice parameter change is associated with the dopant and not with some other extraneous factor. Since the Pauling covalent radii for both tin and tellurium are larger than gallium and arsenic respectively, which they can replace, these increases are qualitatively consistent with the misfit, unlike those obtained by powder techniques (Kolm *et al* 1957), where the effects of precipitation could have been a complicating factor.

On a quantitative level, an attempt has been made to understand the changes in lattice parameter with doping in terms of Vegard's law. Although there is a limited range over which Vegard's law is applicable, it can be used as a first approximation for solute atoms which are larger than the solvent atoms (King 1966). The predicted changes of lattice parameter with doping concentration have been calculated using covalent radii from the formulation of Vegard's law (Willoughby *et al* 1971) given in figure 5 and is shown by the theoretical curve in this figure, where the difference between the covalent radii for solute and solvent atoms is 0·14 Å in both cases, assuming that tellurium ($r = 1·32$ Å) substitutes for arsenic ($r = 1·18$ Å) and tin ($r = 1·40$ Å) substitutes for gallium ($r = 1·26$ Å). For both dopants, with reference to figure 5, the observed lattice parameter change corresponds theoretically to a larger doping concentration than that measured electrically, the discrepancy being apparent above 10^{18}cm^{-3}.

Although the validity of Vegard's law is not proven in this analysis, an explanation of this discrepancy at these high doping levels could be in terms of dopant which was not electrically active but still affecting the lattice parameter. The transmission electron microscopy work of Laister and Jenkins (1968) has indicated that layers doped with tellurium above about 10^{18}cm^{-3} contain plate-like defects considered to be rafts of tellurium substituting for arsenic in (111) planes, which were suggested to produce an expansion in the lattice perpendicular to the gallium—tellurium layer. The tellurium in such stacking fault defects might not be expected to be electrically active and might account for the change in lattice parameter being larger (from theory) than expected from the electrically active doping concentration. The presence of such defects would also be consistent with the increased x-ray broadening observed in these heavily doped samples. Other tentative possibilities are large concentrations of tellurium interstitials or Ga_2Te_3 precipitates (Kressel *et al* 1968).

In support of the proposal that for tellurium doped gallium arsenide the observed change in lattice parameter corresponds to a higher doping level than that measured electrically (the difference being shown by the shaded region in figure 5) Kuznetsov *et al* (1973) have shown that at doping concentrations above 10^{18}cm^{-3} not all tellurium in doped gallium arsenide is electrically active. Their results are in close agreement with those obtained from the present study and are compared in figure 6. The lattice parameter measurements on the selenium doped samples may also have been complicated by precipitation effects. Abrahams *et al* (1967) have shown that in selenium-doped samples with carrier concentrations of $2 \times 10^{18} \text{cm}^{-3}$, the major part of the total selenium content is in the form of precipitate particles of Ga_2Se_3 and thus the electrically active selenium concentration is less than the total selenium content.

Different factors, similar to those already discussed for silicon and germanium doped gallium arsenide, may be involved in considering the effects of tin doping on the lattice parameter of gallium arsenide. Although at concentrations of the order of 10^{17}cm^{-3} one may expect Sn_{Ga} donors to predominate electrically, at higher concentrations one might expect neutral $Sn_{Ga}-Sn_{As}$ pairs, analogous to silicon in gallium arsenide, and this may explain the discrepancy between the theoretical and electrical curves of figure 5 at high doping levels.

4.3. The growth of an undoped layer on a heavily tellurium doped substrate

The measurements summarized in figure 4 present a unique opportunity to examine the influence of a substrate on an epitaxial layer when the two have different lattice

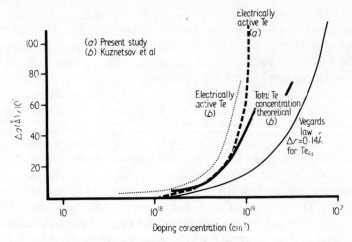

Figure 6. The effect of tellurium doping on the lattice parameter of gallium arsenide: comparison of experimental variations (*a*) present study, (*b*) Kuznetsov *et al* (1973) with theory.

parameters caused by differences in doping. Firstly, it is important to note that the effect of tellurium on lattice parameter is confirmed in this *direct* comparison, as opposed to measurements on different samples reported above. Secondly, the uniformity of lattice parameter through the thickness of the layer (apart from the 27 micrometre layer) is noteworthy and represents one of the first such studies.

The significance of the measurement on the 27 micrometre thick section of the 'step-etched' layer must be discussed. Although only one step, it does indicate, as noted above, that strain in this layer extends to at least 20 micrometres from the interface. The simplest possible analysis to attempt to understand this result is a simple elastic model of lattice mismatch. Assuming that the undistorted lattice parameters are 5·65342 Å for the substrate and 5·65325 Å for the layer, and that this mismatch is accommodated by elastic strain only, it can be seen that the layer will be in *tension,* and the substrate in *compression* near the interface, as illustrated in figure 7. As also illustrated in that figure, it is clear that this will produce a bend, such that the composite is concave on the layer side, and convex on the substrate side. As pointed out by Reinhart and Logan (1973), the stress situation is identical to that of a bimetallic strip except that the lattice mismatch in the latter situation arises from differences in thermal expansion coefficients. Employing the stress analysis of Brotherton *et al* (1973) stress distributions were calculated for each of the layer thicknesses measured, assuming Young's modulus for gallium arsenide to be $8·5 \times 10^{11}$ dyn cm^{-2} (Bateman *et al* 1959) and the substrate thickness to be 430 micrometres. A typical stress distribution is shown in figure 7, and it can be seen that while there is a neutral axis in the substrate well away from the interface, the stress in the layer is *tensile* throughout the whole thickness. Since the lattice parameter measurement on the 27 micrometre layer shows an extension perpendicular to the interface, this implies a *compressive strain* parallel to the interface, and this elastic model is clearly not in accord with the measurements.

Having ruled out a simple elastic bending model of lattice mismatch, other possible explanations will be considered. The possibility of incorporation of tellurium in the

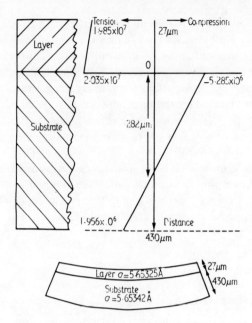

Figure 7. Elastic model of stress distribution (schematic) in gallium arsenide layer and substrate (stress in dynes cm^{-2}).

layer has already been mentioned, and it is clear that, although incorporation might have taken place to a small extent, the concentration needed to explain the change in lattice parameter certainly would have been detected by the capacitance profiling. It is also felt to be unlikely that incorporation of other impurities could have occurred to this extent.

While the elastic bending model considered above is the simplest that could be considered, there is a good deal of evidence that such behaviour is rarely realized in practice. The accommodation of misfit across the interface between a single crystal film grown epitaxially on a single crystal substrate has been considered by Frank and van der Merwe (1949), van der Merwe (1963, 1964), Jesser and Kuhlmann Wilsdorf (1967) and Jesser and Matthews (1967, 1968). It has been found that if the misfit is small and the interfacial bonding strong, then all misfit is expected to be eliminated by elastic strain until the thickness of the film reaches a critical value. When this critical thickness is exceeded it is energetically favourable for part of the misfit to be accommodated by misfit dislocations. Simple calculations (Mroczkowski *et al* 1968) indicate this crystal thickness will be about 10 to 15 micrometres for the present system. These conclusions have been confirmed in a wide range of systems, and the direct observations of G R Booker (1973 private communication) employing homoepitaxially grown gallium arsenide layers containing different doping levels from the substrate, appear also to confirm this model for the present system. Misfit dislocations in homojunctions have also been reported (Abrahams *et al* 1969) for zinc doped gallium arsenide layers on tellurium doped gallium arsenide substrates, these being close to the interface (Kishino *et al* 1972, Kishino and Iida 1972).

Further confirmation of such a model lies in the calculations of stress in figure 7, which show that the maximum stress in the strained layer, which is 2×10^7 dynes cm^{-2}, is great enough to cause dislocation movement at the growth temperature (Laister and Jenkins 1973), and hence introduce a misfit array, and may also cause large scale slip, which has been observed in some layers by etching. If, as seems likely, such a situation exists in the sample under discussion, the elastic stress distribution is now modified to a three region system, rather than two, (ie, the substrate, the elastically strained part of the layer near the substrate and the rest of the layer where strain is partially relieved by dislocations). This latter part of the layer will still, however, be partially elastically strained and the concave surface will be most likely to be in compression, as observed in this work. These measurements thus clearly favour such a model as opposed to one of simple bending, where the elastic strain is opposite in sign to that measured, or where misfit dislocations occur at the interface, where strain is relieved entirely.

5. Conclusions

The principle conclusions of this work are as follows:

(i) Lattice parameter measurements on undoped gallium arsenide layers and substrates show very close agreement despite differences in growth temperature, and indicate that there are no residual point defect or point defect-impurity effects on the lattice parameter above the 1 part per million level.

(ii) The effects of doping with group IV and VI elements on the lattice parameter has been measured in the 10^{17}–10^{19} cm^{-3} concentration range. Some elements show behaviour close to that predicted from expected misfit factors, but, in others, the lattice parameter change above the 10^{18} cm^{-3} concentration level is larger than that predicted, probably due to the effects of precipitation or non-electrically active dopant.

(iii) The growth of an undoped layer on a heavily tellurium doped substrate shows evidence for strain in the layer extending up to 20 μm from the interface.

Acknowledgments

The authors wish to thank Dr D T J Hurle and Dr R M Logan (RRE, Malvern) for helpful and searching discussions on many aspects of this work: D J Ashen (RRE, Malvern) for the growth of v PE material; Miss A Royle and S Benn (RRE, Malvern) for electrical and profiling measurements; J C Brice and M J King (Mullard Laboratories, Redhill), J R Knight (Allen Clark Research Centre, Plessey Company), M C Rowland (SERL, Baldock) and P D Greene (STL, Harlow) for the supply of gallium arsenide single crystals and layers; Mrs J Cooke (RRE, Malvern) for the gallium arsenide step-etching process; T G Read (Department of Electronics, Southampton University) for assistance in the stress analysis; and Professor R L Bell (Southampton University) for provision of laboratory facilities and useful discussion. The work was carried out on a CVD contract and is published by permission of the Ministry of Defence (Procurement Executive). Financial support from the Science Research Council is also acknowledged.

References

Abrahams M S, Buiocchi C J and Tietjen J J 1967 *J. Appl. Phys.* **38** 760
Abrahams M S, Weisberg L R, Buiocchi C J and Blanc J 1969 *J. Mater. Sci.* **4** 223
Allen G A 1968 *Brit. J. Appl. Phys.* **1** 593
Baker T W, George J D, Bellamy B A and Causer R 1968 *Adv. x-ray Anal.* **11** 359
—— 1969 *AERE Harwell Res. Rep.* No 5152
Bateman T B, McSkimmin H J and Whelan J M 1959 *J. Appl. Phys.* **30** 544
Black J and Lublin P 1964 *J. Appl. Phys.* **35** 2462
Bond W L 1960 *Acta Cryst.* **13** 814
Brotherton S D, Read T G, Willoughby A F W and Lamb D R 1973 *Solid St. Electron.* **16** 1367
Bullough R and Newman R C 1963 *Prog. Semicond.* **7** 100
Driscoll C M H and Willoughby A F W 1972 *Defects in Semiconductors* (London: Institute of Physics) p377
Frank F C and van der Merwe J H 1949 *Proc. R. Soc.* **A198** 216
Friedel J 1964 *Dislocations, International Series of Monographs on Solid State Physics* ed R Smoluchowski and N Kurti vol 3
Hasegawa H and Ito H 1972 *Appl. Phys. Lett.* **21** 107
Jesser W A and Kuhlmann-Wilsdorf K 1967 *Phys. Stat. Solidi* **19** 95
Jesser W A and Matthews J W 1967 *Phil. Mag.* **15** 1097
—— 1968 *Phil. Mag.* **17** 461, 595
King H W 1966 *J. Mater. Sci.* **1** 79
Kishino S, Ogirima M and Kurato K 1972 *J. Electrochem. Soc.* **119** 617
Kishino S and Iida S 1972 *J. Electrochem. Soc.* **119** 1113
Kolm C, Kulin S A and Averbach B L 1957 *Phys. Rev.* **108** 965
Kressel H, Hawrylo F Z, Abrahams M S and Buiocchi C J 1968 *J. Appl. Phys.* **39** 5139
Kurtz A D, Kulin S A and Averbach B L 1956 *Phys. Rev.* **101** 1285
Kuznetsov G M, Pelevin O V, Barsukov A D, Olenin V V and Savel'eva I A 1973 *Izvest. Akad. Nauk SSSR–Neorgan. Mater.* **9** 759
Laister D and Jenkins G M 1968 *J. Mater. Sci.* **3** 584
—— 1973 *J. Mater. Sci.* **8** 1218
Logan R M and Hurle D T J 1971 *J. Phys. Chem. Solids* **32** 1739
Mullin J B, Straughan B W and Brickell W S 1965 *J. Phys. Chem. Solids* **26** 782
Madelung O 1964 *Physics of III–V Compounds* (New York: Wiley)
Mroczkowski R S, Witt A F and Gatos H C 1968 *J. Electrochem. Soc.* **115** 750
Pauling L 1960 *The Nature of the Chemical Bond* (Oxford)
Petroff P and Hartman R L 1973 *Appl. Phys. Lett.* **23** 469
Pierron E D and McNeely J B 1968 *Adv. x-ray Anal.* **12** 343
Potts H R and Pearson G L 1966 *J. Appl. Phys.* **37** 2098
Reinhart F K and Logan R A 1973 *J. Appl. Phys.* **44** 3171
Spitzer W G and Panish M B 1969 *J. Appl. Phys.* **40** 4200
van der Merwe J H 1963 *J. Appl. Phys.* **34** 117
—— 1964 *Single Crystal Films* (New York: Pergamon) p139
Whelan J M, Struthers J D and Ditzenberger J A 1960 *Properties of Elemental and Compound Semiconductors, Metall. Soc. Conf.* (New York: Interscience) vol 5 p275
—— 1961 *Proc. Int. Conf. Semiconductors* (Prague: Czech. Acad. of Sci.) p966
Willoughby A F W, Driscoll C M H and Bellamy B A 1971 *J. Mater. Sci.* **6** 1389

Mobility, dopant and carrier distributions at the interface between semiconducting and semi-insulating gallium arsenide

Kurt Lehovec

University of Southern California, Los Angeles, California 90007, USA

Rainer Zuleeg

McDonnell Douglas Astronautics Company, Huntington Beach, California 92547, USA

Abstract. Dopant and mobility distributions in an epitaxial semiconducting layer are determined by capacitance and Q-factor measurements between a gate electrode and a conducting channel. Degradation of mobility and carrier concentration by neutron bombardment is measured and the observed deterioration of the saturation current of junction field effect transistors is accounted for.

It is shown that the $C-V$ method for determining impurity distribution becomes invalid when the depletion layer adjacent to the gate electrode expands to a depletion layer adjacent to the substrate interface. Examples of artificial impurity distributions obtained by the $C-V$ method are calculated for various model substrates and compared with measurements.

1. Introduction

High frequency gallium arsenide junction field effect transistors comprise an epitaxial semiconducting layer on a semi-insulating chromium doped substrate (Hower *et al* 1969, Turner and Wilson 1969, Zuleeg 1969). This paper describes $C-V$ and $C-Q$ measurements to obtain the impurity, carrier and mobility distributions in the epitaxial semiconducting layer adjacent to the semi-insulating substrate.

It is well known that the depletion layer between gate and channel widens with increased reverse bias. As the conducting channel width shrinks, the channel resistance increases. The dependence of channel resistance on gate-to-channel bias can be used to obtain the low field channel mobility of carriers, as is well known from the Shockley analysis (Shockley 1952) of the $I-V$ characteristics of junction field effect transistors at small drain-to-source voltages. Alternatively, the low field mobility can be obtained from measurements of the Q factor of the gate-to-channel capacitance (Lehovec 1974), the Q factor arising from the distributed $R-C$ network involving the channel resistance and the gate capacitance. Application of the $C-V$ and $Q-V$ methods to obtain information on the degradations of carrier concentration and of mobility by neutron radiation will be discussed in this paper and compared with published data (McNichols and Ginell 1970, Behle and Zuleeg 1972) obtained by Hall measurements. Degradation of the $I-V$ characteristics of junction field effect transistors will be accounted for.

The gate depletion layer penetrates to the substrate as the gate bias voltage increases, almost pinching off the channel. A theoretical investigation of the $C-V$ method for a nearly pinched-off channel (Lehovec 1975) has shown that the impurity distribution so obtained is an artefact. Experimental 'impurity distributions' obtained by the $C-V$ method for epitaxial n-type layers on semi-insulating substrates will be compared with theoretical models.

2. Mobility determination by the $C-Q-V$ method

The device used for determination of the carrier mobility in an epitaxial layer is similar to a junction field effect transistor, except that only one ohmic contact to the channel is required (figure 1(a)). Capacitance and Q-factor are measured between the contact to the gate (Schottky barrier or shallow p^+–n junction) and a laterally spaced ohmic contact. Preferably a geometrical arrangement is used which allows one-dimensional linear analysis and utilizes a close spacing between the ohmic contact and the gate edge to suppress a parasitic series resistance. The dotted lines in figure 1(a) signify the boundaries of the gate depletion layer and of a depletion layer at the substrate interface, the origin of which will be discussed in section 4.

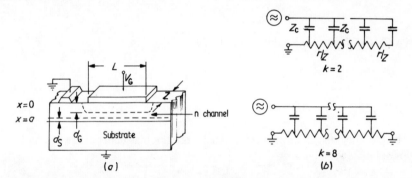

Figure 1. (a) Test structure; (b) Small signal equivalent circuit.

As the negative bias voltage increases, the channel width (a-d_G-d_S) shrinks and the low field resistance per unit channel length and unit channel width r increases. The equivalent circuit for a small signal AC voltage applied at the gate is the distributed r–c network shown in figure 1(b) with c the capacitance per unit gate area.

Resistance r and capacitance c can be derived from measurements of Q-factor and equivalent series capacitance. Transmission line theory provides equivalent series resistance R_S, equivalent series capacitance C_S and

$$Q = (\omega R_S C_S)^{-1}$$

as functions of the reduced frequency

$$\omega^* = \omega r c L^2/k \tag{1}$$

(figure 2). The numerical constant k is 2 for the single ohmic contact of figure 1(b) and 8 for the symmetric double contact of a conventional junction field effect transistor

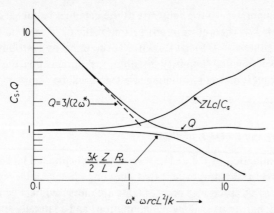

Figure 2. Equivalent series capacitance C_S, resistance R_S, and Q-factor as functions of angular frequency ω for the distributed networks shown in figure 1(b) (after Lehovec 1974).

structure of gate length L with source and drain connected. In the range $Q \gtrsim 3$, the Q factor decreases inversely with frequency:

$$Q = 3/2\omega^* = 3k/(2\omega rcL^2) \tag{2}$$

while C_S and R_S become independent of frequency

$$C_S = ZLc \tag{3}$$

and

$$R_S = 2rL/(3Zk). \tag{4}$$

The mobility distribution is determined from the bias dependence of r and c as follows: let

$$r^{-1} = -\bar{\mu}Q_c \tag{5}$$

where Q_c is the mobile channel charge per unit area and $\bar{\mu}$ is its average mobility. Thus

$$\delta r^{-1} = -\mu\delta Q_c - Q_c\delta\bar{\mu} \tag{6}$$

where μ instead of $\bar{\mu}$ has been used, since the change in channel charge occurs at the boundary of the gate depletion layer and $\mu(d_G)$ is the mobility of the carriers at that location. The capacitance is by definition

$$c = \delta Q_G/\delta V_G \tag{7}$$

where Q_G is the charge per unit area of the gate and V_G is the applied gate-to-channel voltage. Equations (6) and (7) show that

$$\mu = c^{-1}\delta r^{-1}/\delta V_G \tag{8}$$

if (i) $\delta\bar{\mu}=0$, and (ii) $\delta Q_G = -\delta Q_c$. The condition $\delta\bar{\mu}=0$ is not strictly valid because of increased scattering at the channel walls with decreasing channel width (Schrieffer effect, Schrieffer 1955) but it is believed to be a satisfactory first approximation. The condition $Q_G = Q_c$ is not satisfied if deep lying impurities in the depletion layer become ionized as the depletion layer expands, or if the peak carrier concentration in the channel becomes less than the local impurity concentration in an almost pinched-off channel. The latter case will be discussed in more detail in section 3.

It can be inferred from equations (8) and (2)–(4) that

$$\mu = \frac{2\omega L^2}{2kC_S} \frac{\partial (QC_S)}{\partial V_G}. \tag{9}$$

The corresponding distance is obtained by

$$d_G = \epsilon\epsilon_0/c = \epsilon\epsilon_0 ZL/C_S. \tag{10}$$

3. Determination of the impurity concentration by the *C–V* method

The *C–V* method (Schottky 1942) defines a distance

$$X = \epsilon\epsilon_0/c \tag{11}$$

and a concentration

$$N = -2(\epsilon\epsilon_0 q)^{-1}\partial V_G/\partial c^{-2} \tag{12}$$

and identifies $N(X)$ with the donor distribution $N_D(d_G)$ where d_G is the width of the gate depletion layer.

Assuming that the charges of opposite sign compensating δQ_G occur at the depletion boundary, one has

$$\delta(V_G + V_{BG}) = \delta Q_G d_G/\epsilon\epsilon_0 \tag{13}$$

where V_{BG} is the built-in potential between gate and channel. It follows from equations (7) and (11)–(13) that

$$X = d_G - \epsilon\epsilon_0 \partial V_{BG}/\partial Q_G \tag{14}$$

and

$$N(X) = -\partial Q_G/q\partial X. \tag{15}$$

Thus $N(X) \equiv N_D(d_G)$ only if

(i) $\quad \delta V_{BG} = 0 \tag{16}$

and

(ii) $\quad \delta Q_G = -N_D(d_G)q\delta d_G. \tag{17}$

Figure 3 shows the changes in carrier distribution in the channel with gate voltage for two cases: (A) an undepleted channel where carrier concentration equals impurity concentration at the boundary of the depletion layer, and (B) an almost pinched-off channel where peak carrier concentration n_0 is less than impurity concentration N_D.

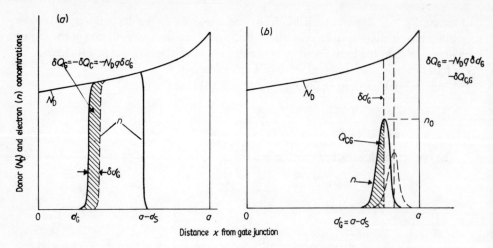

Figure 3. Donor (N_D) and electron (n) distributions in an epitaxial layer between a Schottky barrier gate contact at $x = 0$ and a substrate at $x \geqslant a$. Dashed curve is the electron distribution shifted by applied voltage increment. Left hand figure (a): an electrically neutral channel $n = N_D$ exists within $d_G \leqslant x \leqslant a - d_S$. Right hand figure (b): the channel is almost pinched-off with peak carrier concentration $n_0 < N_D$ at $x = d_G = a - d_S$. The $C–V$ method gives the donor concentration $N_D(d_G)$ in case (a), but not in case (b). δQ_G is the change in gate electrode charge per unit area and $\delta Q_{C,G}$ is the change of mobile channel charge per unit area at left of the position $a - d_S$ of channel maximum in case (b).

Equation (17) is satisfied in case (A). However, in case (B) we have

$$\delta Q_G = -q\delta \int_0^{d_G = a - d_S} N_D \, dx - \delta Q_{C,G} \tag{18}$$

where the first term on the right hand side results from the shift of the centre of the conducting channel with gate bias voltage and the second term is the change of the channel charge located at left from the peak value n_0. Only the left half of the channel carriers (dashed area under the curve n^B) enters into the charge balance with the gate since the electric field perpendicular to the channel is zero at $d_G^B = a - d_S^B$. The condition (17) is not valid in case (B) because of $\delta Q_{C,G} \neq 0$. The condition (16) is also not valid because

$$\delta V_{BG} = -V_T \delta n_0 / n_0 \neq 0 \tag{19}$$

where V_T is the voltage equivalent of temperature. The negative built-in potential increases, $\delta V_{BG} > 0$, with increasingly negative gate voltage, $\delta Q_G < 0$, so that $X > d_G$. In fact, the artificial distribution $N(X)$ may extend deep into the substrate.

The channel potential has a maximum in case (B) at $d_G = a - d_S$ where $\partial^2 V / \partial x^2 = -(N_D - n_0)q/\epsilon\epsilon_0$. Assuming a Boltzmann distribution of channel carriers one obtains for $n_0 \ll N_D$

$$Q_{C,G} \simeq -n_0 q L_D \sqrt{\pi/2} \tag{20}$$

where

$$L_D = (2\epsilon\epsilon_0 V_T / N_D q)^{1/2} \tag{21}$$

is the Debye length.

It follows from equations (14) and (15) with the aid of equations (18), (19) and (20) that

$$X = a - d_S + L_D \frac{n_0^2 \sqrt{\pi} + L_D N_D n_0'}{n_0(2N_D + L_D\sqrt{\pi}n_0')} \tag{22}$$

and

$$N = (n_0^2/2) [2N_D + L_D\sqrt{\pi}n_0']^3$$
$$\times [N_D(2n_0^2 + L_D^2n_0'^2)(2N_D + \sqrt{\pi}L_Dn_0') - L_D^2n_0''n_0(2N_D^2 - \pi n_0^2)]^{-1} \tag{23}$$

where $n_0' = \partial n_0/\partial d_S$. The small term $Q_{CG}\delta d_G/\epsilon\epsilon_0$ which has been neglected on the right hand side of equation (13) is included in equations (22) and (23).

At this point we shall distinguish between the 'high frequency' case where the substrate charge is not modulated, so that

$$\delta \left[q \int_{a-d_S}^{a} N_D \, dx + Q_{C,S} \right] = 0 \tag{24}$$

and the 'low frequency' case where the substrate charge can equilibrate with the AC variation of the channel charge.

In the 'low frequency' case the carriers assume a Boltzmann distribution across the substrate interface so that

$$n_0 = N_D \exp(V_{BS} - V_S)/V_T \tag{25}$$

where N_C is the equivalent density of states in the conduction band, V_T is the voltage equivalent of temperature, $V_{BS} < 0$ is the built-in potential between substrate and electrically neutral ($n_0 = N_D$) channel, and V_S is the electrostatic potential between substrate and channel. Evaluation of equations (22) and (23) with (25) requires detailed assumptions on the space charge distribution in the substrate to derive $V_S(d_S)$.

4. Special interface cases

Figure 4 shows energy level diagrams, and electron and space charge distributions near the substrate interface for various types of substrates. Model cases of particular interest are:

(a) Abrupt p–n junction interface between an n-channel and a p-substrate of constant dopant concentration (figure 4(a)). For $N_D < N_A$, the larger portion of V_S extends over the depletion region in the epitaxial layer and the channel remains substantially within the epitaxial layer and is symmetric around its maximum. The potential between substrate and channel is then

$$V_S \simeq - V_T(1 + N_D/N_A)d_S^2/L_D^2. \tag{26}$$

The channel charge moves closer to the interface with increasing N_D/N_A, and a contribution to V_S arises from the tail of carriers extending into the substrate (Gummel and Scharfetter 1967). This contribution is dominant in case of an intrinsic substrate $N_A = 0$.

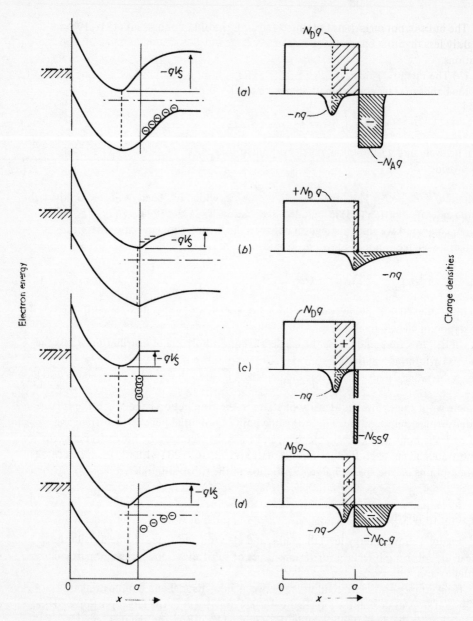

Figure 4. Energy level diagrams (left column) and charge distributions (right column) for the following special cases: (*a*) p-substrate; (*b*) intrinsic substrate; (*c*) large negative interface charge; (*d*) Cr-doped compensated GaAs substrate. Constant dopant concentrations, N_D, N_A, N_{Cr} have been assumed. Location of substrate and interface charges are indicated in the energy level diagram. The vertical dotted lines mark the positions $(a - d_S)$ of the potential maximum in the channel.

The built-in potential between the p-substrate and the undepleted channel is only slightly less than the bandgap for reasonably large but non-degenerate dopant concentrations.

(b) The intrinsic substrate (figure 4(b)). The built-in potential is only about one-half of the bandgap. The condition of charge neutrality for $x \geqslant a - d_S$,

$$(N_D - n_0)\, d_S = \int_a^\infty n\, \mathrm{d}x,$$

can be evaluated with the known electron distribution (Dousmanis and Duncan 1958) to provide

$$n_0 = N_D d_S^2 / L_D^2. \tag{27}$$

The tail of the electron distribution extending into the intrinsic substrate from an undepleted channel $n = N_D$ is

$$n = N_D L_D^2 / (x - a)^2 = 2\epsilon\epsilon_0 V_T / q (x - a)^2 \tag{28}$$

for $x - a \gg L_D$.

JFET on intrinsic substrate show a poor saturation of drain current with drain voltage (Reiser 1970) because of channel conduction extending into the substrate, in contrast to JFET on heavily Cr-doped substrate which exhibit well saturated drain currents and in which channel conduction remains restricted to the epitaxial layer (figure 4(d)).

(c) Large negative interface charge (figure 4(c)). The potential between epitaxial layer and substrate extends only across the depletion layer in the epitaxial film and is obtained from equation (26) for $N_A = \infty$. The built-in potential V_{BS} depends on the energy distribution of the interface charges.

(d) Cr-doped semi-insulating GaAs substrate (figure 4(d)).

The Fermi level in the compensated Cr-doped substrate lies about midgap and the built-in potential is thus somewhat less than half of the bandgap. In other respects the space charge and potential distributions are similar to a p-substrate of $N_A = N_{Cr}$, provided that the chromium concentration N_{Cr} is not large compared to the donor concentration in the epitaxial layer. The fraction of $|V_S|$ extending across the space charge layer in the substrate is then large compared to V_T, and most of the Cr impurities in the space charge layer are ionized. However, if $N_{Cr} \gg N_D$ only a small fraction of the potential V_S extends across the space charge layer and a significant portion of the chromium impurities are then electrically neutral.

Figure 5 shows $N(X)$ distributions calculated by equations (22), (23), (25), and (26) for various N_A/N_D ratios.

The solid curves pertain to $V_{BS}/V_T = -52.5$ for $N_D/N_A = 0$, and

$$V_{BS}/V_T = -50 + \ln N_D/N_A \quad \text{for} \quad N_D/N_A > 0.$$

The value of -52.5 corresponds to 80% of the bandgap of GaAs and the logarithmic relation is the built-in potential of a GaAs p–n junction. The dotted curves pertain to $V_{BS}/V_T = -25$. The dotted curve for $N_D/N_A = 0$ may be considered to arise either from a large negative interface state density which clamps the Fermi level near midgap, or else from application of a forward bias to the p–n substrate junction. Note that the position of infinity of N/N_D is not affected by the value of V_{BS}/V_T in the case of $N_D/N_A = 0$.

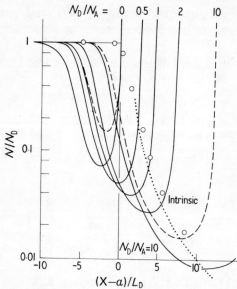

Figure 5. Distributions $N(X)$ calculated by equations (22) and (23) for the abrupt boundary between an n-type semiconducting channel and p-substrate of various dopant ratios N_A/N_D. These curves are artefacts which do not represent the true impurity distribution. The dashed curves pertain to $V_{BS}/V_T = -25$, while for the full curves $V_{BS}/V_T = -52.5$ in the case of $N_D/N_A = 0$ and $V_{BS}/V_T = -50 + \ln N_D/N_A$ for $N_D/N_A > 0$. The dotted curve pertains to an intrinsic substrate and coincides with the carrier distribution extending into the substrate from an undepleted channel. The circles are experimental points obtained by the $C-V$ method, equations (11) and (12), for a lightly compensated substrate.

However, this is not true for finite values of N_D/N_A, in other words the dotted curve for $N_D/N_A = 10$ and the corresponding value for the solid curve for which the infinity occurs at 18·37, outside the range shown in figure 5.

Clearly, these distributions which depend on temperature and on substrate bias are artefacts and do not represent the actual impurity distributions which are step functions.

The dotted line in figure 5 is the $N(X)$ relation for the intrinsic substrate, obtained by equations (22), (23) and (27). For $N \ll N_D$, this relation turns out to be identical to equation (28), that is $n(x) \equiv N(X)$. Thus our analysis proves the contention of Smith (1970) that the tail of the distribution obtained by the $C-V$ method is not the actual impurity distribution, but that it is the carrier distribution in the intrinsic substrate near the interface. Cases where the $C-V$ method provides the carrier distribution rather than the impurity distribution in the presence of space charges have been described by Kennedy *et al* (1968). However, the other examples of figure 5 show that this is not generally true.

The experimental points shown in figure 5 will be discussed in the next section.

5. Experimental results

5.1. GaAs junction field effect transistor degradation by neutron exposure

An n-channel GaAs junction field effect transistor with semi-insulating Cr-doped

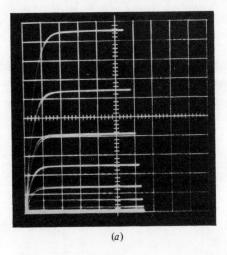

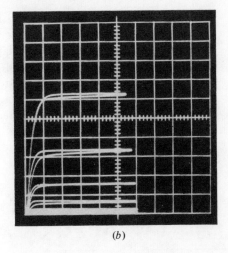

(a)   (b)

Figure 6. *I–V* characteristics of a GaAs JFET (*a*) before and (*b*) after irradiation by 1.6×10^{15} neutrons/cm² of $E > 10$ keV. (*a*) Vertical scale I_D in $20 \, \mu A$/division, horizontal scale V_D in 1 V/division, negative V_G in 0·2 V/step. (*b*) Vertical scale I_D in $10 \, \mu A$/division, horizontal scale V_D in 1 V/division, negative scale V_G in 0·2 V/step.

GaAs substrate and shallow p⁺-gate of dimensions $L = 8.8 \times 10^{-3}$ cm and $Z = 2.0 \times 10^{-2}$ cm was investigated. Figure 6 shows the degradation of the $I-V$ characteristic by neutron bombardment with 1.6×10^{15} neutrons/cm² $(E > 10 \, \text{keV})$. Since $\mu V_p / v_S L \ll 1$, drift velocity does not reach its saturation value v_S over most of the channel (Lehovec and Zuleeg 1970), and the Shockley (1952) theory can be used to obtain the saturation current:

$$I_{SAT} = Z a N_D q \mu V_p \left[1 - 3U + 2U^3 \right] / (3L) \tag{29}$$

where

$$U = \left(\frac{|V_{BG}| + |V_G|}{V_p} \right)^{1/2} \tag{30}$$

and

$$V_p = N_D q a^2 / 2 \epsilon \epsilon_0. \tag{31}$$

The current degradation thus becomes

$$\frac{\delta I_{SAT}}{I_{SAT}} = \frac{\delta \mu}{\mu} + \frac{\delta N_D}{N_D} \left[\frac{2 - 3U^2 + U^3}{1 - 3U^2 + 2U^3} \right] \tag{32}$$

where we used

$$\delta U \simeq -\tfrac{1}{2} U \delta N_D / N_D \tag{33}$$

on account of the change of V_p with N_D, the change of V_{BG} with N_D being negligible.

For devices with small pinch-off potential V_p, so that U is not small with respect to unity, the factor in brackets can become significantly larger than 2, enhancing the effect of $\delta N_D/N_D$ on transistor current degradation.

The $C-Q-V$ technique was used to determine mobility and carrier concentration in the epitaxial layer before and after neutron exposure. The ω^{-1} dependence of Q (equation (2)) was confirmed for the range $3 < Q < 100$ (figure 7) and C_S was found to be independent of frequency in this range, as predicted by equation (3). Q-values

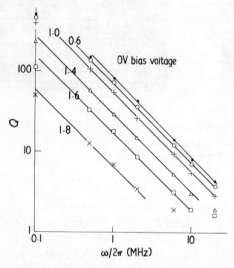

Figure 7. Dependence of the Q-factor on frequency and gate bias voltage. The slopes of the straight lines indicate an inverse frequency relation.

somewhat lower than those expected from the ω^{-1} relation were generally observed in the range $Q > 100$, which is attributed to the shunt leakage resistance of the gate depletion layer becoming significant as the junction impedance increases with lower frequency.

Figure 8 shows the distribution of impurity concentration and mobility derived from C_S and Q by equations (9), (10) and (12) before and after irradiation with $1 \cdot 6 \times 10^{15}$ neutrons/cm^2 ($E > 10$ keV). The impurity distribution has a flat region at $d \leqslant 0 \cdot 2 \, \mu$m and decays with increasing distance at $d \geqslant 0 \cdot 2 \, \mu$m. The tail of the impurity distribution beyond $0 \cdot 2 \, \mu$m is an artefact; it is discussed in section 5.2.

The impurity concentration up to $0 \cdot 2 \, \mu$m is $N_D = 3 \cdot 7 \times 10^{16}$ cm^{-3}; before the Zn gate diffusion, an average concentration in the epitaxial layer of $5 \cdot 9 \times 10^{16}$ cm^{-3} was determined by Hall effect measurements.

The impurity concentration in the flat region for $d < 0 \cdot 2 \, \mu$m decreases from $3 \cdot 7 \times 10^{16}$ to $3 \cdot 0 \times 10^{16}$ cm^{-3} and the peak mobility decreases from 3500 to 2500 cm^2 by neutron exposure. Table 1 shows that these degradations agree well with published data (Behle and Zuleeg 1972). Parameters pertinent to the transistor degradation are shown in table 2. The value before radiation U corresponds to $V_p = 2 \cdot 8$ V, and is obtained by adding the gate voltage for pinch-off ($1 \cdot 6$ V) of figure 6 to the built-in voltage

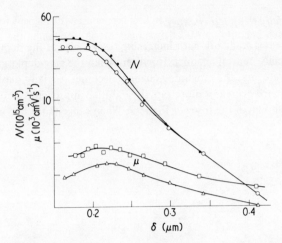

Figure 8. Mobility and carrier distributions as functions of position obtained by $C-Q-V$ measurements before and after irradiation by $1\cdot6 \times 10^{15}$ neutrons/cm² of $E > 10$ keV. The tail of the $N(d)$ curve beyond $0\cdot2 \mu$m is an artefact. Full circles and squares – not irradiated, open circles and triangles – neutron irradiated.

Table 1. Degradation of net impurity concentration N_D and of mobility μ by exposure to $1\cdot6 \times 10^{15}$ neutrons/cm² of $E > 10$ keV

	Experiment	Theory	
ΔN_D	-7×10^{15} cm⁻³	$-7\cdot1 \times 10^{15}$ cm⁻³	Behle and Zuleeg (1972)
μ^R/μ	0·71	0·76	McNichols and Ginell (1970)
I^R_{sat}/I_{sat}	0·288	0 294	Shockley (1952)

Table 2. Analysis of radiation effect on the saturation current at zero gate bias.

	U	N_D	μ	$(1-3U^2+2U^3)$	$(1-3U^2+2U^3)N_D^2\mu$
Non-irradiated device	0·66	$3\cdot7 \times 10^{16}$ cm⁻³	$.3500$ cm²V⁻¹s⁻¹	0·27	13
After irradiation by $1\cdot6 \times 10^{15}$ neutrons/cm²	0·74	$3\cdot0 \times 10^{16}$	2500	0·17	3·8

$V_b = 1\cdot2$ V. The value after irradiation U^R, is obtained by multiplying U by the ratio $(N_D/N_D^R)^{1/2}$, as is to be expected from the decrease in pinch-off potential V_p with N_D, namely equations (30) and (31). Mobility μ, $N_D V_P \simeq N_D^2$ and the factor

$$1 - 3U^2 - 2U^3$$

contribute about equally to the degradation of saturation current $I^R_{sat}/I_{sat} = 0\cdot294$; figure 6 gives 0·288 for this ratio.

5.2. C–V analysis for a partially depleted channel

The gate capacitance decreases gradually with increasing reverse bias as the depletion layer expands into an electrically neutral channel. However, when the gate depletion layer penetrates to the substrate, and the remaining channel becomes partially depleted (ie, for $n_0 < N_D$), the capacitance decreases more rapidly and then approaches a constant value with increasing bias as shown in the inset of figure 9. When capacitance becomes

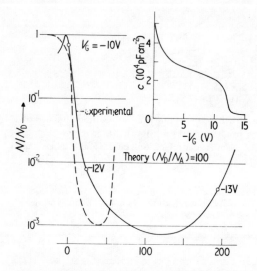

Figure 9. $N(X)$ distribution calculated by equations (11) and (12) from the $C-V$ measurements shown in the inset. The peak value of N_D is $1 \times 10^{17}\,\text{cm}^{-3}$, and $L_D = 1 \cdot 7 \times 10^{-6}\,\text{cm}$. A theoretical curve for an abrupt p–n junction of $N_D/N_A = 100$ is shown for comparison.

independent of bias voltage the artificial impurity distribution $N(X)$ of equations (11) and (12), shown in figure 9, becomes infinity at the position X corresponding to the limiting capacitance.

The accurate determination of the limiting small capacitance at large gate bias voltages poses experimental difficulties due to parasitic effects such as lead stray capacitances and edge capacitances at the gate circumference. Obviously the accuracy of measurements is enhanced by increased gate area. However, errors due to systematic thickness variation of the epitaxial layer along the substrate become then more significant.

Also plotted in figure 9 is a theoretical curve for an abrupt p–n junction of $N_D/N_A = 100$, having about the same minimum concentration N/N_D as the experimental curve. Using the impurity concentration of $10^{17}\,\text{cm}^{-3}$ for the epitaxial layer we obtain $N = 10^{15}\,\text{cm}^{-3}$ for the chromium concentration in the substrate, which is a reasonable order of magnitude for that sample. However, the observed minimum is located at a much larger distance from the metallurgical junction than that which is calculated for the abrupt p–n junction case; that is to say, the Debye length calculated from the impurity concentration of $10^{17}\,\text{cm}^{-3}$ in the epitaxial layer is too small by a factor of three to match the theoretical abrupt p—n junction curve. A possible explanation for this discrepancy is the following: for $N_D/N_A = 10^{-2}$ the width of the depletion layer in the epitaxial film adjacent to the substrate is only 10^{-6} cm, so that impurity diffusion during the epitaxial growth could well have caused a junction grading within the distances d_S. The relation (26) would then become invalid. However, the effect of an error in the limiting small capacitance on the $C-V$ distribution should also be investigated.

The experimental points marked by circles in figure 4 correspond to a high purity, high resistivity substrate. The Debye length required to obtain the abscissa scale factor was calculated from the impurity concentration measured by the $C-V$ method for $(x - a)/L_D \ll -1$. The experimental points have been shifted along the abscissa to match the curve for an intrinsic substrate. A reasonably good fit is obtained. Similar experimental results have been reported by Smith (1970).

Artificial impurity distributions by the $C-V$ method can be expected for narrow ion-implanted channels in high resistivity substrates. Indeed, such distributions are known to exhibit tails extending beyond what would be expected by ordinary ion implant theory (Welch *et al* 1974).

6. Summary and conclusions

(i) Combining $C-V$ with $Q-V$ measurements using a blocking contact to an epitaxial layer on insulating substrate is a useful method for obtaining carrier mobility.

(ii) Degradations of carrier concentration and mobility by neutron bombardment studied by this method agree with published data (Zuleeg and Behle 1972).

(iii) Decrease of the saturation current of JFET by neutron bombardment is explained by these degradations using the Shockley (1952) theory for a long channel device.

(iv) The impurity distribution obtained by the $C-V$ method for nearly pinched-off channel is shown to be an artefact. Distributions obtained for an abrupt p—n junction model and for an intrinsic substrate were calculated and compared with experimental data.

(v) The effect of the impurity distribution in the channel adjacent to a semi-insulating substrate must be investigated in more detail on hand of theoretical models. The theory for the high frequency case, in which substrate or interface charges cannot follow the AC variation of the gate potential, should be developed.

It is hoped that $C-V$ and $Q-V$ measurements as functions of frequency, temperature and substrate bias will eventually provide a powerful tool for analysis of substrate interface properties.

Acknowledgments

This work was sponsored in part by the Joint Services Electronics Program through the Air Force Office of Scientific Research under contract F 44620-71-C-0067 and by Air Force Cambridge Research Laboratories under contract F 19628-74-C-0129.

References

Behle A F and Zuleeg R 1972 *IEEE Trans. Electron Dev.* **ED-19** 993
Dousmanis G C and Duncan R C Jr 1958 *J. Appl. Phys.* **29** 1627
Gummel H K and Scharfetter D L 1967 *J. Appl. Phys.* **38** 2148
Hower P L *et al* 1969 *Proc. 2nd Int. Symp. Gallium Arsenide* (London: Institute of Physics) p187
Kennedy D P, Murley P C and Kleinfelder W 1968 *IBM J. Res. Dev.* **12** 399
Lehovec K and Zuleeg R 1970 *Solid St. Electron.* **13** 1415
Lehovec K 1974 *Appl. Phys. Lett.* **25** 279
—— 1975 *Appl. Phys. Lett.* in print
McNichols J L and Ginell W S 1970 *IEEE Trans. Nucl. Sci.* **NS-17** 52
Reiser M 1970 *Electron. Lett.* **6** 493
Schottky W 1942 *Z. Phys.* **118** 539
Shockley W 1952 *Proc. IRE* **40** 1365
Schrieffer J R 1955 *Phys. Rev.* **97** 641
Smith B L 1970 *J. Phys. D: Appl. Phys.* **3** 1179
Turner J A and Wilson B L H 1969 *Proc. 2nd Int. Symp. Gallium Arsenide* (London: Institute of Physics) p195
Welch B M, Eisen F H and Higgens J A 1974 *J. Appl. Phys.* **45** 3685
Zuleeg R 1969 *Proc. 2nd Int. Symp. Gallium Arsenide* (London: Institute of Physics) p1811

Analysis of metal–GaAs Schottky barrier diodes by secondary ion mass spectrometry

H B Kim, G G Sweeney and T M S Heng

Westinghouse Research Laboratories, Pittsburgh, Penn, USA 15235

Abstract. The degradation characteristics of heat-treated Al–GaAs, Au–GaAs, Au–Ni–GaAs and Pt–GaAs Schottky barriers are investigated by employing secondary ion mass spectrometry (SIMS), x-ray diffractometry and electrical I–V and C–V methods. SIMS data on interdiffusion between metal film and GaAs are presented relating the temperature and diffusion characteristics of these Schottky barrier systems. Of these, only the Al–GaAs system appears to be metallurgically non-reactive and to exhibit stable electrical characteristics up to 500 °C. Recommendations of upper temperature limits are presented for each Schottky barrier system based on electrical and SIMS metallurgical findings.

1. Introduction

The degradation characteristics of the metal–GaAs Schottky barrier interface at high temperatures have been investigated by a number of authors. Ogawa *et al* (1971) using x-ray diffractometry, have observed that Pt–GaAs formed brittle alloys of GaPt, PtAs$_2$ and Ga$_2$Pt after heat treatment which resulted in degradation of the Schottky barrier junction. Auger spectroscopic and x-ray diffractometric measurements have also been carried out by Finn *et al* (1973) on Pt–GaAs junctions. Sinha and Poate (1973) have examined Au–GaAs, Pt–GaAs and W–GaAs barriers by using ^{4}He$^+$ ion Rutherford backscattering measurements and found that only the W–GaAs barrier remained metallurgically non-reactive with stable electrical I–V characteristics after heat treatment temperatures of 350 and 500 °C. Severe interdiffusion was observed to occur in both the Au–GaAs and Pt–GaAs systems.

In this study we have employed secondary ion mass spectrometry (SIMS) and x-ray diffractometry to investigate metallurgical changes in a number of Schottky barrier systems (Al–GaAs, Au–GaAs, Pt–GaAs and Au–Ni–GaAs). The concentration profiles of several elemental species as a function of heat treatment conditions were obtained by SIMS. The detailed electrical degradation characteristics for each system are correlated to SIMS findings and recommendations for upper safe temperature limits are presented in the next sections.

2. Experimental

2.1. Sample preparation

Pt, Al, Au and Au–Ni metal Schottky barriers were formed on epitaxial (100) oriented n on n$^+$ GaAs layers grown by the Ga–AsCl$_3$–H$_2$ vapour transport system.

The epitaxial n layers were doped with Sn to 5×10^{15} cm^{-3} and the substrate material was Si doped to 10^{18} cm^{-3}. The substrate surface was etched in $H_2SO_4 : H_2O_2 : H_2O$: 50 : 1 : 2 for 15 s and rinsed in double-distilled H_2O just prior to metallization. The Al, Au and Au–Ni Schottky metals (1000–3000 Å thick) were evaporated by electron beam deposition, and the Pt junction was formed by RF sputtering of the Pt film (2000 Å thick). The substrate was held at room temperature during metallization. Planar Schottky barrier diodes (0·254 mm in diameter) were delineated in the case of Al, Au and Au–Ni metals while Pt Schottky barrier diodes were of the mesa form. The heat treatment of all the diodes was carried out in an H_2 atmosphere for seven minutes' duration, at temperatures ranging from 200 to 500 °C. For secondary ion mass spectrometric analysis, larger samples (1 × 1 cm) were used and subjected to identical heat treatment conditions in each case.

2.2. Secondary ion mass spectrometry

A Cameca SMI 300 ion mass analyser was employed to assess the mass spectrum and ion image of each element with planar resolution of 1 μm, and to delineate the concentration profile of selected constituent element with an in-depth resolution of 100 in a 2000 Å metal film. Both Ar$^+$ and O_2^+ primary ion beams were used to sputter etch the sample surface over a diameter of 300 μm and at an etch rate of about 600 Å min^{-1}. Reduced apertures of 10 μm and 40 μm were used to enhance sensitivity and resolution of the concentration profile when needed. The etch depth was checked by interference microscope and on the 'Talystep' instrument.

3. Results and discussions

3.1. Au–GaAs system

Figure 1 illustrates a typical set of SIMS concentration profiles for Au, Ga and As obtained for the Au–GaAs Schottky barrier system. The depth profiles of Au and Ga are again displayed in figure 2 in three-dimensional form to describe the effects of heat treatment. These profiles show a well behaved Au–GaAs interface with little or no interdiffusion taking place up to 350 °C. At 400 °C, however, a low concentration of Ga was detected in the Au film. This was followed by a rapid increase in concentration at the Au–GaAs eutectic temperature of 450 °C. At 500 °C, Ga build-up in the Au film was noted. Au was also observed to diffuse into the GaAs quite significantly for temperatures above 450 °C. No diffusion of As was detected up to 450 °C, but at 500 °C a rapid out-diffusion of As into the Au film was observed, as shown in figure 1. The SIMS results have also been confirmed by x-ray diffraction measurements. The pure Au phase remained unchanged after heat treatments.

Figure 3 shows the experimental forward and reverse DC I–V characteristics of Au Schottky barrier diodes subjected to various heat treatments. The parameters n and ϕ_B are derived from these using the thermionic emission current equation (Sze 1969)

$$J = A^*T^2 \exp\left(-q\phi_B/kT\right) \left[\exp\left(qV/nkT\right) - 1\right] \tag{1}$$

where J is the diode current density, A^* the effective Richardson constant, ϕ_B the

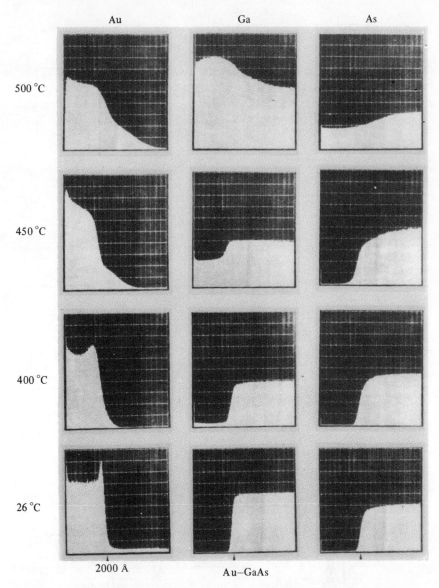

Figure 1. Typical SIMS Au, Ga and As concentration profiles against heat treatment temperature for Au–GaAs Schottky barrier system, Au is 2000 Å thick. Aperture: 40 μm for Au and As; 10 μm for Ga; O_2^+ ion beam.

barrier height, n the forward bias ideality factor (which takes into account the barrier height lowering by the applied voltage, Sze *et al* 1964) and V is the applied voltage. Under forward biased conditions where $V > 3kT/q$, the current may be expressed by $J_f \simeq J_s \exp(qV/nkT)$ where J_s is the saturation current density by $\phi_B = (kT/q) \ln(A^*T^2/J_s)$. The saturation current density is obtained by extrapolating the linear portion of the $I-V$ curve to zero bias. The parameters n and ϕ_B, tabulated in

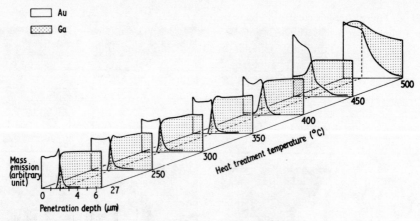

Figure 2. Three-dimensional composite of SIMS Au and Ga profiles against heat treatment temperature; Au–GaAs system.

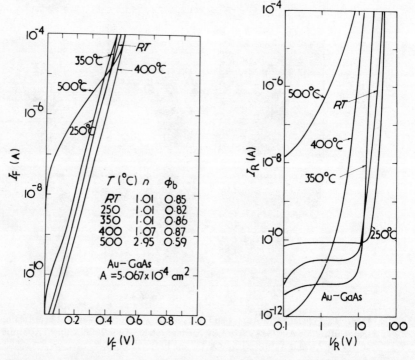

Figure 3. (*a*) Forward and (*b*) reverse DC I–V characteristics of Au Schottky barrier diodes.

figure 3, are calculated respectively from the slope $(q/kT)\,dV/d\ln J$ and the relationship $A^* = (m^*/m_0)A$, where $m^*/m_0 = 0.068$ for n-type GaAs, and the Richardson constant $A = 120\ \mathrm{A\,cm^{-2}\,K^{-2}}$ for thermionic emission into a vacuum. It can be seen from figure 3 that both n and ϕ_B are unchanged (1·01 and 0·86 V respectively) from room temperature to 350 °C. A large step increase in n to 2·95 and ϕ_B down to 0·59 V occurred at

500 °C. These changes clearly indicate degradation in the Au–GaAs Schottky barrier due to heat treatment. The electrical degradation behaviour of n and ϕ_B of the junction agrees with the metallurgical degradation pattern of figures 1 and 2. Figure 3(b) shows the reverse bias I–V characteristics for the Au–GaAs Schottky barrier diodes. A noticeable increase in the leakage current was observed on diodes subjected to heat treatment at temperatures above 400 °C. Although there were some leakage currents from edge breakdown, the observed I–V changes due to heat treatment above 400 °C are readily discernible from the magnitude and 'soft' breakdown characteristics.

The Schottky barrier height ϕ_B was also obtained from differential capacitance against voltage measurements. The barrier height was calculated from the equation (Sze *et al* 1964):

$$\phi_B = V_i + \zeta - \Delta\phi + (kT/q) \tag{2}$$

taking into account image force lowering, $\Delta\phi$ at the interface, and the thermal (kT/q) contribution of the mobile carriers (Goodman 1963). V_i was obtained from $1/C^2$–V extrapolated intercept and the Fermi level; $\zeta = (kT/q)(N_c/N_D)$ where N_c is the effective density of states in the conduction band (Smith 1959) given by

$$N_c = 2M_c(2\pi m_{de}kT/h^2)^{3/2},$$

where m_{de} is the density of state effective mass for electrons and M_c is the number of equivalent minima in the conduction band. The barrier lowering by image force is obtained from

$$\Delta\phi = [q^3 N_D(V_{bi} - V - kT/q)/8\pi^2 \epsilon_d^2 \epsilon_s \epsilon_0^3]^{1/4} \tag{3}$$

where V_{bi} is the built-in potential, ϵ_0 the permittivity of free space, ϵ_d the dielectric constant at optical frequencies, and ϵ_s the static dielectric constant. Table 1 summarizes the results of electrical data obtained for Au–GaAs Schottky barrier diodes in relation to the heat treatment temperatures. The values of ϕ_B obtained from C–V measurements are slightly greater than those deduced from the I–V measurements. No meaningful values of ϕ_B at 500 °C were derived from C–V data because of high leakage currents. In short, the Au–GaAs Schottky barrier exhibits stable characteristics up to 350 °C heat treatment in H_2. Fast out-diffusing Ga tends to build up in Au. Above the eutectic temperature, the significant Au diffusion into GaAs indicated in the SIMS profile may also lead to compensation in the underlying epitaxial layer.

Table 1. Au–GaAs

T (°C)	n	$\phi_B(I$–$V)$ (V)	$\phi_B(C$–$V)$ (V)	V_B (V)	
				10^{-7} (A)	10^{-4} (A)
27	1·01	0·85	0·88	28	42
250	1·01	0·82	0·88	40	44
350	1·01	0·86	0·88	17	24
400	1·07	0·87	0·86	7	13
500	2·95	0·59	0·91†	0·5	12

† High leakage current in C–V measurements.

3.2. *Au–Ni–GaAs system*

Composite SIMS depth profiles of Au, Ni and Ga for the Au–Ni–GaAs Schottky barrier are presented in figure 4. The Au film (1000 Å) was introduced to protect the underlying Ni film (2000 Å) from oxidation. As can be seen, Ni diffuses into both GaAs and Au, beginning slowly at 350 °C and more rapidly at 500 °C. Note that the Ni concentration level at the Ni–GaAs interface surpasses the Ni level at the Au–Ni interface after a heat treatment at 500 °C. Some diffusion of Au into Ni and GaAs was also observed. The unusually high apparent Au concentration level detected in the Ni film at room temperature is due to a perturbation in spectrometric findings (for example, other species of Ni compounds, eg Ni_3OH, have the same mass number 197 as Au). At any rate, the level of Au detected in the Ni film is now used for referencing the room temperature sample from which heat treatment analysis is made. The out-diffusion of Ga into the Ni film at 350 °C is barely noticeable, but increases rapidly at 500 °C through the Ni film and into the Au film as shown in figure 4. SIMS Ga concentration profiles obtained at higher sensitivity are shown in figure 5. The Ga appears to diffuse through the Ni film and accumulate in the Au film even at 200 °C similar to that found for Au–GaAs system at higher temperatures. No significant amount of As out-diffusion into the Ni film was observed at temperatures up to 350 °C, but at 400 °C some As diffusion became noticeable. At 500 °C, the As out-diffusion reaches into the Au film. From figures 4 and 5 it can be assessed that significant metallurgical degradation at the Ni–GaAs interface occurs at heat treatment temperatures above 400 °C. The Au–Ni–GaAs system, however, exhibited an additional problem. A high density of 'bubble' type surface formation was observed on samples heat treated at temperatures greater than 300 °C as shown in figure 6. At 500 °C, the 'bubbles' were reduced to small clusters. The direct ion imaging SIMS mode was utilized to examine the above non-uniform surface formations shown in figure 7. Here the ion images of Au, Ni, Ga and As were taken on a sample heated to 500 °C. The ion images of all the elements were found to be distributed non-uniformly on the surface and through the depth of the alloyed region. The cluster sizes vary from a few micrometres to about 100 μm. These observations were also complemented by x-ray diffraction results which showed that the pure Au and Ni phases have changed to complex phases for the sample heat treated at 500 °C.

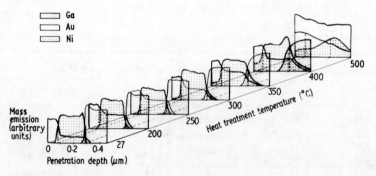

Figure 4. Three-dimensional composite of SIMS Au, Ni and Ga profiles against heat treatment temperature; Au–Ni–GaAs system.

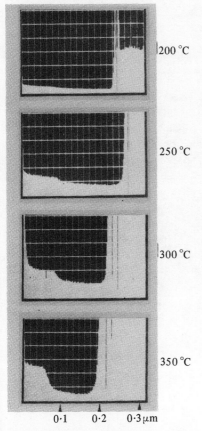

200 °C

250 °C

300 °C

350 °C

0·1   0·2   0·3 μm

Figure 5. SIMS Ga profiles taken at high sensitivity: 20 K counts/second, 10 μm aperture, and O_2^+ ion beam; Au–Ni–GaAs system.

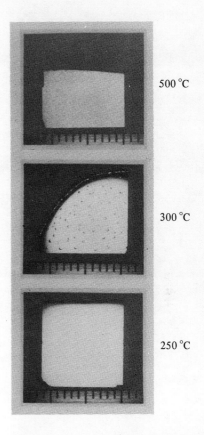

500 °C

300 °C

250 °C

Figure 6. Photomacrograph of 'bubble' type surface due to heat treatment; Au–Ni–GaAs.

The n and ϕ_B parameters obtained from the forward $I–V$ characteristics of the Au–Ni–GaAs Schottky barrier diodes are tabulated in table 2. The room temperature values of $n = 1\cdot01$ and $\phi_B = 0\cdot86$ V remain relatively unchanged up to 350 °C, and rapidly degraded to $1\cdot59$ and $0\cdot65$ V respectively at 400 °C. The Schottky barrier turned ohmic at 500 °C. For heat treatment temperatures up to 350 °C, the reverse biased $I–V$ characteristics of the Au–Ni–GaAs Schottky barrier diodes were similar to those of the Au–GaAs Schottky barrier diodes. The reverse breakdown voltages were less than $0\cdot1$ V for diodes heat treated at 400 °C and the diodes became ohmic at 500 °C. The Schottky barrier heights for the Au–Ni–GaAs diodes from the $C–V$ and $I–V$ measurements are listed in table 2. The results indicate that the double layered Au–Ni–GaAs system exhibits stable electrical and metallurgical characteristics up to heat treatment temperature of 350 °C above which nonuniform distribution of Au, Ni, Ga and As were also observed. There is a rapid out-diffusion of Ga through the Ni film into the Au film, together with Ni and Au diffusion into GaAs which again may compensate the epitaxial GaAs layer.

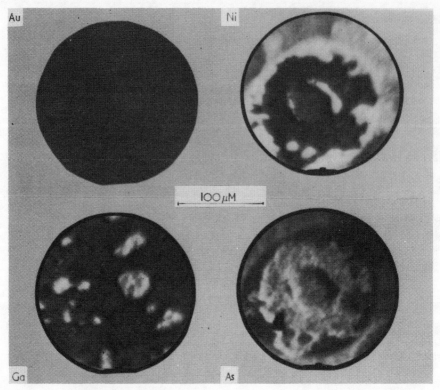

Figure 7. Ion images of Au, Ni, Ga and As cluster formation; 500 °C heat treatment.

Table 2. Au–Ni–GaAs

T (°C)	n	$\phi_B(I-V)$ (V)	$\phi_B(C-V)$ (V)	V_B (V)	
				10^{-7} (A)	10^{-4} (A)
27	1·01	0·86	0·96	16	23
250	1·07	0·85	0·95	32	34
350	1·01	0·89	1·00	28	30
400†	{1·59 Short	0·65}	Short	Short	
500	—	Short	Short	Short	

† Nonuniform.

3.3. Pt–GaAs system

Figure 8 shows the effect of interdiffusion between Pt and Ga due to heat treatment by SIMS measurement. With increasing temperature, there is a systematic increase in the amount and the penetration depth of Ga into the region of the original Pt film layer. No significant amount of Pt diffusion into GaAs was noticed at temperatures up to

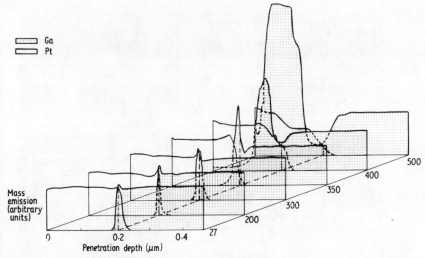

Figure 8. Three-dimensional composite of SIMS Pt and Ga profiles against heat treatment temperature; Pt–GaAs system.

500 °C. Although As out-diffusion into the Pt film was observed at temperatures above 400 °C, the As concentration build-up was contained near the original Pt–GaAs interface region.

Results of x-ray diffraction patterns taken on a heat treated (400 °C) Pt–GaAs sample, show five lines corresponding to a cubic structure with a lattice parameter of approximately 6·00 Å which is associated with $PtAs_2$ (5·967 Å) or Ga_2Pt (5·922 Å). In addition an unindexed line was present at 3·92 Å. PtGa formation was also observed by Ogawa *et al* (1971) and Finn *et al* (1973) in addition to these reaction products. The results of the x-ray analysis provide a cross reference for the SIMS distribution measurements of figure 8. For example, at 350 °C the original Pt film remains unchanged except near the Pt–GaAs interface where a Ga build-up was observed. (The alloy distribution may be a Pt–PtGa (or Ga_2Pt)–GaAs layered structure, since no significant As out-diffusion into Pt film was observed.) At 500 °C, however, the alloy distribution shows a possible Pt–Ga_2Pt–$PtAs_2$–GaAs structure since a high concentration of Ga has diffused throughout the original Pt–GaAs interface. A similar layered structure formation was reported by Sinha and Poate (1973). Under our short-time heat-treatment condition only a negligible amount of the GaAs layer was consumed by reaction at 500 °C. This differs from the reaction rates reported by Coleman *et al* (1974) for the Pt–GaAs system assuming a movement of the Schottky barrier interface to coincide with the $PtAs_2$ layer advancement for temperature ranges from 300 to 400 °C. As shown in figure 9 no significant $PtAs_2$ formation near the original interface was observed up to 400 °C for an alloying time of seven minutes.

The values of n and ϕ_B for the Pt–GaAs diodes were found to be 1·04 and 0·95 V respectively at room temperature. The peak values of 1·34 and 0·85 at 300 °C suggest a progressive departure from the ideal Schottky barrier behaviour with increasing heat treatment temperature. At 500 °C, n is 1·09 and ϕ_B is 1·07 V, exhibiting the effects of alloying (see table 3).

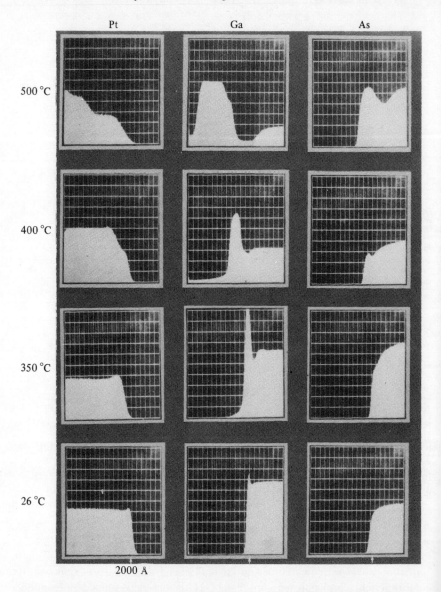

Figure 9. Typical SIMS Pt, Ga and As concentration profiles against treatment temperature for Pt–GaAs Schottky barrier system; Pt is 2000 Å thick. Aperture: 40 μm for Pt and As; 10 μm for Ga; O_2^+ ion beam.

In summary, Ga out-diffuses rapidly into Pt and forms a layered structure of Pt alloys which is dependent on the alloy temperature. It was found also that As out-diffuses slowly, forming $PtAs_2$, and remains near the original Pt–GaAs interface. Since no significant Pt diffusion into GaAs was observed this should eliminate problems associated with compensation by Pt in GaAs in the form of deep acceptors.

Table 3. Pt—GaAs

$T(^{\circ}C)$	n	$\phi_B(I-V)$ (V)
27	1·04	0·95
200	1·17	0·90
300	1·34	0·86
350	1·24	0·90
400	1·31	0·90
500	1·09	1·07

3.4. Al—GaAs system

Among the metal—GaAs systems examined, the Al—GaAs system shows the most stable metallurgical character with heat treatment. Composite Al and Ga SIMS profiles are shown in figure 10. No significant degradation of the Al—GaAs interface, due to interdiffusion of adjacent species, was observed up to heat treatment temperature of 500 °C. Figure 11 shows the SIMS Ga and As profiles examined with high sensitivity. Again no noticeable erosion of the Al—GaAs interface was noted up to 500 °C. Both Ga and As out-diffuses slowly, advancing the Al—GaAs interface a negligible amount at 500 °C, while still maintaining a metallurgically sound barrier with an abrupt Al—GaAs interface. X-ray diffraction results showed only a pure Al phase without any new complex metal phases for the 500 °C sample.

The n and ϕ_B parameters for the Al—GaAs junction are shown in table 4. There is a slight improvement of both parameters over the whole temperature range with increasing heat treatment temperatures up to 400 °C, at which n is 1·04 and ϕ_B is 0·89. A slight increase in reverse leakage current was also observed at 500 °C. This may be a normal variation, since no softening of the reverse breakdown was observed. Table 4 is a summary of the pertinent electrical parameters as a function of heat treatment temperature for the Al—GaAs Schottky barrier diodes.

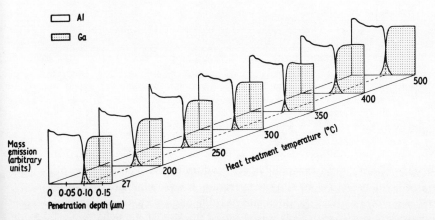

Figure 10. Three-dimensional composite of SIMS Al and Ga profiles against heat treatment temperature; Al—GaAs system.

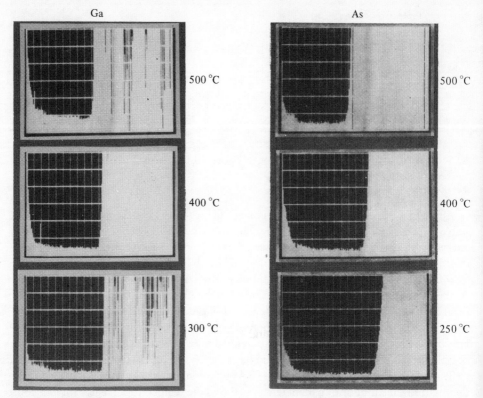

Figure 11. SIMS Ga and As profiles taken at high sensitivity: 2 K counts/second for Ga and As; 10 μm aperture for Ga; 40 μm aperture for As.

Table 4. Al–GaAs

T (°C)	n	$\phi_B(I–V)$ (V)	$\phi_B(C–V)$ (V)	V_B (V)	
				10^{-7} (A)	10^{-4} (A)
27	1·01	0·74	0·80	30	46
250	1·04	0·77	0·80	36	40
350	1·11	0·87	0·84	42	46
400	1·04	0·89	0·96	44	47
500	1·04	0·79	0·87	32	39

4. Summary

Measurements of metallurgical changes by SIMS depth profiles, ion image, x-ray diffraction and optical methods are in a good agreement with electrical measurements for characterizing the Pt–GaAs, Au–GaAs, Au–Ni–GaAs and Al–GaAs Schottky barriers. The effects of heat treatment on these junctions are summarized in table 5. These data suggest a range of safe and stable heat treatment temperatures for device operation and processing for each metal system.

Table 5. Metal–GaAs Schottky barriers

Schottky barrier metal	Heat treatment	T (°C)	n	ϕ_B (V)	Degradation by	T_{max} (°C)
Al	Before	500	1·01	0·74	Stable	500
	After		1·04	0·79		
Au	Before	500	1·01	0·85	Ga and As out-diffusion	350
	After		2·95	0·59	Au in-diffusion	
Pt	Before	400	1·04	0·95	Ga, As–Pt alloy	300
	After		1·31	0·90	Ga out-diffusion	
Ni	Before	400	1·01	0·86	Ga, As–Ni alloy	350
	After		1·59	0·65	Ga out-diffusion Ni in-diffusion	

All the metals examined exhibit acceptable Schottky barrier characteristics within the recommended maximum heat treatment temperature. Of these metals, the Al–GaAs Schottky barrier was found to be the most stable and to have a non-reactive interface with minimum electrical characteristic degradation for temperatures up to 500 °C.

Acknowledgments

The authors wish to acknowledge helpful discussions with Dr S Y Wu, and are grateful to Dr W Takai for carrying out the x-ray diffraction measurements. The able assistance of A F Lovas in making electrical measurements is gratefully acknowledged. We wish to thank Bernard Autier (of Cameca Instruments, Inc) for cross-checking our ion probe measurements.

References

Coleman D J Jr, Wisseman W R and Shaw D W 1974 *Appl. Phys. Lett.* **24** p355
Finn M C, Hong H Y-P, Lindley W T, Murphy R A, Owens E B and Strauss A J 1973 *Preparation and Properties of Electronic Materials Conf. Paper D1 Las Vegas*
Goodman A M 1963 *J. Appl. Phys.* **34** 329
Ogawa M, Shinoda D, Kawamura N, Nozaki T and Asanabe S 1971 *NEC Res. Dev.* **22** 1–7, 7–1971
Sinha A K and Poate J M 1973 *Appl. Phys. Lett.* **23** 666–8
Smith R A 1959 *Semiconductors* (London: CUP)
Sze S M 1969 *Physics of Semiconductor Devices* (New York: Wiley) p20
Sze S M, Crowell C R and Kahng D 1964 *J. Appl. Phys.* **35** 2534

An electrochemical technique for automatic depth profiles of carrier concentration

T Ambridge and M M Faktor

Post Office Research Department, Dollis Hill, London NW2 7DT, UK

Abstract. An electrochemical technique has been developed which provides continuous automatic plots of carrier concentration profiles in semiconductors, with no limitation on depth. This is achieved by using a concentrated electrolyte both to form a Schottky barrier suitable for depletion layer capacitance measurements, and to provide a means for simultaneous controlled anodic dissolution. Highly doped material is easily accommodated, as it is necessary to maintain only a small constant reverse bias (below the avalanche breakdown voltage); thus complete continuous profiles may be obtained through epitaxial layer structures including the entire active layer, buffer layer, and as much as required of the substrate. Profiles have been obtained in n-GaAs, including structures for IMPATT devices, and in n-GaP. Preliminary results, describing extension of the technique to the analysis of p-type material; and to hetero-structures (eg, GaAs–GaAlAs), are also reported.

1. Introduction

The technique to be described was developed at the Post Office Research Department, originally to meet internal requirements for the analysis of epitaxially grown multilayer structures of n-type gallium arsenide. It will be shown, however, that the technique is neither restricted to gallium arsenide, nor to n-type material. The standard technique for obtaining carrier concentration profiles in such layers relies on capacitance–voltage measurements upon reverse-biased Schottky barriers, formed by metal deposition on the surface, or by a mercury probe system (Hughes *et al* 1972). The analysis is most frequently performed using a commercially available instrument, based on the design of Baxandall *et al* (1971), or Copeland (1969). This technique has one major drawback, which is that the maximum depletion depth that can be probed beneath the surface, is limited by the maximum reverse bias that can be applied before electrical breakdown occurs. For gallium arsenide, this depth is about $3\,\mu\text{m}$ for a doping level of $10^{16}\,\text{cm}^{-3}$, but less than $0.1\,\mu\text{m}$ for doping levels about $10^{18}\,\text{cm}^{-3}$. To obtain profiles to greater depths requires chemical step etching. To obtain a complete profile usually requires several steps, and the technique becomes impractical where extended regions of high doping level are to be investigated, in complex multilayer structures. In some cases, structures have been analysed using a combination of capacitance measurements, Hall-effect measurements, and the cleave and stain technique (Rosztoczy and Kinoshita 1974). These procedures could be replaced by the fully automatic measurement provided by the electrochemical technique.

The Post Office electrochemical technique combines the concept of the Schottky barrier capacitance technique with a means of continuous dissolution of the semiconductor; this replaces the step-etches previously required, allowing the procedure to be fully automated. We have found that, over a well-defined range of applied voltage, the capacitance–voltage behaviour of gallium arsenide, in contact with an electrolyte of concentrated potassium hydroxide solution, fits the theoretical Schottky relationship, the electrolyte behaving essentially as a metal (Ambridge and Faktor 1974). The electrolyte also provides the medium for anodic dissolution of the semiconductor.

2. Experimental background

The electrochemical cell used for this work is shown in figure 1. Electrolyte flows through the cell from a reservoir, and out to waste (about 50 ml per hour). The gallium

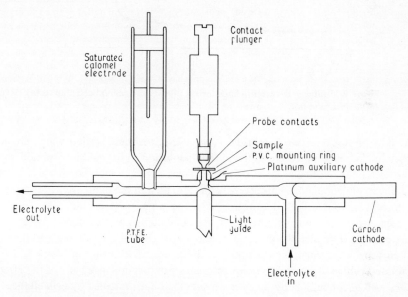

Figure 1. The electrochemical cell.

arsenide anode is held onto a moulded PVC mounting ring by the pressure of two tin-coated wire probes. Current is passed via one contact and a carbon cathode, whilst potential is measured via the other contact with respect to a saturated calomel electrode SCE. Capacitance measurements are made via a back contact, and an auxiliary cathode of platinum wire. The sample may be illuminated from a filtered standard projector lamp source, by means of a light guide; the illumination is essential for even anodic dissolution as described below.

Figure 2 depicts current–voltage curves obtained from an n-type gallium arsenide sample, in the dark and under illumination. Note that the potential scale is given as measured with respect to the calomel electrode. The 'rest' potential marked is that measured with zero current, in the dark. The solid curve represents the current–voltage behaviour under illumination. There is a potential range (from about −1 to 0 V) over which the current has almost reached a saturation value; this value is directly proportional to illumination intensity, and is due to minority carrier current — that of holes

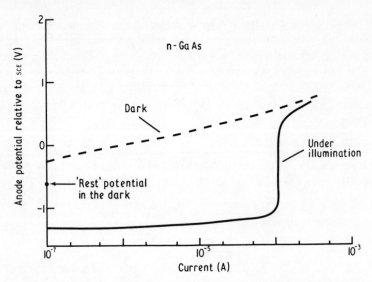

Figure 2. Voltage–lg current characteristics for n-type gallium arsenide ($n \simeq 10^{18}\,\text{cm}^{-3}$) in the dark, and under illumination.

created by photoexcitation (Ambridge *et al* 1973). It is these carriers which are necessary for the anodic reaction which results in the dissolution of the gallium arsenide, essential to our technique. At higher potentials the current begins to increase sharply, corresponding to an 'anomalous current' which is also seen in the dark. We have found that this current is not distributed uniformly over the sample surface, but arises from localized regions where apparently premature breakdown occurs, due to material imperfections. If anodic dissolution is attempted in the dark, by drawing a high current (eg $1\,\text{mA cm}^{-2}$), dissolution occurs only at these regions, as shown in figure 3. Deep pits are formed, which develop a complex internal structure; enhancement of the effective surface area of the sample by up to seven times has been revealed by capacitance measurements, which are thus entirely unsuitable for carrier concentration determination in these circumstances (Ambridge and Faktor 1974). If, on the other hand, dissolution is carried out under uniform strongly absorbed illumination, at low anodic potential (where the dark current is negligible), the dissolution is uniform†, as shown in figure 4.

Referring again to figure 2, a 'safe' working potential range may be defined from about -0.6 to $0\,\text{V}$. At higher potentials the anodic dark current becomes significant, and at lower potentials there is a danger of drawing cathodic current in the dark. It is over this safe range, also, that capacitance–voltage behaviour is in accordance with the Schottky law – the depletion layer capacitance is inversely proportional to the square root of potential barrier height, for uniformly doped material. Also excursions outside

† The dissolution rate has a weak dependence on carrier concentration, which is minimized by using illumination with photon energies much greater than the bandgap energy, so that the absorption width is smaller than the Schottky depletion width. Poor quality substrate material, having gross variations in carrier concentration, may show small variations in dissolution rate (up to 10%), although the effect is not sufficient to enhance the surface area.

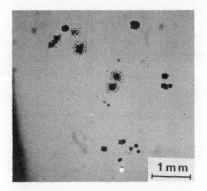

Figure 3. n-type gallium arsenide in which localized regions have been attacked by performing electrochemical dissolution in the dark.

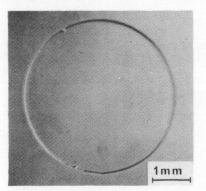

Figure 4. Epitaxial n-type gallium arsenide, showing an area where dissolution has been performed under illumination (at −0·3 V with respect to SCE).

this range in either direction give erroneous measured values of capacitance due to high differential conductance through the depletion layer.

The basis of the automatic profile plotting technique is that dissolution of the material is performed continuously at a 'safe' potential, the value of −0·3 V being used for gallium arsenide. Simultaneous capacitance measurement is made which allows the carrier concentration to be determined, and directly plotted, as a function of depth. The system will now be explained in more detail.

3. The automatic plotting system

The manner in which carrier concentration is derived from capacitance measurements bears some similarity to the method of Baxandall *et al* (1971), although our electronics system performs analogue computation according to a differently written, but equivalent operating formula (Hilibrand and Gold 1960).

The net donor concentration, at the depletion depth is given by

$$(N_D - N_A) = \frac{2\epsilon_0\epsilon\delta V}{q\delta(W_D{}^2)}$$

where ϵ_0 is the absolute permittivity of free space, ϵ is the relative permittivity (dielectric constant), q is the electronic charge, V is the anodic potential, W_D is the depletion width, and

$$W_D = \epsilon_0\epsilon A/C$$

where C is the capacitance and A is the surface area.

A phase sensitive system, operating at 3 kHz, measures the Schottky barrier capacitance C, which gives the depletion width W_D, via an analogue unit. The depletion width is also modulated as a result of a deliberate modulation of the anodic potential at 30 Hz. A second phase sensitive system, and analogue unit, determine $\delta W_D{}^2$, the modulation of $W_D{}^2$. Hence the donor concentration can be obtained.

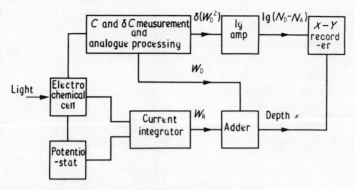

Figure 5. Block diagram of the automatic plotting system.

A very much simplified block diagram of the electrochemical system is shown in figure 5. The capacitance measurement system just described is shown as a single block. (For further information, see Ambridge and Faktor 1975.) The δW_D^2 output is fed to a logarithmic amplifier, to give $\lg(N_D - N_A)$. The other essential parts of the system are the potentiostat, which maintains the 'safe' anodic potential, and a current integrator which determines the thickness removed W_R, according to Faraday's law of electrolysis (Ambridge *et al* 1973). The total measurement depth x is the sum of the thickness removed, and the depletion width, obtained via the adder unit. An $X-Y$ recorder is used for plotting.

4. Application of the profile plotter

4.1. n-type gallium arsenide

Figure 6 shows a profile plot obtained on this equipment from a two-layer n-type gallium arsenide structure, grown by vapour phase epitaxy. Note that the plot is

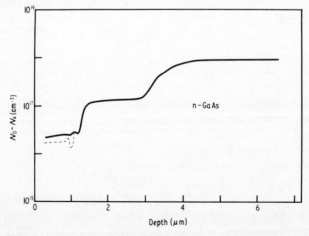

Figure 6. Automatic profile plot: a two-layer n-type epitaxial gallium arsenide structure. The additional plots (broken lines) shown in this and subsequent figures are explained in the text (section 4.1).

continuous through the low-doped 'active' layer, the buffer layer, and the heavy doped substrate. The curve actually consists of individual points, plotted at small intervals of dissolution, although this feature is not reproduced in the figures. The heavily doped material is accommodated as easily as the low doped layer, since the applied reverse bias is about 0·5 V at all times, which is much lower than the breakdown value (except for that of exceptionally poor material which may contain dopant precipitates). The dissolution rate during profile plotting through gallium arsenide is about $2\,\mu$m per hour; thus the active layer was analysed within about half an hour. It should be emphasized that the equipment requires no attention during the plotting so that the slow rate is not a serious disadvantage.

A second example is shown in figure 7. This is a profile of a three-layer n-type gallium arsenide structure, which was intended to be used for the fabrication of modified Read-type IMPATT diodes. It demonstrates the wide range of carrier concentration

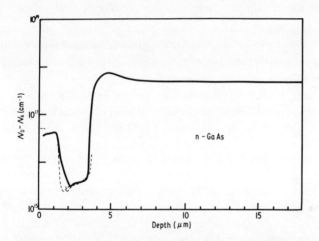

Figure 7. Automatic profile plot: a three-layer n-type epitaxial gallium arsenide structure.

values which may be incorporated in a single plot. Two unusual features are apparent in the structure. One is the dip in carrier concentration at the finish of growth of the active region; a new procedure has since been adopted by the growers to eliminate this. The other feature is the unusual state of affairs, in which the buffer layer is more heavily doped than the substrate. A more heavily doped substrate should have been used; manufacturer's data on substrate doping tends to be unreliable as large variations in doping level are common in bulk-grown ingots.

The above plot is an example of one which has reached full scale, and initiated automatic close-down of the profiling system, including stopping dissolution. It is a useful system, as the machine can, therefore, be left to run overnight, if a sample is loaded before leaving in the evening. Alternatively, if the approximate layer thicknesses are not known, the machine can be left running in the recycle mode. When full scale is reached, the recorder pen returns to the beginning of the scale, and continues to plot, whilst the dissolution continues, extending the depth scale indefinitely.

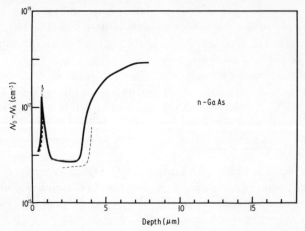

Figure 8. Automatic profile plot: a 'clump' structure of n-type epitaxial gallium arsenide.

Figure 8 is a very recent plot showing a 'clump' structure in epitaxial n-type gallium arsenide, also intended for IMPATT diode fabrication. This demonstrates the increasing complexity of structures which are being presented for analysis.

Also shown in figures 6, 7 and 8 are plots obtained from different regions of the same samples, using a commercial dotter (Baxandall *et al* 1971) in conjunction with smaller area (0·25 to 0·75 mm diameter) evaporated Schottky barriers. In some cases conventional step etching was used. It should be stressed that these do not represent 'control' experiments for quantitative comparison, since all of the structures depicted exhibited significant lateral variations in layer properties. (In separate experiments, on relatively uniform material, close agreement in absolute carrier concentration values has been obtained.) The essential comparison that can be made here between the two types of measurement is that, whilst the electrochemical method provides a continuous plot over a large depth, the alternative method is at present capable of greater resolution of abrupt changes, in this non-uniform material, mainly because the measurement is integrated over a smaller area. A single electrochemical measurement therefore provides the crystal grower with an indication of the lateral uniformity of his product, as judged by the sharpness of interface definition. However, in structures selected for further processing into devices, the final assessment would normally need to be performed on test areas typical of the devices themselves. In principle, it should be possible to use the electrochemical technique in conjunction with such areas, defined by suitable masking.

4.2. n-type gallium phosphide

Gallium phosphide has also been studied using the electrochemical technique. It is a noteworthy feature that none of the n-type samples, that we have studied, has given a measurable dark current in the range of potential covered in figure 2. A profile plot is shown for a thick epitaxial layer in figure 9; this demonstrates the essentially limitless depth range of the technique. The artefacts shown within a depth of about 5 μm from the surface were reproduced on several different areas of the same layer, and correlate with a pre-close-down procedure in the growth technique.

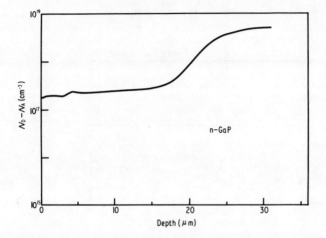

Figure 9. Automatic profile plot: a single thick epitaxial layer structure of n-type gallium phosphide.

4.3. p-type gallium arsenide and p–n junctions

A modification of the profiling technique has been tried, to enable p-type material to be studied. For a p-type semiconductor, one would expect reverse Schottky behaviour with cathodic polarization of the sample. However, Brattain and Garrett (1955) failed to observe cathodic photocurrent saturation in their classic research on germanium electrodes. Green (1959), using a sophisticated approach to achieve and maintain purity of his electrochemical system, did obtain this, which was evidence that the cathodic reaction for the evolution of hydrogen was limited by the supply of electrons from the semiconductor conduction band.

Of course, cathodic reactions tend to contaminate the sample surface, leading to an experimentally intractable situation, as indicated above. We argued, however, that if cathodization could be confined to brief periods of capacitance test measurement (during which little charge flowed) alternating with periods of extensive anodic dissolution, which is in any case required for depth profiling, a 'clean' semiconductor–electrolyte interface might be maintained. Note that no illumination is required for the anodic dissolution, since holes are readily available. The modified profiling technique, which we have therefore tried, allows dissolution to take place at the usual anodic potential, but after each interval of thickness, where a point is normally plotted, the potential is switched to a suitable cathodic value (-1.3 V relative to SCE). The carrier concentration is then automatically plotted from the measurement made at this potential, before the potential is switched back to the anodic value, and dissolution continues. The ratio of anodic to cathodic charge transfer per cycle is arranged to be at least 100:1. Preliminary results show this technique to be successful, with limitation of the cathodization being essential to maintain well-defined Schottky behaviour. The charge ratio of 100:1 apparently contains an adequate safety factor, but this is strictly an empirical figure. Figure 10 shows plots taken from nominally uniformly doped, bulk grown, p-type gallium arsenide samples. The values of carrier concentration obtained agree to within

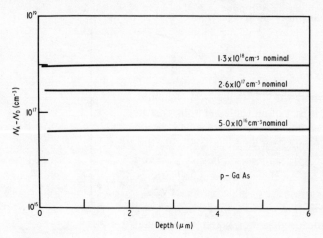

Figure 10. Automatic profile plots: three plots through uniformly doped p-type bulk gallium arsenide. The marked values of carrier concentration are the manufacturer's nominal values, based on Hall measurements.

the usual tolerance levels (about ±20%), with the manufacturers quoted nominal values, derived from Hall-effect measurements.

A profile obtained from a liquid phase epitaxially grown layer of p-type gallium arsenide on an n-type substrate is shown in figure 11. The apparent increase in carrier concentration beyond a depth of 5 μm is not real, but indicates a departure from p-type Schottky barrier behaviour, as the junction is approached. When about 8 μm had been removed, the usual n-type profiling procedure was adopted, although it was not until a total depth of 12 μm had been reached that well-behaved n-type Schottky characteristics were obtained, corresponding to the substrate material. The rather large width over which the measurement was invalid reflects partly a variation in p-layer thickness, as determined by a cleave and strain measurement. More work is required to

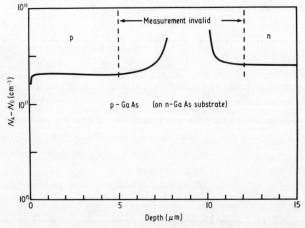

Figure 11. Automatic profile plot: p-type epitaxial gallium arsenide on an n-type substrate: see text for comments on region where measurement is invalid.

evaluate fully the applicability of the electrochemical technique to p–n junction regions.

4.4. Heterolayer analysis

An area of study where the electrochemical technique may well be exploited, with further development, is that of multiple heterolayer structures, such as those of gallium arsenide and gallium–aluminium arsenide. Preliminary results indicate that gallium–aluminium arsenide can be processed by the automatic plotter, although, of course, in heterolayer profiles there will be a small error in carrier concentration values, unless correction is made for the variation in dielectric constant. This depends upon alloy composition, a knowledge of which is also of vital interest in its own right. We hope to explore illumination-wavelength dependence of photocurrent, or open-circuit photovoltage or photocapacitance, to determine the bandgap, and hence alloy composition at different points in the profile.

4.5. Deep level trap analysis

The electrochemical method of depth profiling provides considerable scope for the application of different measurement techniques to study the Schottky barrier. Information on deep-level traps within the semiconductor could be gained from frequency-dependent, or temperature-dependent impedance measurements, or from a study of the wavelength-dependence of one, or more, of the photoeffects mentioned above.

5. Conclusions

To summarize, the electrochemical plotter presently provides fully automatic continuous measurement of carrier concentration profiles, in epitaxial layers of n-type gallium arsenide, on a routine basis. Much wider analytical applications may be predicted, because the electrochemical technique combines the versatility of a Schottky barrier with accurately controlled dissolution.

Acknowledgments

We wish to thank E G Bremner for designing and constructing most of the electronic equipment. Acknowledgment is made to the Director of Research of the British Post Office for permission to present this paper.

References

Ambridge T, Elliott C R and Faktor M M 1973 *J. Appl. Electrochem.* **3** 1–15
Ambridge T and Faktor M M 1974 *J. Appl. Electrochem.* **4** 135–42
—— 1975 *J. Appl. Electrochem.* to be published
Baxandall P J, Colliver D J and Fray A F 1971 *J. Phys.* E: *Sci. Instrum.* **4** 213–21
Brattain W H and Garrett C G B 1955 *Bell. Syst. Tech. J.* **34** 129–76
Copeland J A 1969 *IEEE Trans. Electron Dev.* **ED-16** 445–9

Green M 1959 *Modern Aspects of Electrochemistry* ed J Bockris vol II (London: Butterworths)

Hilibrand R and Gold D R 1960 *R.C.A. Rev.* **21** 245–52

Hughes F D, Headon R F and Wilson M 1972 *J. Phys. E: Sci. Instrum.* **5** 291–2

Rosztoczy F E and Kinoshita J 1974 *J. Electrochem. Soc.* **121** 439–44

The ion implantation of donors for n⁺-p junctions in GaAs

J M Woodcock and D J Clark

Mullard Research Laboratories, Redhill, Surrey, England

Abstract. GaAs has been doped by the ion implantation of silicon, sulphur, selenium and tin. After annealing at $700\,°C$ the layers were n-type with Hall mobilities of $2000\ cm^2\,V^{-1}\,s^{-1}$ in all cases but with the heavier ions, selenium and tin, it was necessary to implant above room temperature. The spatial distribution of active donors was obtained from differential Hall measurements on implanted layers in semi-insulating or epitaxial n on semi-insulating GaAs. With selenium and tin implants the concentration and mobility of free electrons near the surface and the depth of the donor distribution were dose dependent. The group IV impurities, silicon and tin, produced the maximum peak carrier densities. The tin implants in particular had carrier densities at the peak slightly in excess of $10^{18}\ cm^{-3}$ with very abrupt transitions to the background concentration of the substrate. Implanted layers of tin and silicon were used to make n⁺-p diodes which were electrically isolated by proton bombardments of suitable energies. These diodes exhibited low reverse leakage currents and reverse breakdown voltages close to the theoretical value for GaAs of the carrier concentration used but detailed studies of the current–voltage characteristics showed that they were not ideal. The electroluminescence spectra were dominated by low energy emission bands. Identical spectra were obtained from just below the junction with measurements combining cathodoluminescence and chemical stripping. Capacitance measurements suggested the presence of an insulating layer less than a micrometre thick.

1. Introduction

It is difficult to diffuse dopants into GaAs and its related compounds. This is particularly true of donor impurities which tend to have much lower diffusion coefficients than acceptor impurities (Goldstein 1961, Fane and Goss 1963). Temperatures around $1000\,°C$ are necessary for donor diffusion, well above that at which GaAs freely dissociates. Impurities can be introduced at a much lower temperature by ion implantation, a doping technique fairly well developed for silicon but not for semiconducting compounds. However a high temperature stage is required to anneal out radiation damage produced by the energetic ions and to enable the implanted atoms to reach substitutional sites. If electrical activity could be obtained at temperatures below those required for diffusion, ion implantation would be a useful alternative to thermal diffusion. It would be particularly useful, for example, in any planar technology requiring donor regions in GaAs.

In recent years a number of publications have appeared concerned with the ion implantation of donors into GaAs (Sansbury and Gibbons 1970, Whitton and Bellavance 1971, Foyt *et al* 1969a, Eisen *et al* 1973). Although implants have given layers with a reasonably high electron mobility (Sansbury and Gibbons 1970), the maximum carrier concentrations have been at least an order of magnitude below those obtained with material doped during growth in all cases except one. In that case (Eisen *et al* 1973), high electrical activity was measured after a tellurium implanted layer had been annealed at 950 °C, but the results were not reproducible owing to the difficulty of maintaining an effective encapsulating layer at that temperature. Furthermore, nothing was reported concerning the quality of the substrate material following an anneal at temperatures where donor diffusion begins to become significant.

In this paper we report on investigations into implanted donor layers encapsulated and annealed at temperatures (700 °C) where the degradation of the bulk material as determined by photoluminescence is negligible. Implants of both group IV and group VI donors have been examined (Woodcock *et al* 1975) but we are presenting the results of only the group IV donors, silicon and tin, since these produced the highest electrical activities.

2. Experimental procedure

2.1. Implantation details

A Cockroft Walton accelerator (Goode 1971) was used to implant samples with 400 keV silicon and tin ions. They were implanted 8° off the [100] axis to minimize channelling and the substrate temperatures were varied between 300 and 600 K. Before any electrical activity could be detected the samples had to be annealed.

Before annealing the samples were coated with a silicon nitride encapsulating layer to prevent the GaAs surface from dissociating. This layer was found to maintain the photoluminescence efficiency of unimplanted n-type GaAs for anneal temperatures up to 750 °C. The samples were annealed in evacuated quartz ampoules.

2.2. Measurement details

The carrier concentration and Hall mobility of the implanted layers were obtained from Van der Pauw measurements (Van der Pauw 1958) on implants in epitaxial n on semi-insulating GaAs. Epitaxial layers were chosen which had thickness—carrier concentration products of less than $5 \times 10^{11} \, \mathrm{cm}^{-2}$ to enable accurate measurements of the implanted layers to be made without interference from the epitaxial layers. In order to find the depth distribution of implanted donors, differential Hall measurements (Petritz 1958) were carried out while progressively etching clover leaf samples. The etchant used was $H_2O_2 : H_2SO_4 : H_2O$ in the volume ratio 2 : 2 :100. Experiments showed that this removed GaAs at a rate of $500 \pm 50 \, \text{Å} \, \text{min}^{-1}$ uniformly throughout the implanted region and the rate was independent of the implanted dose in the range 10^{13} to 3×10^{15} ions/cm². The amount of material removed was measured with a Rank Taylor Hobson talystep machine and the etch rate calculated after each completed set of measurements and sometimes at intermediate stages to check the uniformity of the etch rate.

Cathodoluminescence spectra were obtained as a function of depth through the implanted layers and beyond by progressively etching layers with the 2:2:100 etchant. The spectra were excited with a low energy electron beam (2 keV) for minimum electron penetration and detected with a Bausch and Lomb monochromator and either a cooled photomultiplier with a S1 photocathode or a lead sulphide cell.

Electroluminescence spectra were obtained with a similar optical set up.

3. Results

3.1. Electrical measurements

Two temperature stages have been reported (Carter *et al* 1970, Harris and Eisen 1971) for the annealing of implantation damage in GaAs: lightly damaged GaAs anneals at temperatures between 100 and 300 °C whereas annealing temperatures between 600 and 700 °C are required for GaAs which is damaged so heavily that it is amorphous. With silicon implants into GaAs at room temperature it was found necessary to anneal at temperatures above 600 °C to obtain any electrical activity, in agreement with the high temperature annealing stage. The activity increases sharply with anneal temperatures between 600 and 700 °C but shows insignificant increase between 700 and 750 °C. However with room temperature implants of the heavier ion, tin, negligible electrical activity was obtained even after a 700 °C anneal. Heavy ions result in higher damage densities than light ions of the same energy and it is concluded that very heavily damaged GaAs does not anneal completely at 700 °C. On the other hand implants of tin into GaAs held at 200 °C result in high levels of electrical activity following a 700 °C anneal. This is related to the lower temperature annealing stage; damage annealing during the implantation maintains a density low enough for electrical activity to be subsequently obtained.

Unless otherwise stated all the samples described in this section were annealed at 700 °C for 20 min.

The results of differential Hall measurements on 400 keV silicon implants into epitaxial GaAs at room temperature and 300 °C are shown in figure 1. Both were implanted with a dose of 10^{15} ions/cm². The projected range calculated from LSS theory (Johnson and Gibbons 1970) and the theoretical tail of the atomic distribution (Brice 1971) are indicated on the figure. The peak concentration of the theoretical distribution is about 3.7×10^{19} atoms/cm³ indicating that the doping efficiency is low. Doping efficiencies around 1% are typical for implants with doses of 10^{15} ions/cm². Figure 1 shows that the free electron concentration at the peak of the distribution is higher for the higher temperature implant and the electron mobilities are high throughout the implanted region for both implants. The peak positions are both close to the predicted range and the tails of the distributions are similar and do not extend far beyond the predicted distribution; this means that diffusion is not an important process for silicon implants at this temperature and dose.

Results of the differential Hall measurements on 400 keV tin implants into epitaxial GaAs at 200 °C are shown in figure 2. Tin ions are heavier than silicon ions and lead to shallower implanted distributions; the depth scale in figure 2 has been expanded compared with figure 1. Profiles for two different doses are shown and it is evident that the implant depth is dose dependent; the higher dose implant is at twice the predicted depth and the tail of the active distribution extends beyond the theoretical distribution. This is not an unusual result (Sansbury and Gibbons 1970) and is thought to be caused by the diffusion of the implanted atoms in the wake of migrating defects generated during the implantation. The higher dose the larger is the number of defects generated. Attractive features of the high dose implant are the peak electron concentration slightly in excess of 10^{18} cm⁻³ and the abrupt transition from the peak to the background concentration with no sign of carrier removal or reduced electron

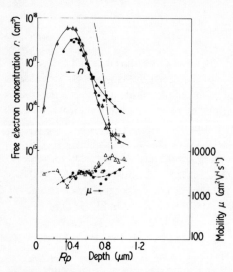

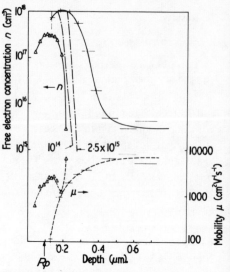

Figure 1. Profiles of carrier concentration and electron mobility for 10^{15} 400 keV silicon ions/cm² implanted at 20 and 300 °C. △ implant at 300 °C, ● implant at room temperature, chain curve theoretical tail of distribution.

Figure 2. Profiles of carrier concentration and electron mobility for doses of 10^{14} and $2 \cdot 5 \times 10^{15}$ 400 keV tin ions/cm² implanted at 200 °C. △ 10^{14} ions/cm², + $2 \cdot 5 \times 10^{15}$ ions/cm², chain curve theoretical tails of distributions.

mobility below the implant. However, the electron mobility at the surface has been reduced as a result of defects which neither migrate into the bulk nor anneal out at 700 °C. These defects increase the density of scattering centres which reduces the electron mobility.

3.2. Cathodoluminescence measurements

Cathodoluminescence measurements as a function of depth indicate the way in which the luminescence efficiency changes through the layer and give some information about the distribution of defects.

Figure 3 shows the variation with depth of the spectra at 77 K from an epitaxial n-type layer implanted with 10^{15} 400 keV silicon ions/cm². This is a similar implant to that of figure 1 with the peak of the distribution at about $0 \cdot 4 \,\mu$m and the tail extending to approximately $1 \cdot 2 \,\mu$m. Through most of the implanted region the spectra are dominated by broad low energy emission bands. The most intense is a band at $1 \cdot 18$ eV at a depth close to the peak of the implant. Detailed study of the intensity and half width variation of this band with temperature (Woodcock *et al* 1975) suggests that it is the result of recombination at a gallium vacancy—silicon complex. This defect is considered to be a deep acceptor (Williams and Bebb 1972) and so it will add to any compensation of the implanted donor layer.

Emission near the band gap energy does not appear until $0 \cdot 74 \,\mu$m of material has been removed; hence this emission comes from the tail of the implant well below the

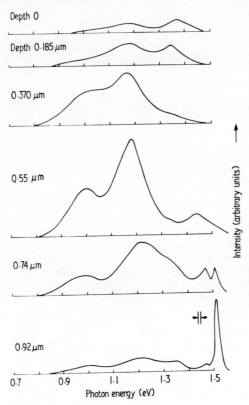

Figure 3. The variation with depth of the cathodoluminescence spectra at 77 K from an epitaxial n-type GaAs layer implanted with 10^{15} 400 keV silicon ions at 20 °C and annealed at 750 °C.

peak (figure 1). At this depth two bands can be resolved near the band gap energy: 1·48 eV compatible with a conduction band to a shallow acceptor recombination and 1·51 eV compatible with a shallow donor to valence band recombination. Since the 1·48 eV band is much more intense than that obtained from unimplanted control samples it is thought that the shallow acceptor may be the implanted silicon atoms on arsenic rather than gallium sites. If this interpretation is correct compensation due to implanted silicon on both sublattices is occurring. At a depth of 0·92 μm and deeper the near band gap emission has recovered to the same intensity as that from unimplanted control samples.

The cathodoluminescence spectra from implanted p-type GaAs exhibit the same broad low energy bands in the implanted region as the spectra from implanted n-type. However there is one important difference between the spectra from the two materials: emission close to the band gap energy from the implanted p-type GaAs does not recover to the same intensity as that from unimplanted control samples until several micrometres have been removed by etching. This is illustrated in figure 4 where the percentage recovery of the p-type band edge emission is plotted against depth for the two donor implants. Unannealed melt grown p-type GaAs ($p = 2 \times 10^{17}\,\mathrm{cm^{-3}}$) implanted with

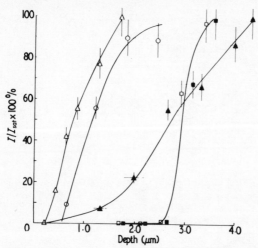

Figure 4. The variation with depth of the intensity of the band edge luminescence from implanted p-type GaAs. △ 200 °C Sn 700 °C anneal LPE material, ▲ 200 °C Sn 700 °C anneal melt-grown, ○ room temperature Si 700 °C anneal melt grown, ▫ 200 °C Si unannealed melt grown, ■ room temperature Si unannealed.

silicon recovers the luminescence efficiency at a depth between $2 \cdot 5$ and $3 \cdot 5 \mu m$; on annealing at 700 °C the luminescence recovers at a depth around $2 \mu m$. Tin implanted into melt grown p-type GaAs degrades the efficiency of the band edge emission to a depth of $4 \mu m$. However the depth of the luminescence degradation depends on the substrate material; tin implanted liquid epitaxial material of the same doping level as the melt grown material recovers at a depth of $1 \cdot 5 \mu m$.

These results indicate that defects migrate into the p-type material during the implantation rather than during the anneal and either the depth of migration or the type of defect finally formed depends on the implanted ion and the substrate material. There is no evidence for the migration of defects for implants in n-type GaAs; neither the carrier concentration nor the luminescence below implants are affected. However if the final defects formed after the migration in both p- and n-type GaAs were deep donors, these results might be expected.

3.3. Diodes

The fabrication of n^+–p junction diodes by the ion implantation of donors is shown schematically in figure 5. After annealing a non-localized implant at 700 °C, Ag/Sn contacts are evaporated and alloyed to the implanted donor layer. The diodes are electrically isolated by bombardment with protons of an appropriate energy. The lattice damage produced by proton implantation compensates GaAs resulting in a semi-insulating layer stable up to about 300 °C (Foyt *et al* 1969b, Pruniaux *et al* 1971). The diode areas are masked with photoresist and silicon dioxide during the proton implantation. The latter acts as a buffer layer to enable the masks to be easily removed when the photoresist is polymerized by the high energy proton beam.

A non-detailed current voltage characteristic from a tin implanted diode is shown in figure 6. The implant was 10^{15} 400 keV tin ions/cm^2 into Cd-doped GaAs ($p = 2 \times 10^{17}$

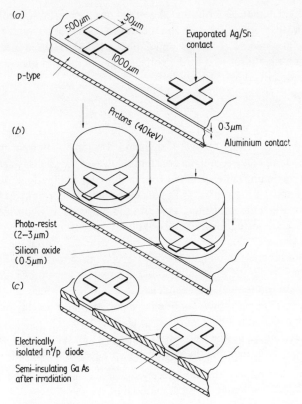

Figure 5. Diode fabrication (*a*) Metallization (*b*) Proton isolation (*c*) Diode array before dicing.

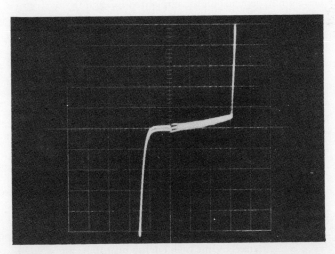

Figure 6. Current–voltage characteristic from a tin implanted n–p diode. Vertical scale $10\,\mu A$/division. Horizontal scale: first and second quadrants 0.5 V/division; third and fourth quadrants 5 V/division.

cm^{-3}) at 200 °C. The characteristic exhibits a well defined reverse breakdown voltage around 15 V which is in close agreement with the theoretical value for an abrupt junction in GaAs of the doping level used (Sze and Gibbons 1966). Detailed measurements showed that over a significant part of the voltage range in forward bias these characteristics could be described by the equation

$$I = I_0 \exp \frac{eV}{nkT}$$

where I is the current, e the electronic charge, V the applied voltage, k Boltzmann's constant, T the absolute temperature, and n a constant. However n was always greater than two, which is the maximum value expected from a simple p–n junction. Measured values of n between 2·5 and 3·5 are typical. The possibility that the current was limited by an insulating layer between the implanted layer and the substrate was considered since such a layer had been reported for the implantation of acceptors into n-type GaAs (Hunsperger and Marsh 1970). In order to investigate this, the capacitance of reverse biased diodes as a function of bias and frequency was measured. Figure 7 shows the results plotted as $1/C^2$ against V with the theoretical curve for an abrupt junction in $p = 2 \times 10^{17}$ cm^{-2} GaAs. The frequency dependence is due to traps which contribute to the capacitance at low frequencies but cannot charge and discharge quickly enough at high frequencies (Sah and Reddi 1964), in this case above 1 MHz. The capacitance at high frequencies is determined by the free carrier density. The experimental capacitance at high frequency is lower than the theoretical capacitance which suggests that

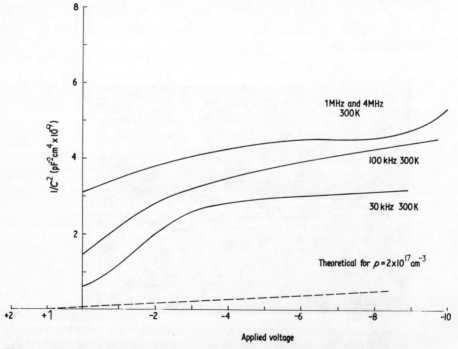

Figure 7. The variation with applied voltage of $1/C^2$ from a tin implanted n–p diode.

there is in fact an insulating layer below the implant. The capacitance at zero bias is effectively due to the i-layer alone but as the reverse bias increases the depletion layer spreads into the p-type substrate and contributes to the measured capacitance. An i-layer $0.6\,\mu$m thick with $p = 2 \times 10^{15}\,\mathrm{cm}^{-3}$ would produce both capacitance and voltage breakdown results similar to our observations.

The migration of defects to form a compensated layer below implants is not surprising in view of the luminescence results from implanted p-type GaAs presented above. Similarly as might be expected from those results the electroluminescence efficiency of the diodes depends on the p-type substrate material. Tin implants into liquid epitaxial material result in diodes with efficiencies of 10^{-5} at 300 K and 5×10^{-4} at 77 K. These efficiencies were measured with a calibrated silicon solar cell. The spectrum from one of these diodes at 77 K is shown in figure 8(a). The near band gap emission is dominant but at a lower energy than expected, 1·46 eV rather than 1·48 eV. This is thought to be due to absorption (Carr and Biard 1964) since the emission is screened by the contact and could undergo several reflections before escaping. Diodes made from implanted melt grown material are less efficient, typically 10^{-6} at 300 K and 10^{-4} at 77 K. In this

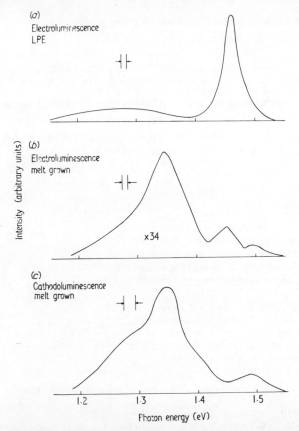

Figure 8. (a) and (b) Electroluminescence spectra at 77 K from forward biased tin implanted diodes ($I = 30$ mA). (c) Cathodoluminescence spectra at 77 K $0.7\,\mu$m below the surface of a tin implanted diode.

case the spectra at 77 K are dominated by a low energy band at 1·35 eV (figure 8(b)). The origin of this band is not known but recombination at a deep acceptor such as copper, a notorious impurity in GaAs, seems a likely possibility. The near band gap emission from the diode of figure 8(b) consists of bands at 1·45 eV and 1·48 eV; it is thought that the 1·48 eV band escapes directly and the other after reflection and absorption. A cathodoluminescence spectrum after 0·7 μm of the diode had been etched from the surface is shown for comparison in figure 8(c). A similar spectrum is observed but with only one recombination line near the band gap energy.

4. Conclusions

N-type layers have been produced with doping levels up to 10^{18} cm^{-3} by the implantation of group IV donors into GaAs. There is evidence that damage induced defects remain after a 700 °C anneal, namely reduced electron mobility at the surface of heavy ion implants and degraded luminescence below implants in p-type GaAs. However there is no evidence of either carrier removal or reduced electron mobilities below implants in epitaxial n-type GaAs.

Diodes made from these donor implants into p-type GaAs appear to be n–i–p devices with i-layers somewhat less than a micrometre thick.

Acknowledgments

The implants for this work were carried out at AERE, Harwell, England, under the supervision of Mr P D Goode. The high purity epitaxial layers were grown by Messrs B Dobson, B C Easton and M J King. The authors would also like to thank Mr J M Shannon for invaluable discussions.

References

Brice D K 1971 *Sandia Labs. Res. Rept.* No 5C-RR 71 0599
Carr W N and Biard J R 1964 *J. Appl. Phys.* **35** 2776
Carter G, Grant W A, Haskell J D and Stevens G A 1970 *Rad. Effects* **6** 277
Eisen F H, Harris J S, Welch B, Pashley R D, Sigurd D and Mayer J W 1973 *Proc. Conf. Ion Implantation in Semiconductors and Other Materials* (New York: Plenum) p631
Fane R W and Goss A J 1963 *Solid St. Electron.* **6** 383
Foyt A G, Donelly J P and Lindley W T 1969a *Appl. Phys. Lett.* **14** 372
Foyt A G, Lindley W T, Wolfe C H and Donelly J P 1969b *Solid St. Electron.* **12** 209
Goldstein B 1961 *Phys. Rev.* **121** 1305
Goode P D 1971 *Nucl. Instrum. Methods* **92** 447
Harris J S and Eisen F H 1971 *Rad. Effects* **7** 123
Hunsperger R G and Marsh O J 1970 *Metall. Trans.* **1** 603
Johnson W S and Gibbons J F 1970 *Projected Range Statistics in Semiconductors* (Stanford University Bookstore)
Petritz R L 1958 *Phys. Rev.* **110** 1254
Pruniaux B R, North J C and Miller G L 1971 *2nd Int. Conf. Ion Implantation in Semiconductors* p212
Sah C T and Reddi V G K 1964 *IEEE Trans. Electron. Dev.* **ED-11** 345
Sansbury J D and Gibbons J F 1970 *Rad. Effects* **6** 269

Sze S M and Gibbons G 1966 *Appl. Phys. Lett.* **8** 111
Van der Pauw L J 1958 *Philips Res. Rep.* **13** 1
Whitton J L and Bellavance G R 1971 *Rad. Effects* **9** 127
Williams E W and Bebb H B 1972 *Semicond. Semimet.* **8** 181
Woodcock J M, Shannon J M and Clark D J 1975 *Solid St. Electron.*

Transport properties of O⁺ implanted in GaAs

M I Abdalla†, J F Palmier and C Desfeux

Centre National d'Etudes des Télécommunications CPM/PMT, 22301 Lannion, France

Abstract. Ion beam implantation of oxygen in n-type GaAs resulted in a high resistivity layer (P N Favennec, G T Pelous and M Boudet 1972 *Paper presented at Int. Conf. Ion Implantation in Semiconductors and Other Materials, Yorktown Heights*). The implantation was carried out at selected ion beam energies and different doses. Two types of devices were fabricated. The first represent those with two ohmic contacts deposited on to the two opposite faces. The second represent those with a Schottky barrier with the metal contact deposited on to the implanted face. All devices were subjected to isochronal anneal for 30 min at 470 °C before metal deposition.

DC conductivity against temperature measurements for devices with ohmic contacts enabled us to determine the position of the Fermi level as well as the mobility against ion dose. Measurements of the AC conductivity established that conduction at low temperatures was due to hopping.

The forward and reverse current–voltage characteristics of these devices were not symmetric. The forward characteristics were due to space charge limited conduction. The reverse characteristics were due to Poole–Frenkel mechanisms modified by Ohm's law at low fields.

The capacitance–voltage characteristics of the Schottky barrier devices displayed clearly that the implanted devices have an MIS structure. From the measurements of the parallel conductance of these devices we were able to determine the density of surface states. Depending on the temperature at which measurements were carried out, the current–voltage characteristics were due to thermionic emission, thermal assisted tunnelling and field assisted tunnelling. The thermionic emission mechanisms enabled us to determine the barrier height as a function of the implanted dose.

1. Introduction

Recent years have witnessed an intensive investigation of ion beam implantation in semiconductors as a unique method of introducing a specific impurity species of well controlled concentrations and device dimensions. This technique when used to implant oxygen in GaAs resulted in the formation of a high resistivity layer. The formation of such a layer could be regarded as due to:

(i) Radiation damage defects. This is considered to be a major drawback of this technique as a result defects (point, line, planar, and perhaps three dimensional) are inherently produced. Such defects will act as sites at which majority carriers are trapped leading to the formation of a depleted space charge region (Gossick 1959). This damage can be partially removed by annealing.

(ii) Chemical effect due to high electronegativity of oxygen. This term refers to the

† Present address: CNET, 196 Rue de Paris, 92220 Bagneux, France.

attractive power or in other words the affinity of an atom for electrons. The electro-negativity of oxygen = 3·5 compared to 1·5 for Ga and 2 for As (Gordy and Thomas 1956).

(iii) Electronic effect due to incorporation of many impurity levels in the forbidden energy gap. It is known that oxygen in GaAs gives rise to many impurity levels in the forbidden gap, some of these levels have been described as deep donors and some as deep acceptors.

After describing the experimental procedures, the experimental results are divided into two parts. Section 3 discusses the results obtained on ohmically contacted devices. This includes their DC conductivity against temperature, AC conductivity against frequency at different temperature and their current–voltage characteristics. Section 4 concerns the results obtained on Schottky barrier devices. This includes their current–voltage, capacitance–frequency, capacitance–voltage, conductance–frequency. All the results were investigated as a function of implanted dose.

2. Experimental procedures

2.1. Sample preparation

Wafers of n-type unintentionally doped GaAs of (111) orientation were used. The carrier concentration, electron mobility and resistivity were stated by the supplier as $2·7 \times 10^{16} \, cm^{-3}$, $3300 \, cm^2 \, V^{-1} \, s^{-1}$ and $0·07 \, \Omega \, cm$ respectively. The as received wafers were mirror polished on one side, $(\bar{1}\bar{1}\bar{1})$ face. The wafers after being chemically polished in bromine–methanol solution were bombarded with ^{16}O ion beam incident $7°$ off the $[\bar{1}\bar{1}\bar{1}]$ axis. Successive implantation energies with a maximum of 1·3 MeV were used in order to produce an almost flat implanted region extending from a position close to the bombarded face to a position determined by the maximum beam energy. The implantation was carried out at room temperature and the integrated dose level ϕ used in this study was between 10^{10} and $10^{15} \, ions/cm^2$. After implantation, all devices were subjected to isochronal anneal at $470 \, °C$ for 30 min in an argon atmosphere.

2.2. Metal contacts

All metal contacts were evaporated in an oil pumped vacuum system provided with a liquid nitrogen trap and operating in the low $10^{-6} \, Torr$ (background pressure), raised by an order of magnitude during evaporation. The ohmic contact used consists of alloying at $400 \, °C$ a $1000 \, Å$ Au–Ge eutectic film with a thin chrome metal layer about $100 \, Å$ thick deposited before (Au–Ge).

2.3. Schottky barrier contacts

Schottky barrier diodes were prepared by depositing an Al contact on to the implanted face and an ohmic contact on to the back face. In order to obtain reproducible results on Schottky barriers, special care was adopted in depositing the Al contact. This consists of heating the GaAs substrate for 1 min at $200 \, °C$ and then the temperature was allowed to drop and maintained at $100 \, °C$ during metal deposition. Aluminium contacts prepared by this method were found to adhere well to the GaAs surface in contrast to the bad adherence and irreproducible results when deposition was carried out on a cold substrate.

Figure 1 is a proposed equilibrium energy band diagram of both ohmic and Schottky barrier implanted devices. Shown in the same figure is a scanning electron photomicrograph taken in the cathodoluminescent mode with the implanted region appearing as a dark band.

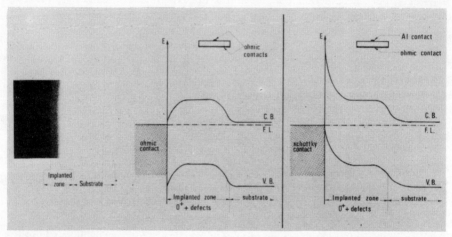

Figure 1. Equilibrium energy band diagram for: (*a*) ohmic devices, (*b*) Schottky barrier devices. (*c*) is a scanning electron micrograph taken in cathodoluminescence mode.

3. Devices with ohmic contacts

3.1. Temperature dependence conductivity.

DC conductivity measurements were carried out by applying a DC voltage across the device and recording the current. This voltage was kept at a minimum, usually about 1 mV, so that it falls in the linear portion of the $I(V)$ characteristics. The logarithm of the measured DC conductivity as a function of the reciprocal of the temperature is shown in figure 2 for different doses.

Generally the DC conductivity σ_O can be written as:

$$\sigma_O = e(n\mu_n + p\mu_n) + \sigma_{\text{hopping}} \tag{1}$$

where σ_{hopping} can be neglected towards high temperatures. Inspection of figure 2 reveals the following remarks:

(i) If $N_O < N_D$, where N_O is the density of implanted oxygen and N_D is the shallow donors density, the conductivity is almost constant with temperature.
(ii) If $N_O \geqslant N_D$, the observed conductivity becomes thermally activated and a compensated high resistivity layer results.

We shall see later from the $I-V$ characteristics that even if N_O is substantially higher than N_D we never obtain a p–n junction. Therefore it is not unreasonable to neglect

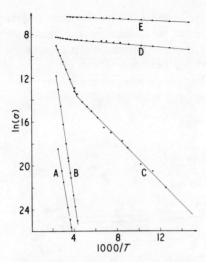

Figure 2. DC conductivity plotted against the inverse of temperature for different implanted devices. ϕ-values (in ions/cm^2) are as follows: A $6\cdot9 \times 10^{12}$, B $2\cdot7 \times 10^{12}$, C $6\cdot9 \times 10^{11}$, D $2\cdot7 \times 10^{11}$, E $1\cdot4 \times 10^{11}$.

hole's conduction. Assuming that conduction is due to only one type ie electrons, we can deduce the position of the Fermi level below the conduction band from the high temperature slope of figure 2 and this is shown in table 1. Values of the mobilities obtained from equation (1) are also listed in table 1. The striking feature is the tremendous reduction in mobility. In the literature impurities which cause large reduction in mobility are termed mobility killers (Weisberg 1962) and are attributed to the presence of deep lying traps in the forbidden energy gap with large capture cross section. It must be stated here that the mobility can not be measured by the conventional direct methods such as Hall effect due to the very low substrate resistivity compared to that of the implanted layer.

At low temperatures a second activation of lower value is observed. Some samples do not even exhibit any activation energy at the very low temperature end (not shown in figure 2), usually extending from 50 to 4·2 K, which might be due to hopping conduc-

Table 1. Fermi level measured from DC conductivity and mobility measured from DC conductivity and from I–V characteristics as a function of implanted dose

ϕ (ions/cm^2)	Slope (eV)	μ (cm^2 V^{-1} s^{-1}) from σ_O	μ (cm^2 V^{-1} s^{-1}) from I–V
$6\cdot9 \times 10^{10}$			5·330
$1\cdot38 \times 10^{11}$	0·004	0·400	1·220
$2\cdot7 \times 10^{11}$	0·018	0·400	0·664
$6\cdot9 \times 10^{11}$	0·202	0·280	0·481
$2\cdot7 \times 10^{12}$	0·412	0·477	
$6\cdot9 \times 10^{12}$	0·491	0·109	0·049
$1\cdot38 \times 10^{13}$			0·197
$6\cdot9 \times 10^{15}$			$5\cdot4 \times 10^{-3}$

tivity. This has been confirmed by making AC conductivity measurements. Figure 3 displays the AC conductivity (inversely proportional to resistance) against temperature measured at different frequencies for a typical device implanted at a dose = $6\cdot9 \times 10^{11}$ ions/cm^2. The figure shows that towards the high temperature end, the experimental AC points appear to approach the straight line DC curve while at low temperatures the conductivity is proportional to $\omega^{0\cdot9}$.

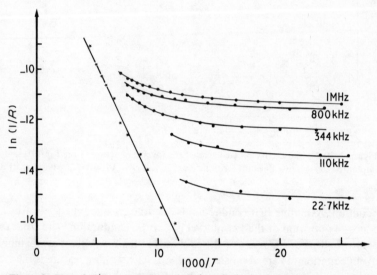

Figure 3. Plot of ($I/R \propto$ conductivity) against the inverse of temperature at different frequencies for a typical device implanted at $6\cdot9 \times 10^{11}$ ions/cm^2.

3.2. Current–voltage characteristics

3.2.1. Positive bias applied to the implanted face. The dependence of current on voltage is shown in figure 4(*a*) plotted on a log–log scale for devices subjected to different implanted doses. The characteristics have an ohmic region at low voltages followed by a V^n dependance law where n is close to 2. These characteristics correspond to the well known space charge limited (SCL) current flow developed by Mott and Gurney (1940) and modified by Lampert (1956) to include the effect of traps on the observed current. The current density in the ohmic region is given by:

$$J = ne\mu V/L \tag{2}$$

and in the SCL region is given by:

$$J = \tfrac{9}{8}\epsilon\epsilon_0\mu\theta\, V^2/L^3 \tag{3}$$

where θ is the coefficient representing trapping of the injected carriers and the other symbols have their usual meaning.

The transition from the ohmic region to the SCL region occurs at a voltage V_{tr} given by:

$$V_{tr} = \frac{8}{9}\frac{enL^2}{\epsilon\epsilon_0\theta} = \frac{3kT}{e}\left[\frac{L}{L_D}\right]^{3/2} \tag{4}$$

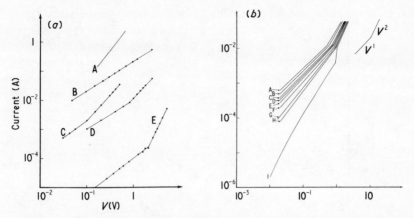

Figure 4. A log–log plot of I against V in the forward direction: (*a*) for devices subjected to different doses. A V^2, B V^1 no implantation, C $\phi = 6\cdot9 \times 10^{10}$ ions/cm², D $\phi = 1\cdot38 \times 10^{13}$ ions/cm², E $\phi = 6\cdot9 \times 10^{15}$ ions/cm². (*b*) for a typical device measured at different temperatures, $\phi = 2\cdot7 \times 10^{11}$ ions/cm². Temperatures (K): A 290·5, B 250, C 190, D 159·8, E 128·9, F 90, G 60, H 31, I 4·6.

where L_D is the Debye length given by:

$$L_D = \left(\frac{4\pi\epsilon\epsilon_0 kT}{e^2 n}\right)^{1/2}.$$

Therefore from equations (2), (3) and (4) we can calculate the mobility μ and this is shown in table 1. The values of μ obtained by this method are comparable to those obtained from conductivity measurements.

These values of μ are surprising and we thought that the anneal temperature was not sufficient to eliminate the defects. However, annealing at 570 °C for 1 h resulted in a minor improvement. Unfortunately we could not investigate the effect of higher annealing temperature on the mobility. This is due to the dissociation of GaAs at higher anneal temperatures.

An overall forward current–voltage characteristic for a typical device plotted at different temperatures from 4·6 to 290 K is shown in figure 4(*b*).

The liquid helium temperature characteristics are difficult to explain and they could be due to junction effects at low voltages (a straight line is obtained when $\ln I$ is plotted against V) and SCL effects at higher voltages.

3.2.2. Negative bias applied to the implanted face. The current–voltage characteristics in the reverse direction were found to obey Ohm's law at low fields followed by a characteristic which obeys a $(\ln I - V^{1/2})$ law at higher fields. This is shown in figure 5(*a*) for devices subjected to different doses. In practice such dependence is normally interpreted as due to the Schottky effect or the Poole–Frenkel effect. Both effects are characterized by the reduction of the potential barrier due to the applied electric field. The characteristics presented in figure 5(*a*) are due to Poole–Frenkel effects as the following argument might prove:

(i) In order to decide whether the characteristics are due to Schottky or Poole–Frenkel

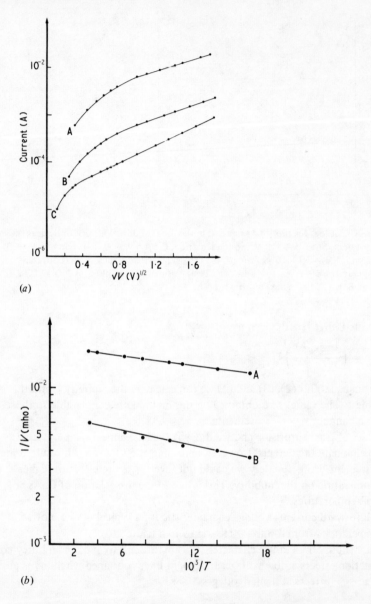

Figure 5. (*a*) Reverse characteristics plotted as $\ln I - V^{1/2}$. ϕ-values (ions/cm^2): A $6\cdot9 \times 10^{14}$, B $6\cdot9 \times 10^{13}$, C $6\cdot9 \times 10^{11}$. (*b*) $\ln I/V - 1/T$ plot establishing that the characteristics are due to the Poole–Frenkel effect. ϕ-values (ions/cm^2): A $2\cdot8 \times 10^{11}$, B $1\cdot4 \times 10^{11}$.

effects, it is necessary to carry out the measurements at different temperatures. A plot of $\ln I/V$ against $1/T$ will be linear for the Poole–Frenkel effect while $\ln I/T^2$ against $1/T$ is linear for the Schottky effect. Figure 5(*b*) is a plot of $\ln I/V$ against $1/T$ for two typical implanted devices which confirm that the reverse characteristics are due to Poole–Frenkel effects.

(ii) The characteristics are dominated by bulk ohmic conductivity at low fields, contrary to the Schottky effect which is due to electrode limited conduction at such fields.

(iii) For a given impurity concentration N_t, Poole–Frenkel effects will limit the current above a field strength given by Voolmann (1974)

$$ E = \frac{N_t^{2/3} e}{\pi \epsilon \epsilon_0}. \tag{5} $$

Generally this relation was obeyed in all of our implanted devices, (eg, if $N_t = 10^{15}$ cm^{-3} the Poole–Frenkel effect would be observed at voltages greater than 100 mV.

Returning to figure 5(*a*), the following remarks can be made:

(i) The slopes of the curves were almost the same for devices implanted at different doses. This is because the implanted devices have almost the same thickness since they were implanted at the same energy. However the measured thickness (obtained from the slope of the curves) was very small compared to the thickness either measured from capacitance data or by scanning electron microscopy.

(ii) The intercept on the current axis increases with increasing dose. The magnitude of the intercept is proportional to the density of Poole–Frenkel centres.

4. Results obtained on Schottky barrier devices

4.1. Current–voltage characteristics

The $I-V$ characteristics of Schottky barrier diodes are complicated and lengthy and due to the available short space, only a short summary will be given. Detailed analysis of these characteristics will be the subject of a further publication.

4.1.1. Forward characteristics. The observed $I-V$ characteristics can be classified according to the temperature ranges at which measurements were carried out.

At temperatures greater than or equal to 400 K, the characteristics were found to follow the familiar diode equation

$$ I = I_0 \exp\left(eV/nkT\right) \tag{6} $$

where n, the ideality factor is approximately unity. Therefore, the high temperature characteristics are governed by thermionic emission where electrons gained enough energy to surmount the existing barrier height ϕ_B given by:

$$ \phi_B = \frac{kT}{e} \ln \frac{AT^2}{J_s} \tag{7} $$

where A is the Richardson constant and J_s is the saturation current density.

Barrier heights measured as a function of dose are shown in table 2. The interesting point is that the barrier passes through a maximum when the density of the implanted oxygen atoms is approximately equal to the density of shallow donors. After that, the barrier decreases very rapidly.

Measurements carried out in the temperature range from 60 to 390 K were found to be governed by thermal assisted tunnelling (Padovani and Stratton 1966), at high

Table 2. Barrier height, density of surface states and percentage of cathodoluminescence as a function of implanted dose

ϕ (ions/cm^2)	ϕ_B (eV)	N_{ss} at low frequency (states/cm^2/eV)	Cathodoluminescence percentage
6.9×10^{10}	1.05 ± 0.18	8.4×10^8	17
2.7×10^{11}	1.30 ± 0.15	1.21×10^9	15
6.9×10^{11}	1.30 ± 0.22	2.52×10^9	12
2.7×10^{12}	1.38 ± 0.16	3.22×10^{10}	10
6.9×10^{12}	0.95 ± 0.21	6.43×10^{11}	8
2.4×10^{13}	0.73 ± 0.16	3.68×10^{13}	7
6.9×10^{13}	0.54 ± 0.08	9.23×10^{13}	
6.9×10^{14}	0.41 ± 0.08	8.81×10^{15}	
6.9×10^{15}	0.20 ± 0.09	9.32×10^{16}	

voltages and by recombination/generation of charge carriers in the space charge region at low voltages ($n \simeq 2$). Consequently this might explain the observed low cathodoluminescence. The percentage of the cathodoluminescence in the implanted layer compared to that of the substrate is shown in table 2.

At very low temperature, below 60 K, the $I-V$ characteristics are dominated by field assisted tunnelling.

For the sake of space, the above mentioned mechanisms are illustrated in figure 7 for the reverse direction (notice that if these mechanisms are seen in forward bias, they must as well be seen in reverse bias).

4.1.2. Reverse characteristics. Figure 6 is the room temperature reverse current–voltage characteristics for diodes subjected to different doses. This figure shows clearly that, the higher the dose level, the more pronounced the reduction of the breakdown voltage. In

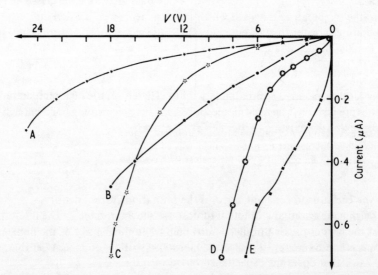

Figure 6. Reverse $I-V$ characteristics for devices subjected to different doses. ϕ-values (ions/cm^2): A 6.9×10^{10}, B 6.9×10^{13}, C 6.9×10^{11}, D 6.9×10^{14}, E 6.9×10^{15}.

all of the examined devices, the reverse characteristics showed a soft breakdown. In the literature soft breakdown is usually attributed to electrons tunnelling from metal to semiconductor. However, the appearance of soft breakdown cannot be taken only as due to tunnelling and the temperature dependence of the breakdown voltage must be examined. Our results showed that the breakdown voltage increases upon decreasing temperature (ie, consistent with tunnelling theory (Mahadevan *et al* 1971)). As in the forward case, the characteristics are classified according to the temperatures at which measurements were carried out.

The high temperature characteristics ($T \geqslant 400$ K) were due to thermionic emission since a straight line was obtained when $\ln I$ was plotted against $V^{1/4}$. The intermediate temperature characteristics ($T = 60$–390 K) were due to thermal assisted tunnelling since a straight line was obtained when $\ln I$ was plotted against V.

The low temperature characteristics ($T < 60$ K) were due to field assisted tunnelling since a straight line was obtained when $\ln I/V$ was plotted against $V^{-1/2}$. These mechanisms are illustrated in figure 7. In all of the examined devices, tunnelling was accomplished through trapping states, that is, electrons will take the route metal to trapping states to semiconductor or vice versa (Parker and Mead 1969).

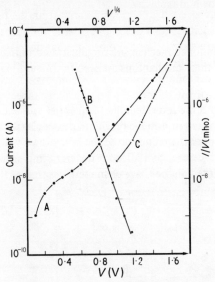

Figure 7. Reverse I–V characteristics at different temperatures. A thermal assisted tunnelling, $\ln I$ against V, 300 K. B field assisted tunnelling, $\ln I/V$ against $V^{-1/2}$, 20 K. C thermionic emission at 413 K $\ln I$ against $V^{1/4}$, $\phi = 1.4 \times 10^{10}$ (ions/cm²).

4.2. Capacitance measurements

Capacitance was measured as a function of frequency and voltage; figure 8 shows the C–f dependence for different doses. This figure shows that for low dose levels (ie, for $\phi = 6.9 \times 10^{10}$ and 6.9×10^{11} ions/cm²) the capacitance was independent of frequency. At higher doses the capacitance varied with the frequency and such dependence was found to be greatly increased with increasing the dose. Moreover capacitance was found to saturate at higher frequencies, usually of the order of 1 MHz.

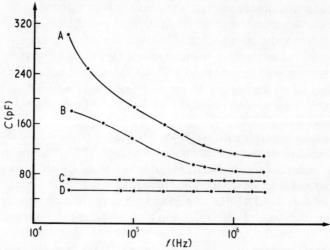

Figure 8. Capacitance as a function of frequency for different implanted devices. ϕ-values (ions/cm^2): A $6\cdot9 \times 10^{15}$, B $6\cdot9 \times 10^{13}$, C $6\cdot9 \times 10^{10}$, D $6\cdot9 \times 10^{11}$.

At present the frequency dependent capacitance is known to arise from the time constant associated with charging and discharging of deep impurities in the space charge region. Sah and Reddi (1964) have shown that the time constant decreases exponentially with depth in the semiconductor (ie, deep levels have shorter time constant than shallow ones). Therefore capacitance decreases with increasing frequency. At frequencies higher than the response of any of the deep levels, capacitance is constant. Moreover, inspection of figure 8 shows that devices implanted at $6\cdot9 \times 10^{11}$ ions/cm^2 are the more insulating (have lower capacitance). This can be understood since such a device is closer to total compensation.

Figure 9 is the capacitance–voltage measured at 1 MHz for different implanted devices. It is clearly seen that the implanted devices have characteristics similar to that

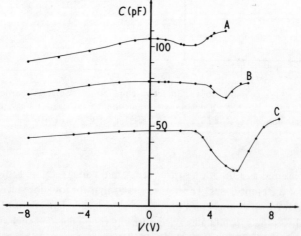

Figure 9. $C-V$ characteristics measured at 1 MHz. The characteristics are similar to those of MIS. ϕ-values (ions/cm^2): A $6\cdot9 \times 10^{15}$, B $6\cdot9 \times 10^{13}$, C $2\cdot7 \times 10^{12}$.

of an MIS device. Increasing the implanted dose resulted in a shift of the minimum capacitance towards the left of the voltage axis.

In order to determine the density of interface states we found it much simpler to measure the parallel conductance of the devices against frequency at zero bias. The parallel conductance G_p is given by Sze (1969):

$$\frac{G_p}{\omega^2} = C_s \tau - G_p \tau^2. \tag{8}$$

This relation has been derived on the assumption that the device capacitance consists of its geometrical capacitance in series with the parallel combination of the space charge capacitance and conductance. Figure 10 is a plot of G_p/ω^2 against G_p for two devices implanted at $6 \cdot 9 \times 10^{11}$ and $2 \cdot 4 \times 10^{13}$ ions/cm². The first gave a straight line with a single time constant while the second has two time constants. From the intercept on

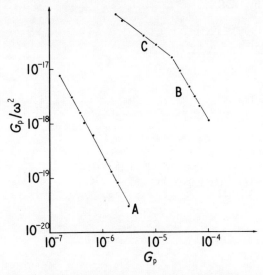

Figure 10. Parallel conductance measured at different frequencies. A $\phi = 6 \cdot 9 \times 10^{11}$ ions/cm², $\tau = 0 \cdot 428 \, \mu s$, $C_s = 3 \cdot 17 \, pF$, $N_{ss} = 2 \cdot 52 \times 10^9$ states/cm²/eV. B $\phi = 2 \cdot 4 \times 10^{13}$ ions/cm², $\tau_2 = 0 \cdot 28 \, \mu s$, $C_s = 0 \cdot 72 \, pF$, $N_{ss} = 5 \cdot 34 \times 10^9$ states/cm²/eV. C $\phi = 2 \cdot 4 \times 10^{13}$ ions/cm², $\tau_1 = 2 \cdot 57 \, \mu s$, $C_s = 46 \cdot 3 \, nF$, $N_{ss} = 3 \cdot 68 \times 10^{13}$ states/cm²/eV.

the G_p axis we can determine the surface states capacitance C_s and hence the surface states density $N_{ss} = C_s/eA$ can be calculated. These measurements were carried out for the other devices and the results are shown in table 2.

5. Conclusion

In the first part we have shown that the implanted devices have a compensated insulating region established from the DC conductivity and the SCL mechanisms. The very low mobility (due to the high density of scattering centres) might explain the observed Poole–Frenkel effect seen in the reverse current–voltage characteristics.

In the second part we have shown that the Schottky barrier devices have an MIS structure. The observed low density of surface states suggests that they are lying at the interface between the implanted region and the substrate rather than at the surface. Depending on the measured temperature range, the current–voltage characteristics were found to be due to thermionic emission, thermal assisted tunnelling and field assisted tunnelling.

With regard to technological applications, the implanted region might be of help for device insulation.

Acknowledgments

The authors would like to thank Dr Combet, Head of the department of PMT for providing research facilities and for his encouragement during the course of this work. Thanks are also due to Mr Pelous and Mr Favennec for carrying out the implantation. This work is partly supported by DGRST.

References

Gordy W and Thomas D E 1956 *J. Chem. Phys.* **24** 439
Gossick B R 1959 *J. Appl. Phys.* **30** 1214
Lampert M A 1956 *Phys. Rev.* **103** 1468
Mahadevan S, Mardas S M and Suryan G 1971 *Phys. Stat. Solidi* (a) **8** 355
Mott N F and Gurney R W 1940 *Electronic Processes in Ionic Crystals* (New York: OUP)
Padovani F A and Stratton R 1966 *Solid St. Electron.* **9** 695
Parker G H and Mead C A 1969 *Phys. Rev.* **184** 780
Sah C T and Reddi G K 1964 *IEEE Trans. Electron Dev.* **ED-11** 345
Sze S M 1969 *Physics of Semiconductor Devices* (New York: Wiley) p453
Voollmann W 1974 *Phys. Stat. Solidi* (a) **22** 195
Weisberg L R 1962 *J. Appl. Phys.* **33** 1817

Automated Czochralski growth of III—V compounds

W Bardsley, G W Green, C H Holliday, D T J Hurle, G C Joyce,
W R MacEwan and P J Tufton

Royal Radar Establishment, Malvern, Worcestershire, England

Abstract. The weighing method of automatic diameter control has been extended to grow bulk crystals of gallium arsenide, indium phosphide and gallium phosphide. An outline of the control theory is given, and the apparatus and servo loop described.

1. Introduction

Large single crystals are usually grown from the melt, and the favoured method is that of pulling due to Czochralski. Dissociable compounds, such as gallium arsenide, indium phosphide and gallium phosphide, have been grown this way on a commercial scale since the development of the liquid encapsulation technique. However, although the basic technological problems have been solved, and bulk crystals of these materials are now widely available, the grower is still unable to pull to order a crystal of accurately specified size and shape.

Automatic Czochralski growth holds the promise of rectifying this situation. It implies closed-loop servo control of crystal diameter from seed-on to completion of pull, from the grow-out through the cylindrical portion to the tapered tail end. Conceptually the easiest way is to measure the diameter directly and use a servo loop to correct deviations from the desired profile as they occur. This technique has been tried, and indeed it has had limited success. It is possible to measure diameter by optical means (Gross and Kersten 1972) or by x-rays (Van Dijk 1974), but these systems are neither easy to design nor easy to operate where the apparatus is pressurized and the melts encapsulated. Weighing a crystal, on the other hand, is a simple and accurate operation even during pulling. From the rate of change in weight, the growth speed and a knowledge of the density the diameter can in principle be calculated. This method has also been tried and considerable success obtained with certain materials such as oxides (Zinnes *et al* 1973, Bardsley *et al* 1974), but as far as we are aware no-one previously has been able to extend this technique to grow semiconductors; the principal reason is that, as we shall see later, there is a class of materials for which it proves impossible to derive the actual diameter of the growing crystal by monitoring its weight signal, and the major semiconductor materials belong to this class. Nevertheless the force experienced by the load cell is still some measure of the diameter, so this does not preclude us from exploiting the signal for controlling the shape, even if the exact diameter is indeterminable.

2. Theoretical considerations

To understand the difficulty peculiar to the semiconductor-like materials it is necessary to consider the mechanics of the weighing process. A crystal being withdrawn from the melt will appear to weigh more than its actual weight because of the pull of the meniscus. Provided that the crystal stays the required diameter, so that the shape and consequent pull of the meniscus remains constant, then the weight increase will be as expected and the weight error zero. If, however, the crystal diameter varies for any cause, then the contribution of the changed pattern of forces due to the altered shape of the meniscus must be taken into account. We have shown previously (Bardsley *et al* 1974) that, if a is the radial deviation, then the normalized weight error is given by the expression:

$$H = \int_0^t a \, dt + \eta a - \lambda \dot{a}$$

where the two parameters η and λ are rather complicated functions of the crystal radius containing, in particular, terms involving the surface tension and the density difference between the solid and liquid states of the pulled material.

In previously described weighing methods of control it is this weight error signal H which is used as the basic measure of the amount of correction needed. Thus H must increase monotonically for increasing a for satisfactory control. Whether it does so depends on the sign of the two parameters. It turns out that η is always positive, but λ is negative (making the third term in the expression positive when a is increasing) only for materials that are denser in the solid state than in their molten form and also the melt wets the growing crystal with a low enough angle of contact. This appears to be always true for oxides and some other materials, but not true for semiconductors. The group IV and III–V compounds have the very open diamond or zinc blende crystal structures and consequently the solid floats on the melt; moreover, the angle of contact is about 15° (Antonov 1965). We thus have two classes of materials: those with negative λ for which H is a well-behaved function and as a consequence the crystal shape is in principle easy to control, and those with positive λ which show anomalous behaviour. Unfortunately it is the very materials which show anomalous behaviour for which the crystal diameter cannot be calculated from measuring the time dependence of the weight signal. When the expression for H is transformed to evaluate a then the integral contains an exponential which is strongly divergent for a positive index, and the index takes the sign of λ.

The effect of a change in diameter on the weight can be seen in figure 1. The initial reaction to a developing bulge is for the power to be lowered; this, of course, can cause an instability. In physical terms we can see what is happening by considering a rise in temperature. This will ultimately decrease the diameter, but initially the melting-point isotherm is raised to give a taller meniscus with the same crystal interfacial area. The apparent weight rises on two counts. Firstly the surface tension forces are swung round towards the vertical while acting on the same line length, and secondly the taller meniscus generates a higher pressure to pull on the crystal interface. While it is true that the crystal loses some weight as the tip melts, for anomalous materials, where the melt density exceeds that of the crystal, it is the increased pull of the meniscus which is the larger effect. The increase in apparent weight is interpreted as an increase in diameter, in spite of the physical result which will be to reduce the diameter. The power is duly raised by the servo. As the diameter decreases so the area is reduced, the

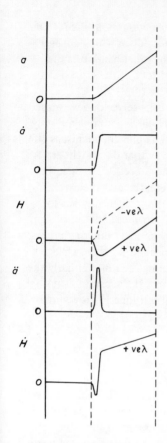

Figure 1. Schematic representation of the time evolution of various parameters following a power change.

peripheral line length shortened and the meniscus height lowered. The weight error signal therefore reverts to the correct sign. Nevertheless, for a space of time, dependent on the growth speed, the correction to an unwanted change in diameter only makes matters worse. Of course, if the growth speed is fast enough then the crystal may scarcely respond to the wrong power adjustment in the short time in which it is applied, but for slow pull speeds the anomalous behaviour must be eliminated.

If the weight error signal is differentiated we obtain:

$$\dot{H} = a + \eta \dot{a} - \lambda \ddot{a}.$$

We can use this differential equally well for control provided that we can generate the signal sufficiently free from noise. As can be seen from the bottom trace in figure 1 $\dot{H}$ suffers from the same anomaly, but our calculations reveal that the integrated power change of the wrong sign is less than if we were using the straight weight error H. This is therefore a better method of control. Several other advantages accrue. Leaving aside the anomaly, the differential is a direct measure of the diameter, which apart from making it easier for the grower to appreciate what is happening to the crystal during a pull, also makes it a more simple operation to change the diameter deliberately, as we have to during the grow-out and tapering procedures. Further, if the crystal diameter is altered and we are using the weight error signal to effect the correction, then it can only be the deviation from the total weight that can be corrected for which appears as a damped oscillation in the diameter. If we work with $\dot{H}$, on the other hand, it is the diameter change itself which is corrected and the previous history of the crystal is ignored.

If we are to use the differential $\dot{H}$ as the control signal for the servo loop we must generate it from the output of the load cell. This is a force F which contains not only the actual weight of the growing crystal at any moment in time plus the meniscus effects but also a contribution from the pull rod, etc. In the same way that the weight error is obtained by subtracting the expected weight from F (Bardsley *et al* 1972), then the differential weight error is produced by subtracting the differential expected weight, $\dot{F}_{ref}$, from $\dot{F}$. The subtlety of using $\dot{H}$ is now apparent. $\dot{F}_{ref}$ is merely a constant for diameter growth, and a linear ramp electronically generated provides a reference signal which can be used to control the shape during both the grow-out and tapering processes. By altering the slope of the ramp the angle at which the crystal grows out is changed, and this is an adjustment which can be made at any time during the run. Similarly the diameter will be fixed by the final holding voltage of the ramp. If the crystal is not fat enough then the final level can be raised and the crystal will continue to grow out at the set angle till it reaches the new required diameter.

3. Practical considerations

How do we weigh the crystal in a hot pressurized environment? The load cell, a commercial linear variable differential transformer (LVDT), can be attached to the pull rod in a re-entrant cavity above the pressure vessel, but we prefer to put it below and weigh the crucible assembly and melt. The crystal weight can be readily obtained by subtraction, and it has the advantage that the load cell becomes a bolt-on attachment to the standard 5-inch RRE crystal puller. Any noise picked up from the pressure changes or convection in the pressure vessel is too high in frequency and too low in amplitude to affect the control signal. To help keep the cell clear of phosphorus, or other volatile constituents, a stream of inert gas is continuously passed through the load cell and on up the narrow connecting stem to the pressure vessel itself. We have been using one cell in this sort of environment for about a year now without corrosion becoming too serious. We find we do not need crucible lift and rotation but suitable engineering can overcome the problems which they present. The other difficulty is that of levitation. With RF heating the induced currents in the susceptor are repelled by those that caused them in the coil. At the centre of the coil there exists a point of neutrality, but if the susceptor is displaced from this position there is a change in apparent weight corresponding to each power change. As the crystal grows the thermal environment alters and therefore the power must be raised or lowered accordingly to keep the diameter constant. If the levitation is not allowed for the crystal usually grows with a slight taper. There is also the possibility that levitation can produce an instability at high loop gains if the corrective power change to an apparent weight change itself causes an apparent weight change of the same sign (ie, the feedback is positive). It is best to remove the levitation effect completely. The magnitude of the repulsion is a function of the RF current, and so is the output of a current transformer placed round one of the

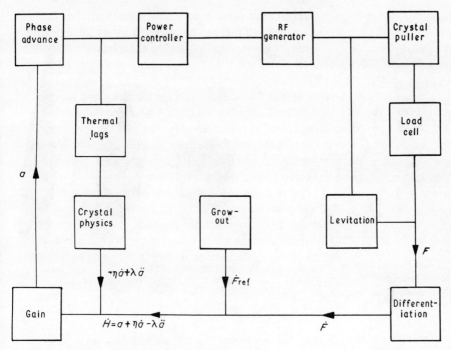

Figure 2. A block diagram of the servo loop.

leads. If we select the right fraction of this and feed it back in the correct phase to the output of the load cell then the levitation is neutralized.

We are now in a position to construct the basic servo loop, and it is shown diagrammatically in figure 2. The output of the load cell, after the levitation effect has been removed, is differentiated and the reference signal from the grow-out unit subtracted. Now we have to remove this anomaly. We have shown that

$$\dot{H} = a + \eta \dot{a} - \lambda \ddot{a},$$

and we wish the drive signal to the power controller to be a alone. If we therefore take this drive signal, differentiate it twice, and feed fractions of each differential back into the $\dot{H}$ signal we can extract a and complete the loop to the power controller. Now we have to introduce a phase advance network. In a large resistance-heated system the time lag between a corrective power change being applied and it becoming effective as a temperature change at the growth interface may be of the order of several minutes, and phase advance is the standard way of speeding up the implementation. It so happens that the use of the $\dot{H}$ control signal automatically introduces a phase correction of the right sign and we find that the network is unnecessary for our smaller R F-heated pullers, though we require it for the larger models.

4. Results

We have used our automatic control gear for a great variety of materials grown in several different pullers. We have pulled small diameter crystals (2 mm) and large

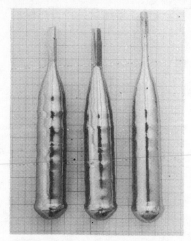

Figure 3. Three germanium crystals grown with the same control settings (but with different orientations). The apparatus was pressurized but no boric oxide used.

Figure 4. A shadowgraph showing the change in grow-out angle that can be accomplished by altering the slope of the grow-out ramp.

diameter crystals (40 mm) of these III–V compounds, and we have pulled crystals successfully at speeds in the range of $1-200$ mm h^{-1}.

We have demonstrated the reproducibility of the process, as shown in figure 3, and our control over shape, as in figure 4. We have even gone so far as to grow in a controlled manner within some thermal environments which have defied our manual attempts to produce a reasonable shape. The germanium crystals in figure 3 have residual surface oscillations showing a loop instability brought about by too high a gain. The minimum control setting at that time was not low enough to compensate for the extremely high contribution to the overall gain produced by the very flat isotherms at the growth interface. If the entire melt is drained to produce a crystal then the tail end will not be tapered but bulged because the melt surface changes shape as it reaches the bottom corner of the crucible.

Although we are confident that we can grow any crystal which can be pulled by the Czochralski technique better than it can be grown manually, we can still envisage thermal situations from which we cannot extract a crystal. Nor is this our only problem. The three III–V compounds, gallium arsenide, indium phosphide and gallium phosphide, are pulled through an encapsulating layer of boric oxide. Not only do we have to allow for the viscous drag and surface tension of the encapsulant for perfect control, but more importantly when the crystal emerges from the top of the boric oxide the additional weight increase will be compensated for by a diameter reduction at the growing interface. Similarly an unwanted bulge will be repeated by a neck the depth of the encapsulant lower down the crystal. As the depth of the boric oxide changes throughout the growth this is an extremely difficult effect to compensate for, and this is an area where a digital computer comes into its own. Nevertheless we are able to minimize the effect by ensuring that the grow-out is not finished by the time the top of the crystal emerges from the encapsulant. In other words we accept a limitation imposed by the thickness of the boric oxide on the grow-out angle. We find all the other effects of the boric oxide unimportant provided that the demanded diameter is kept below 90% of that of the crucible. The final two figures, 5 and 6, demonstrate the success we have on indium phosphide and gallium phosphide respectively with this cheap, simple and reliable apparatus.

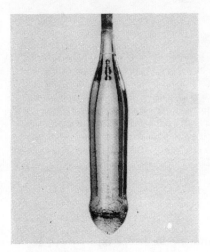

Figure 5. A 200 g crystal of indium phosphide.

Figure 6. A 100 g crystal of gallium phosphide.

References

Antonov P I 1965 *Sb. Rost. Kristallov* **6** 158
Bardsley W, Cockayne B, Green G W, Hurle D T J, Joyce G C, Roslington J M, Tufton P J and Webber H C 1974 *J. Cryst. Growth* **24/25** 369
Bardsley W, Green G W, Holliday C H and Hurle D T J 1972 *J. Cryst. Growth* **16** 277
Gross U and Kersten R 1972 *J. Cryst. Growth* **15** 85
Van Dijk H J A, Jochen C M G, Scholl G J and Van der Werf P 1974 *J. Cryst. Growth* **21** 310
Zinnes A E, Nevis B E and Brandle C D 1973 *J. Cryst. Growth* **19** 187

An in-situ etch for the CVD growth of GaAs: the 'He-etch'

J V DiLorenzo

Bell Laboratories, Murray Hill, New Jersey 07974, USA

Abstract. An 'in-situ' etching procedure for CVD growth of GaAs using the $AsCl_3/Ga/H_2$ system has been developed. The etch is effected by using He as the carrier gas rather than H_2 and optimizing the flow rate of $AsCl_3/He$ over the Ga source to effect free Cl_2 in the deposition zone. Mirror-like surfaces are obtained after etching. Etch rates ranging from $0.1\,\mu m\,min^{-1}$ to greater than $1\,\mu m\,min^{-1}$ can be obtained. The etch rate is controllable by three parameters: the substrate temperature, the H_2/He ratio at the substrate and the mole fraction of $AsCl_3/He$. The application of this technique to the continuous growth of a n/n^+/semi-insulating structure with n-layer thickness of $0.05-0.1\,\mu m$ and a n^+ layer thickness of $4\,\mu m$ is described.

1. Introduction

The necessity for an 'in-situ' etch of a GaAs substrate prior to epitaxial growth has been recognized by many workers as an essential growth sequence to improve the quality of the epitaxial deposit (Fairman and Solomon 1973, Hirao and Nakashima 1970, Nozaki and Saito 1972, Wolfe *et al* 1968). This paper describes results on an 'in-situ' etching technique, the 'He-etch', which results in greatly improved growth procedures.

For purposes of better understanding of the technique a schematic profile of one of our typical CVD reactors is shown in figure 1. The system consists of a high purity quartz reactor tube containing two inlets. The inlets under growth conditions contain H_2 and $AsCl_3$ and H_2 and H_2S as shown in figure 1. During the 'He-etch', one inlet contains He and $AsCl_3$, while the other inlet has the capability of carrying H_2 flow as well as He flow. This inlet provides a means of controlling the H_2/He ratio at the substrate. The hydrogen is purified by diffusion through a Pd—Ag membrane, while the He is used directly from a 99·99995% pure tank filtered through a $0.2\,\mu m$ membrane. The Ga, $AsCl_3$ and cleaning of the wafers and general growth procedures are as given in previous papers (DiLorenzo *et al* 1971, DiLorenzo 1972).

The two general techniques reported in the literature for 'in-situ' etching of GaAs by other workers involve: (i) elevation of the substrate temperature to or above the Ga source temperature (Wolfe *et al* 1968, Fairman and Solomon 1973) or (ii) to remove the Ga source completely and etch by an $AsCl_3/H_2$ flux only (Effer 1965, Jida and Hirose 1965, Bakin *et al* 1972). In the case of elevation of the seed temperature the procedure is cumbersome requiring a change of the furnace profile during the pre-growth cycle and long lag times waiting for substrate temperature equilibration. In the case of etching by an $AsCl_3/H_2$ flux only, the ability to obtain mirror-like surfaces is very sensitive to the substrate temperature and on $AsCl_3/H_2$ molar ratio; after etching matte or rough surfaces can occur.

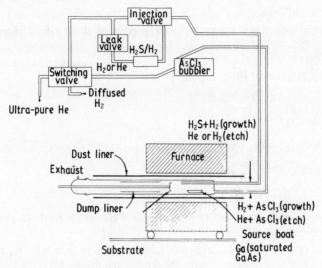

Figure 1. Schematic diagram of CVD reactor.

Recently Nozaki and Saito (1972) found that by introducing a supply of $AsCl_3/H_2$ which did not contact the Ga source (it would enter the reactor from the H_2S/H_2 tube in figure 1) mirror-like surfaces could be produced after 'in-situ' etching at a rate of $0.4 \, \mu m \, min^{-1}$. While their work represents a significant advance in 'in-situ' etching, the complexity of a second bubbler system was required.

The 'He-etch' is simply effected by introducing a flow (described below) of $He/AsCl_3$ at the onset of the growth process.

2. Chemistry of the system

In order to understand why etching takes place in a He carrier gas it is necessary to consider the chemistry of the $AsCl_3/Ga/H_2$, He system.

The formation and transport of GaAs at the Ga boat can be represented by the following equations:

$$2AsCl_3(g) + 3H_2(g) \rightleftharpoons \tfrac{1}{2}As_4(g) + 6HCl(g) \tag{1}$$

$$2Ga(l) + \tfrac{1}{2}As_4(g) \rightleftharpoons 2GaAs(s) \tag{2}$$

$$2GaAs(s) + 2HCl(g) \rightleftharpoons 2GaCl(g) + \tfrac{1}{2}As_4(g) + H_2(g). \tag{3}$$

Equation (3) represents the transport of the GaAs 'skin' from the boat but also is the reaction of deposition at the substrate, in other words,

$$2GaCl(g) + \tfrac{1}{2}As_4(g) + H_2 \rightleftharpoons GaAs(s) + 2HCl(g). \tag{4}$$

Therefore, as pointed out clearly (Boucher and Hollan 1970), the amount of GaAs 'thermodynamically' available for deposition at the substrate is simply the degree of 'supersaturation' of the gas. This degree of 'supersaturation' is simply the source-to-substrate temperature difference. For reaction (3) above, the equilibrium constant, setting $a_{GaAs} = 1$ for equilibrium conditions, K_1, has a value of about 1 for temperatures in the range of 750–800 °C (Boucher and Hollan 1970).

In a gas other than H_2, for example N_2 or He, the dominant reactions at the boat are (Ihara *et al* 1974, Dazai *et al* 1974, Kirwan 1970):

$$2AsCl_3(g) \rightleftharpoons \tfrac{1}{2}As_4(g) + 3Cl_2(g) \tag{5}$$

$$2Ga(l) + \tfrac{1}{2}As_4(g) \rightleftharpoons 2GaAs(s) \tag{6}$$

$$2GaAs(s) + Cl_2(g) \rightleftharpoons 2GaCl(g) + \tfrac{1}{2}As_4(g) \tag{7}$$

$$2GaCl(g) + 2Cl_2(g) \rightleftharpoons 2GaCl_3(g) \tag{8}$$

$$3Cl_2(g) + 2GaAs(s) \rightleftharpoons 2GaCl_3(g) + \tfrac{1}{2}As_4(g). \tag{9}$$

At equilibrium, deposition via reaction (9) would occur in He similar to H_2 since the gas at the substrate is supersaturated.

However, as Ihara *et al* (1974) point out clearly in their experiments with H_2 as a carrier gas and in work at this laboratory with He and H_2 as a carrier gas (Cox and Luther 1974) the transport of GaAs by either HCl or Cl_2 via reactions (9) or (3) is extremely sensitive to flow over the source. Using Ihara's and Dazai's definition of reaction transport kinetics of free Cl_2 or HCl to the deposition zone:

$$R = \frac{F(Cl_2 \text{ or HCl}) \text{ free in deposition zone}}{F(Cl_2 \text{ or HCl}) \text{ feed into source zone}}$$

where F = molar concentration/unit time, equation (9) can be rewritten as:

$$2(1-R)GaAs(s) + 3Cl_2(g) \rightleftharpoons 2(1-R)GaCl_3(g) + \tfrac{1}{2}(1-R)As_4 + 3RCl_2 \tag{10}$$

and equation (3) can be rewritten as:

$$2(1-R)GaAs(s) + 2HCl(g) \rightleftharpoons 2(1-R)GaCl(g) + \tfrac{1}{2}(1-R)As_4(g)$$
$$+ (1-R)H_2(g) + 2RHCl(g). \tag{11}$$

The importance of the Japanese work and ours is that by proper choice of flow over the source, the quantity of free Cl_2 or HCl in the deposition zone, R, can have a value of the partial pressure leading to etching of the substrate rather than growth.

However, in practice the flow rate of $AsCl_3/H_2$ necessary to effect etching is very high (more than $4\cdot0$ cm s^{-1} and R(HCl) is very sensitive to this flow. In He, etching occurs at normal flow rates and R(Cl_2) is relatively insensitive to flow.

We have carried this concept one step further by introducing H_2 downstream of the boat so that competing growth reactions takes place via the H_2 system of reactions (1)–(4) and the He system (5)–(9). In this manner, the etch rate can be readily controlled.

3. Results and discussions

A series of etching experiments were performed using either n$^+$ or semi-insulating substrates as samples with a quartz plate as an etching mask and determining the etch rate from the height difference between the etched and masked areas.

The experimental conditions used in the etch experiments were as follows:

Ga source temperature = 820 °C
Substrate temperature = 748–711 °C
$\%H_2/He = 0–20\%$
Reactor tube = 31 mm ID
Flow rate $(He/AsCl_3) = 407 \text{ cm}^3 \text{ min}^{-1}$ (25 °C)
Flow velocity over source $(He/AsCl_3) = 3\cdot3 \text{ cm s}^{-1}$ (820 °C)†.

In figure 2, the highly specular nature of the surfaces of 'He-etched' substrates having an orientation of $2\cdot5°$ off {100} are shown. (Note the reflection of the ruler from

Masked (unetched) surface

Masked (unetched) surface

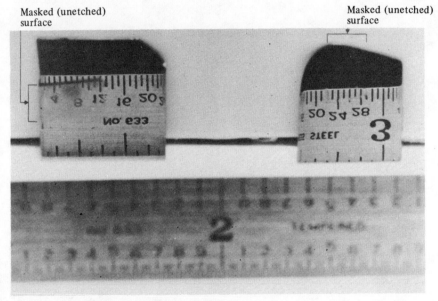

Figure 2. 'He-etched' surfaces of boat-grown GaAs substrates.

the etched surface relative to the masked surface.) In figure 3, an interference contrast photomicrograph of the surface of a boat-grown n^+ substrate is shown. The surface is free of defects having smooth surface morphology, with no obvious faceting. This is to be expected since for small departures from equilibrium (low etch rates in this case), a crystal in quasi-equilibrium with its vapour phase has a featureless surface (Cabrera and Coleman 1963, Burmeister 1971).

In figure 4, the variation of the etch rate with the reciprocal substrate temperature is shown for three different etching conditions. The etch rate decreases with decreasing substrate temperature. The data, while limited, suggest a linear fit with a range of activation energy of 20–40 kcal mol^{-1} depending on etching conditions.

From equilibrium thermodynamics alone (Seki *et al* 1968) one would predict that the etch rate should decrease with decreasing substrate temperature, suggesting that this is the case in the He system.

† Except for curve C, figure 4, with a flow-velocity of $2\cdot5 \text{ cm s}^{-1}$.

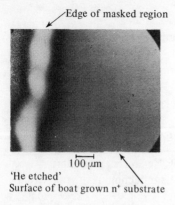

Edge of masked region

|← 100 μm →|

'He etched'
Surface of boat grown n⁺ substrate

Figure 3. Interference contrast photograph of a 'He-etched' surface in the vicinity of the masked region.

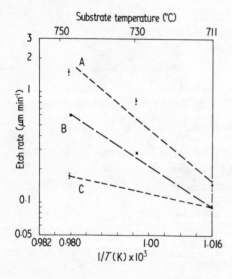

Substrate temperature (°C)

Figure 4. Dependence of the etch rate (μm min^{-1}) on reciprocal substrate temperature $1/T$ °F for different etching conditions. A: X (AsCl$_3$) = 10^{-2}, % H$_2$ = 20. B: X (AsCl$_3$ = $3 \cdot 5 \times 10^{-3}$, % H$_2$ = 5. C: X (AsCl$_3$) = $3 \cdot 5 \times 10^{-3}$, % H$_2$ = 5.

Note also in figure 4, the difference in etch rate between curves B and C at the same values of AsCl$_3$ mole fraction and %H$_2$. Data presented in curve C were obtained at a lower flow velocity over the Ga source (about $2 \cdot 5$ cm s^{-1} compared to $3 \cdot 3$ cm s^{-1}). This is consistent with the chemistry of equation (11) since at lower flow velocities R is smaller (Ihara *et al* 1974), resulting in less free Cl$_2$ in the deposition zone and thus a lower etch rate.

In figure 5 the variation of the etch rate with increasing addition of H$_2$ for two different mole fractions of AsCl$_3$/He is shown. As can be seen from the figure, a dramatic decrease in the etch rate occurs with a small addition of H$_2$ ($1 \cdot 5\%$), increases again with further addition of H$_2$ (5%) and then decreases with further addition out to 20% H$_2$. The reasons for this are unclear but further work is underway to gain insight into the chemistry causing the variation. As can be seen from figures 4 and 5, etch rates at temperatures between 740 and 750 °C can be obtained from $0 \cdot 1$ to about $1 \cdot 0 \, \mu$m min^{-1}. This temperature range is significant since it is the usual temperature for growth of epitaxial films for device applications.

An important point not previously brought out concerning the 'He-etch' is that it

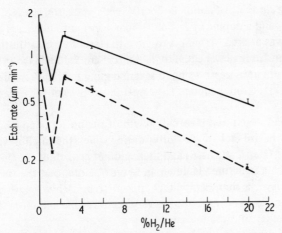

Figure 5. Dependence of the etch rate on % H_2/H_2 and $AsCl_3$ mole fraction in He X ($AsCl_3$) at a substrate temperature of 748 °C. Full curve: T = 748 °C, X ($AsCl_3$) = 10^{-2}. Broken curve: T = 748 °C, X ($AsCl_3$) = $3\cdot5 \times 10^{-3}$.

insures source saturation at the onset of growth, since an arsenic flux is present over the source during the etch cycle and insures complete 'GaAs skin' formation. Further, growth is commenced immediately by a simple change to H_2, insuring that a stable growth skin exists during the entire growth cycle. The results of growing from an unsaturated source are disastrous as has been recently pointed out (Shaw 1971).

An example of the application of the 'He-etch' to growth of a device structure is shown in figure 6. For applications as a microwave mixer diode, a n/n^+/semi-insulating

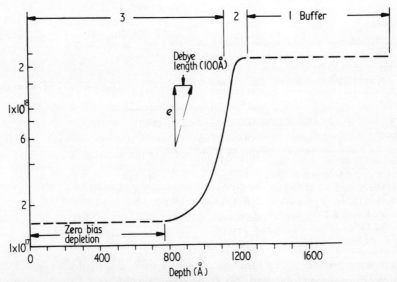

Figure 6. Doping profile plotted against depth of a buffer (1), n-layer (3) structure with growth procedure indicated in steps 1, 2, 3. Region 1, $N \simeq 3 \times 10^{18}$ cm^{-3}, region 2, He-etch doping changed in vapour to give 1×10^{17} cm^{-3} (in solid), region 3, growth initiated, thin-film-grown.

(substrate) structure, having a doping of about $1-2 \times 10^{17} \mathrm{cm}^{-3}$ and a thickness of $0.05-0.1\mu m$ for the n layer, and a doping of $2-5 \times 10^{18} \mathrm{cm}^{-3}$ with a thickness of $2-10\,\mu m$, for the n^+ layer, was desired. Attempts to grow the structure by adjusting the pressure of H_2S in the vapour from a level equal to $2-5 \times 10^{18} \mathrm{cm}^{-3}$ doping (solid) to $1-2 \times 10^{17} \mathrm{cm}^{-3}$ (solid) were unsuccessful since a severe doping tail between the n and n^+ layer were obtained (the time required to grow $0.05\,\mu m$ was of the order of 0.5 min).

A technique was developed which first involved growth of the buffer (step 1, figure 6) followed by a slow in-situ etch of the buffer during which the sulphur was changed to give $1 \times 10^{17} \mathrm{cm}^{-3}$ (step 2, figure 6) and then another growth of the thin film. A doping profile of such a structure† is shown in figure 6. Note, that the transition between the n and n^+ layer is sharp and in fact appears to be Debye length limited, without any interfacial anomalies. This procedure is very reproducible and more than 50 films have been grown similar to those in figure 6.

In summary the 'He-etch' provides a simple, controllable means of in-situ etching of substrates in the $AsCl_3/Ga/H_2$ system. The surfaces obtained after etching are specular, non-faceted and chemically clean. The etch insures a means of stabilizing the GaAs 'skin' on the Ga source, which is an essential part of growth in the $AsCl_3$ system. Its use for growth of device structures is established, and it is a simple means of achieving very low growth and/or etch rates for growth of thin films.

Acknowledgments

The author wishes to thank L C Luther, H M Cox and A R Von Neida for many helpful discussions concerning this work. In particular the use of the unpublished data of L C Luther and H M Cox is greatly appreciated.

References

Bakin N N, Dedkov V D and Porokhonvichenko L P 1972 *Inorg. Mater.* **8** 1206
Boucher A and Hollan L 1970 *J. Electrochem. Soc.* **117** 932
Burmeister J 1971 *J. Cryst. Growth* **11** 131
Cabrera N and Coleman R V 1963 *The Art and Science of Growing Crystals* ed J J Gilman (New York: Wiley)
Cox H M and Luther L C 1974 *J. Electrochem. Soc.* to be published
Dazai K, Ihara M and Ozeki M 1974 *Fujitsu Sci. Tech. J.* **10** 125
DiLorenzo J V 1972 *J. Cryst. Growth* **17** 189
DiLorenzo J V and Moore G E 1971 *J. Electrochem. Soc.* **118** 1823
Effer D 1965 *J. Electrochem. Soc.* **112** 1020
Fairman R D and Solomon R 1973 *J. Electrochem. Soc.* **120** 541
Hirao M and Nakashima H 1970 *Proc. 2nd Conf. Solid State Devices* p6
Ihara M, Dazai K and Ryuzan O 1974 *J. Appl. Phys.* **45** 528
Jida Sh and Hirose S 1965 *Japan. J. Appl. Phys.* **4** 1025
Kirwan D J 1970 *J. Electrochem. Soc.* **117** 1572
Nozaki T and Saito T 1972 *Japan. J. Appl. Phys.* **11** 110
Seki H, Moriyama K, Asakama I and Korie S 1968 *Japan. J. Appl. Phys.* **7** 1324
Shaw D W 1971 *J. Cryst. Growth* **8** 117
Wolfe C M, Foyt A G and Lindley W T 1968 *Electrochem. Tech. J.* **6** 208

† The doping profile was obtained by a reverse biased 20 mil Au Schottky barrier and an automatic $C-V$ profiler.

Nouvelle technique de croissance épitaxiale de InP

N Sol, J P Clariou, N T Linh, G Bichon and M Moulin

Laboratoire des Matériaux Monocristallins et Couches Minces, Thomson CSF LCR, 91401 Orsay, France

Abstract. A new technique for liquid phase epitaxy of InP is proposed: saturation of the indium melt is performed *in situ* by PCl_3.

Correlated characterizations (ion mass analysis, electric measurements, photoluminescence) have shown that epitaxial layers prepared by this technique are of high purity.

Sommaire. Une nouvelle technique d'épitaxie en phase liquide est proposée: la saturation du bain d'indium est réalisée *in situ* par PCl_3.

Des caractérisations corrélées (analyse ionique, mesures électriques, photoluminescence) ont montré que les couches préparées par cette technique sont de haute pureté.

1. Introduction

L'épitaxie en phase liquide connaît depuis quelques années un développement important dans de nombreux domaines, entre autres pour la réalisation de dispositifs hyperfréquences pour lesquels on recherche en premier lieu un faible taux de dopage.

De nombreux travaux (Rosztoczy *et al* 1970, Astles *et al* 1973, Williams *et al* 1973) sur le phosphure d'indium ont été réalisés. De bons résultats ont été obtenus en particulier par Joyce et Williams (1970), Hales *et al* (1970) en phase vapeur, et récemment par Astles *et al* (1973) en phase liquide.

La présente étude a pour but de mettre au point une nouvelle technique d'épitaxie en phase liquide. On comparera ensuite les propriétés des couches ainsi obtenues à celles des couches préparées par la méthode d'épitaxie en phase liquide classique (Miller *et al* 1972).

Les différentes méthodes de caractérisation sont les suivantes:

(i) détection d'impuretés chimiques par la microsonde ionique,
(ii) mesures électriques selon la méthode de Van der Pauw,
(iii) photoluminescence à 77 et 4 K (laser HeNe $\lambda = 6328$ Å).

2. Resultats experimentaux

2.1. Mise au point d'une nouvelle technique d'épitaxie

La technique classique d'épitaxie en phase liquide de InP consiste à faire croître une couche à partir d'une solution d'indium préalablement saturée par InP polycristallin

(Miller *et al* 1972). Récemment Astles *et al* (1973) ont développé une méthode de saturation *in situ* par PH_3. Nous proposons ici une saturation par PCl_3. Le principe du réacteur d'épitaxie est schématiquement montré sur la figure 1. Cette technique de saturation par PCl_3 présente, par rapport à celle de PH_3, l'avantage d'une possibilité de décapage *in situ* du creuset de graphite par HCl provenant de la décomposition de PCl_3.

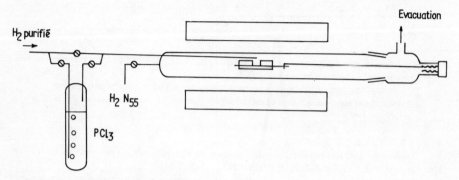

Figure 1. Schéma du réacteur d'épitaxie.

2.2. *Etude comparative des couches épitaxiales*

Dans ce qui va suivre nous comparerons les propriétés des couches épitaxiales obtenues par saturation avec:

InP polycrystallin (couches α)
InP monocristallin pur de provenance RRE (couches β)
PCl_3 (couches γ).

2.2.1. Analyse à la microsonde ionique. L'analyse chimique par la microsonde a mis en évidence l'existence de Si, O et Zn dans les couches épitaxiales α. Leurs concentrations sont respectivement de l'ordre de 4×10^{17}, 5×10^{16} et 5×10^{18} atom/cm^3. Ces mêmes impuretés se retrouvent en grande quantité et de façon inhomogène dans l'InP poly-cristallin de saturation. Par contre, dans les couches β et γ, ces impuretés se trouvent en plus faible concentration: [Si] $\sim 5 \times 10^{16}$ atom/cm^3, [O] $\sim 10^{16}$ atom/cm^3 et [Zn] $\sim 10^{18}$ atom/cm^3. Les limites de détection sont respectivement de 5×10^{15}, 5×10^{15} et 10^{18} atom/cm^3. Au stade actuel de nos analyses il n'est pas possible de trouver de grandes différences entre les échantillons β et γ.

2.2.2. Mesures électriques. Les mesures de Hall ont été effectuées en utilisant la méthode de Van der Pauw, avec un champ magnétique de 1100 G.

Les couches α ont une concentration de porteurs libres variant de 5×10^{17} à 2×10^{18} cm^{-3} à 300 K.

Avec une saturation par InP monocristallin nous avons obtenu des couches épitaxiales ayant une concentration de 10^{16} porteurs libres par cm^3 à 300 K.

L'utilisation de la saturation par PCl_3 nous a permis d'améliorer la pureté des couches émitaxiales préparées ($n = 8 \times 10^{15}$ cm^{-3}, $\mu = 3800$ cm^2 V^{-1} s^{-1} à 300 K et $n = 5 \times 10^{15}$ cm^{-3}, $\mu = 20\,000$ cm^2 V^{-1} s^{-1} à 77 K).

Cette amélioration de la pureté des couches épitaxiales préparées a été corrélée avec les mesures de photoluminescence.

2.2.3. Photoluminescence. Le spectre de photoluminescence à 77 K de différentes couches épitaxiales est représenté sur la figure 2.

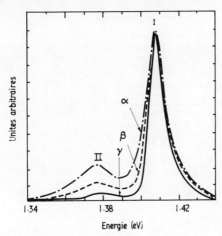

Figure 2. Spectre de photoluminescence à 77 K: α: couche obtenue par saturation avec InP polycristallin (n (77 K) = 4 × 10^{17} cm^{-3}, μ (77 K) = 3800 cm^2 V^{-1} s^{-1}). β: couche obtenue par saturation avec InP monocristallin pur (n (77 K) = 4 × 10^{16} cm^{-3}, μ (77 K) = 16 000 cm^2 V^{-1} s^{-1}). γ: couche obtenue par saturation avec PCl$_3$ (n (77 K) = 5 × 10^{15} cm^{-3}, μ (77 K) = 20 000 cm^2 V^{-1} s^{-1}).

On observe deux pics, le pic I (1·41 eV) est attribué à la transition bande de conduction–bande de valence, le pic II (1·37–1·38 eV) est associé à la recombinaison sur un niveau accepteur situé à 45 meV de la bande de valence. Ce niveau accepteur correspondrait au zinc (Mullin *et al* 1972) déjà détecté par analyse chimique. Pour faciliter la représentation des spectres, la hauteur du pic I a été prise comme référence.

Nous pouvons remarquer que la hauteur du pic II varie énormément, prouvant que le taux de compensation du matériau diminue des échantillons α aux échantillons γ, donc que la nombre d'accepteurs (Zinc) présent dans le matériau est considérablement réduit.

D'autre part, nous constatons en même temps un affinement du pic I donc la largeur à mi-hauteur peut être reliée au nombre de porteurs libres dans le matériau (Williams *et al* 1973).

Cette évolution est confirmée sur les spectres de luminescence réalisés à 4·2 K (figure 3). La hauteur du pic situé à 1·416 eV (transitions excitoniques associées aux donneurs) croît par rapport au pic situé à 1·378–1·382 eV (transition de paire – Williams *et al* 1973).

3. Conclusion

Les résultats des différentes méthodes de caractérisation (microanalyse ionique, mesures électriques, photoluminescence) montrent que les couches épitaxiales obtenues

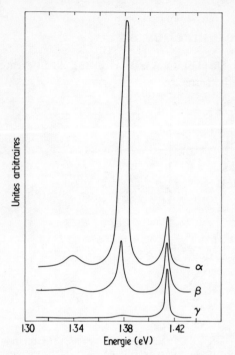

Figure 3. Spectres de photoluminescences à 4·2 K. Les échantillons α, β et γ sont les mêmes que ceux décrits à la figure 2.

par la technique au PCl_3 proposée sont de plus grande pureté que celles préparées par les techniques classiques.

Remerciements

Les auteurs remercient Mademoiselle Huber pour sa participation à la caractérisation des épitaxies par le microanalyseur ionique et Monsieur J B Mullin (Royal Radar Establishment) pour la fourniture d'InP monocristallin de haute pureté.

Cette étude a bénéficié du soutien de la DGRST.

Bibliographie

Astles M G, Smith F G M et Williams E W 1973 *J. Electrochem. Soc.* **120** 1750

Hales M C, Knoght J R et Wilkins C W 1970 *Proc. 3rd Int. Symp. Gallium Arsenide and Related Compounds* (London: Inst. Phys. Phys. Soc.) p50

Joyce B D et Williams E W 1970 *Proc. 3rd Int. Symp. Gallium Arsenide and Related Compounds* (London: Inst. Phys. Phys. Soc.) p57

Miller B I, Pinkas E, Hayashi I et Capik R J 1972 *Appl. Phys.* **43** 2817

Mullin J B, Royle A, Straughan B W, Tufton P J et Williams E W 1972 *Proc. 4th Int. Symp. Gallium Arsenide and Related Compounds* (London: Inst. Phys.) p118

Rosztoczy F E, Antypas G A et Casau C J 1970 *Proc. 3rd Int. Symp. Gallium Arsenide and Related Compounds* (London: Inst. Phys. Phys. Soc.) p86

Williams E W, Elder W, Astles M G, Welb M, Mullin J B, Straughan B et Tufton P J 1973 *J. Electrochem. Soc.* **120** 1741

A (GaAl) As–GaAs heterojunction structure for studying the role of the cathode contact on transferred electron devices

B W Clark, H G B Hicks, I G A Davies and J S Heeks

Standard Telecommunication Laboratories Ltd, Harlow, Essex, England

1. Introduction

Much work has been carried out in recent years on the influence of the cathode contact on transferred electron devices (Colliver *et al* 1972). This includes theoretical treatments which point out the advantages of having an injection limited contact in order to make the field profile within a device more uniform (Hariu *et al* 1970). Also there have been reports (Rees 1974) describing the advantage of injecting hot electrons into a transferred electron device in order to overcome the initial low field dead space normally associated with ohmic contacts in which electrons are heated before they can transfer to an upper conduction band valley. Most startling results of improved oscillator efficiency have been reported with InP devices but improved bandwidth of amplifiers and changes in oscillator efficiency have been reported for GaAs devices as well.

Practically these contacts are produced by alloying a mixture of metals into the active layer surface in order to form a low potential Schottky barrier. The recipe used is usually the result of a series of inspired experiments and the actual structure of the resulting contact is almost impossible to analyse physically or electrically. This paper describes a practical method of obtaining an injection limited contact with known properties and by which the injected electron temperature can be electronically controlled via a third terminal.

2. Device operation

The basic concept of the device was suggested to us by Dr Hilsum of R R E and is similar to that proposed by Atalla and Moll (1969) but incorporates a heterojunction transistor-like contact for control of injection level into the supercritically doped active region of the device. A schematic representation of the device structure is shown in figure 1. The DC potential across the active device is controlled by reverse bias of the base–collector terminals and the current level in the device is controlled by injection of carriers from the forward biased emitter–base terminals. No R F modulation of the emitter–base junction is required and thus a microwave transistor type construction is not necessary. Either the base–collector or emitter–collector terminals (depending on device geometry) are used as the R F connections with the third terminal simply for DC bias control.

The cathode conditions in the active part of the device are controlled by the injected current level as shown qualitatively in figure 2. At low current levels there is a large

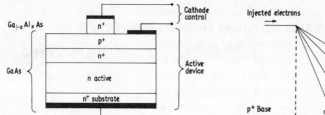

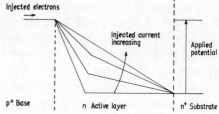

Figure 1. Three-terminal device configuration.

Figure 2. Three-terminal device energy band diagram.

potential drop close to the cathode as in a standard reverse biased p–n junction. For an increasing injected current, at a fixed bias, the potential gradient in the device becomes more uniform. When the current reaches the maximum that can be carried by drift of the conduction electrons in the collector the field profile is flat throughout the device. If the current exceeds this value then there is no potential drop close to the cathode and the effective device length is reduced. The purpose of the n^+ control layer under the base layer, which was shown in figure 1, is to control the maximum width of the high field depletion region at the cathode and also to be able to maintain a high field at the cathode with a high level of injection. This high field then serves to accelerate the electrons rapidly and provides a source of hot electrons for injection into the remainder of the device. If the n^+ region is grown slightly thicker than the depletion width of the p–n junction then it is possible to produce a contact that will inject hot electrons into a low-field region of the device (Rees 1974).

3. Material and device preparation

The material for these devices is grown by the sliding boat liquid epitaxy method. This enables us to grow the multilayer structure while maintaining uniform doping and control thickness of the submicron layers. Figure 3 is a photomicrograph of a cleaved section through a crystal which has been stained to expose the various layers. The layers grown in sequence on an n^+ GaAs substrate are an n^+ GaAs buffer layer, an n GaAs

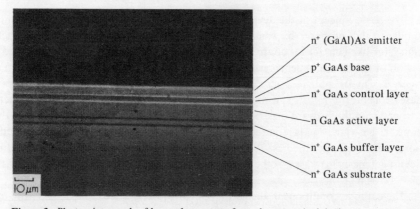

Figure 3. Photomicrograph of layered structure for a three-terminal device.

active layer, an n^+ GaAs control layer, a p^+ GaAs base layer and an n^+ (GaAl)As emitter layer. The dopants used were Sn in the n^+ GaAs layer and Ge in the p layer. The (GaAl)As layer was doped with Se up to a level of approximately 10^{18} donors/cm^3. Se was used rather than Sn, despite its more volatile nature, in the light of Panish's (1973) results. He has shown that it is difficult to obtain high ionized doping levels using Sn when the AlAs content is at our required level of approximately 50%.

The devices are produced by first contacting to the GaAs substrate and to the (GaAl)As layer with an evaporated and alloyed AgSn film. Emitter contact areas are then defined by a standard photolithography process. Using the AgSn contact as a mask the (GaAl)As around the emitter is etched away to expose the p-type base layer. This is a particularly easy process as (GaAl)As with a suitable AlAs content can be etched preferentially to GaAs. This property was first pointed out by Dumke *et al* (1972) who used hydrochloric acid as the preferential etch for layers with an AlAs concentration greater than 30%. We have found that hydrofluoric acid gives a more uniform etch and can be used at lower temperatures than hydrochloric, although we also found it necessary to have an AlAs concentration greater than 40% to achieve satisfactory preferential etching. Once the base layer has been exposed the base contact, which is a ring around the emitter, is made by a further evaporation and photolithography process. Finally, the devices are separated and encapsulated in a three-terminal coaxial package.

4. Preliminary results

Preliminary results which demonstrate the feasibility of the three-terminal approach are encouraging. Firstly, figure 4 shows the DC characteristics of a device that has been processed into a standard transistor. As can be seen from the common emitter characteristic a DC current gain of 1200 has been obtained at an emitter current of 35 mA, which corresponds to a current density of approximately 1000 A cm^{-2}. Current density through this device can be increased to approximately 3000 A cm^{-2} which corresponds to the valley current of GaAs with a doping density of 2.4×10^{15} donors/cm^3. This demonstrates the potential usefulness of the structure we have produced for operation as a transistor and represents the best reported gains for a (GaAl)As–GaAs heterojunction

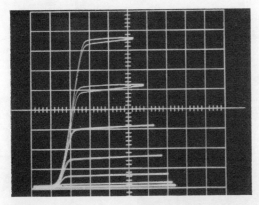

Figure 4. DC characteristics of device IGT 224/2. Vertical scale 5 mA/division. Horizontal scale 0·5 V/division. I_b in 0·01 mA steps.

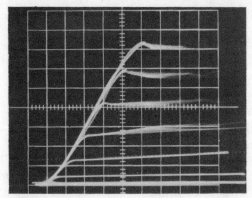

Figure 5. DC characteristics of device IGT 222/4. Vertical scale 20 mA/division. Horizontal scale 1 V/division. I_b in 1 mA steps.

transistor. A suitable geometry for the structure could produce a very high frequency device with the advantages over Si transistors that a GaAs transistors can have.

Secondly, devices have also been produced which include an active transferred electron layer. Figure 5 shows the effect that this layer has on the DC characteristics of the device. Again, the common emitter characteristics are shown and as can be seen, low frequency bias oscillations, characteristic of two-terminal Gunn devices, occur for a range of injected current levels. Suitable loading in the bias circuit can suppress these oscillations and the output characteristic is then seen to have a region of negative differential resistance.

Devices of this type have been operated in microwave circuits and have emitted microwave power, at J-band frequencies, although only as yet at a very low level. It is anticipated, however, that by elimination of microwave power leakage through the non-active part of the device our objective of achieving an operational device with electronic control of the cathode will have been achieved.

Acknowledgment

This work has been carried out with the support of Procurement Executive, Ministry of Defence UK sponsored by CVD.

References

Atalla M M and Moll J L 1969 *Solid St. Electron.* **12** 619–29
Colliver D J, Gray K W, Jones D, Rees H D, White P M and Gibbons G 1973 *Proc. 4th Int. Symp. Gallium Arsenide and Related Compounds* (London and Bristol: Institute of Physics) pp286–94
Dumke W P, Woodhall J M and Rideout V L 1972 *Solid St. Electron.* **15** 1339–43
Hariu T, Ono S and Shibata Y 1970 *Electron. Lett.* **6** 666–7
Panish M B 1973 *J. Appl. Phys.* **44** 2667–75
Rees H D 1974 *Metal–Semiconductor Contacts* (London and Bristol: Institute of Physics) pp105–15

Author Index